香港

觀鳥系列
**01**

香港

**觀**

**鳥**

全圖鑑

捕捉 267 種雀鳥姿態 · 觀鳥入門準備

# A PHOTOGRAPHIC GUIDE TO THE BIRDS OF HONG KONG

萬里機構

HKBWS
香港觀鳥會 編著

## 香港觀鳥會《香港觀鳥全圖鑑》製作組
## HKBWS "*A Photographic Guide to the Birds of Hong Kong*" Working Group

| | |
|---|---|
| 顧問 | 林超英 |
| 攝影 | 文超凡、孔思義、黃亞萍、文權溢、王煌容、王維萍、王學思、古愛婉、甘永樂、任永耀、任德政、伍昌齡、伍耀成、朱祖仁、朱詠兒、朱翠萍、朱錦滿、何志剛、何建業、何國海、何瑞章、何萬邦、何維俊、何錦榮、余日東、余柏維、吳思安、吳璉宥、呂德恆、宋亦希、李君哲、李佩玲、李炳偉、李振成、李偉仁、李啟康、李逸明、李雅婷、李鶴飛、杜偉倫、周家禮、林文華、林釗、林鳳兒、柯嘉敏、洪敦熹、胡明川、英克勁、夏敖天、容姨姨／李啟源、高俊雄、高偉琛、崔汝棠、張玉良、張勇、張振國、張浩輝、梁釗成、深藍、許兆杰、許淑君、郭子祈、郭加祈、郭匯昌、陳土飛、陳巨輝、陳志光、陳志明、陳志雄、陳佩霞、陳佳瑋、陳俊兆、陳建中、陳家強、陳家華、陳詠芝、陳燕明、陳燕芳、陶偉意、陸一朝、勞浚暉、彭俊超、森美與雲妮、馮少萍、馮振威、馮啟文、蕭敏晶、馮漢城、黃才安、黃志俊、黃良熊、黃卓研、黃倫昌、黃理沛、江敏兒、黃瑞芝、黃福基、黃寶偉、楊加強、溫柏豪、葉紀江、詹玉明、賈知行、劉柱光、劉振鴻、劉健忠、劉匯文、劉劍明、蔡美蓮、馬志榮、鄧玉蓮、鄧仲欣、鄧詠詩、鄭兆文、鄭偉強、鄭諾銘、黎凱輝、盧嘉孟、霍棟樑、薛國華、謝鑑超、鍾潤德、簡志明、羅文凱、羅瑞華、羅錦文、譚業成、譚耀良、關朗曦、關子凱、關寶權、蘇毅雄、Aka Ho、Geoff Smith、Geoff Welch、John Clough、Michael Schmitz |
| 撰文 | 區俊茵、陳慶麟、陳明明、張勇、周家禮、馮寶基、何文輝、孔思義、黃亞萍、洪維銘、江敏兒、林鳳兒、劉偉民、馬嘉慧、蘇毅雄、黃志俊、王學思、余日東、呂德恆 |
| 圖片搜集及挑選 | 譚業成、張浩輝、何萬邦、江敏兒、黃理沛、李佩玲、盧嘉孟、呂德恆、黃卓研、葉思敏、郭子祈 |
| 文稿編輯、翻譯及校對 | 陳慶麟、陳佳瑋、張嘉穎、張浩輝、孔思義、黃亞萍、林傲麟、李銘慧、勞浚暉、呂德恆、黃志俊、駱俊賓、馬嘉慧、溫翰芝、黃卓研、王學思、楊莉琪、余日東、劉偉民、陳燕明、梁嘉善、左治強、劉家雁、黎淑賢、陳翠楣、方海寧、李鍾海、郭子祈、羅偉仁、陳愷瑩、謝偉麟、簡瑋彤 |
| | |
| Consultant | C.Y. Lam |
| Contributors (photos) | Geoff Carey, Allen Chan, Christina Chan, Daniel Chan, Gary Chan, Helen Chan, Isaac Chan, Jacky Chan, Chan Jun Siu Brian, Chan Kai Wai, Natalie Chan, Sam Chan, Simon Chan, Thomas Chan, Chan Wing Kam, Cheng Nok Ming, Raymond Cheng, Cheng Wai Keung, Andy Cheung, Cheung Ho Fai, Louis Cheung, Cheung Mok Jose Alberto, Owen Chiang, Jimmy Chim, Gary Chow, Chu Cho Yan, Chu Chui Ping, Doris Chu, Francis Chu, Frankie Chu, Chung Yun Tak, John Clough, Stanley Fok, Ken Fung, Maison Fung, Fung Siu Ping, Sonia & Kenneth Fung, Martin Hale, Aka Ho, George Ho, Jan Ho, Ho Kam Wing, Kinni Ho, Danny Ho, Marcus Ho, Pippen Ho, Wan Pak Ho, Jemi & John Holmes, Kami Hui, KK Hui, Tony Hung, Herman Ip, Kam Wing Lok, Kan Chi Ming, Koel Ko, Kitty Koo, Kou Chon Hong, Matthew Kwan & TH Kwan, Kwan Po Kuen, Andy Kwok, Kwok Ka Ki, Kwok Tsz Ki, Dick Lai, James Lam, Kenneth Lam, Shirley Lam, Angus Lau, Benson Lau, Lau Chu Kwong, Jac Lau, Jianzhong, Law Kam Man, Anita Lee, Eling Lee, Lee Hok Fei, Jasper Lee, KY Lee, Lee Kai Hong, Lee Yat Ming, Evans Leung, Geoffery Li, Harry Li, Li Wai Yan, Aaron Lo, Lo Chun Fai, Lo Kar Man, Roman Lo, Henry Lui, Mike Luk, Christine and Samuel Ma, Bill Man, Kennic Man, Felix Ng, Karl Ng, Ng Lin Yau, Ng Sze On, Or Ka Man, Pang Chun Chiu, Sammy Sam & Winnie Wong, Michael Schmitz, Leo Sit, Geoff Smith, Samson So, Sung Yik Hei, Tam Yip Shing, Tam Yiu Leung, Tang Chung Yan, Joyce Tang, Wing Tang, Allen To, Ann To, Wallace Tse, Wee Hock Kee, Geoff Welch, Captain Wong, Cherry Wong, Wong Choi On, Dickson Wong, Wong Hok Sze, Wong Leung Hung, Peter Wong & Michelle Kong, Wong Po Wai, Wong Shui Chi, Wong Wai Ping, Wong Wong Yung, Woo Ming Chuan, Kelvin Yam, Yam Wing Yiu, Yeung Ka Keung, Ying Hak King, Yu Yat Tung, Freeman Yue |
| Contributors (text) | Joanne Au, Alan Chan, Chan Ming Ming, Louis Cheung, Gary Chow, Robin Fung, Ho Man Fai, Jemi & John Holmes, Hung Wai Ming, Michelle Kong, Shirley Lam, Apache Lau, Carrie Ma, Samson So, Dickson Wong, Wong Hok Sze, Yat-tung Yu, Henry Lui |
| Photographic editors | Tam Yip Shing, Cheung Ho Fai, Marcus Ho, Michelle Kong, Peter Wong, Eling Lee, Lo Kar Man, Henry Lui, Cherry Wong, Cecily Yip, James Kwok |
| Text editing, translation & proofread | Alan Chan, Peter Chan, Fion Cheung, Cheung Ho Fai, Jemi & John Holmes, Alan Lam, Maggie Li, Lo Chun Fai, Henry Lui, Dickson Wong, Luk Tsun Pun, Carrie Ma, Judy Wan, Cherry Wong, Wong Hok Sze, Vicky Yeung, Yu Yat Tung, Apache Lau, Christina Chan, Katherine Leung, George Jor, Norma Lau, Lai Suk Yin, May Chan, Helen Fong, Tom Li, James Kwok, Lo Wai Yan, Zoey Chan, Ivan Tse, Vivienne Kan |

# 序

香港山川秀美，城市邊緣不遠處，總有讓人感受自然悠閒之所，稍加留意，更可見到種類繁多的鳥類，與我們一起活在同一天空下。

勞碌匆忙的現代城市生活，往往把我們的視野局限在眼前的瑣碎。以物質享受為主導的價值觀念，使我們與自然生命的距離拉得越來越遠。自然雖近，心中卻無處安放。我們很容易在不經意之中陷入一種渾噩的機械狀態，日子一天一天過去，精神生活卻只剩下一片空白。

觀鳥表面上是一種消閒活動，但是多年經驗告訴我，觀鳥促使我們走近生命，認識生命的多元和微妙，感受生命的美麗和舒泰。我的觀鳥朋友全都開心樂觀，不論環境順逆，總是活在安祥自在之中。

在推廣觀鳥的過程中，我發現人家開始時都很重視分辨鳥種和很倚重照片的幫忙，因此一本囊括香港較多鳥種的照片圖鑑是十分需要的。2003年香港觀鳥會得到鳥友慷慨借出在野外多年努力才拍得的珍貴照片，結集出版了《香港鳥類攝影圖鑑》，是香港歷史上第一本以照片向廣大市民展示香港鳥類美麗一面的圖鑑，加上知識廣博的鳥友以文字提供相關知識，實在難得，更令人欣喜的是攝影圖鑑出版後受到市民熱烈歡迎，遠超想像。

在攝影圖鑑的鼓舞下，很多人愛上鳥類攝影，2009年吸納了大批精彩鳥類照片和增添豐富內容，改名《香港鳥類圖鑑》出版。十年來，鳥類攝影水平大幅提升，《香港鳥類圖鑑》因應重修，易名《香港觀鳥全圖鑑》，並更新照片和增添近年拍攝到的鳥種，使本書既精美又完備。

本圖鑑成書過程中，所有鳥友及攝影師都是義務投入的，實在是我們香港人的福分，謹向他們致以誠心的謝意。漁農自然護理署香港濕地公園於2009年的支持促成本圖鑑廣泛流通，我們銘記心中，是次2020年新版得以實現，有賴本會紅耳鵯俱樂部成員無名氏慷慨解囊襄助，我們衷心感激。

願望此書促進越來越多人親近大自然，並在觀賞鳥類中感受到生命的喜悅。

香港觀鳥會榮譽會長
林超英
二零二零年五月

# Preface

Hong Kong is a land blessed with hills and streams. In spite of urbanization, one is never far away from a place for unwinding oneself in the embrace of Nature. For people who care to look around, they would find a great variety of feathered friends sharing the same sky with us.

While Nature lies virtually at our doorsteps, it escapes our attention most of the time. The hustle and bustle of the modern city confine our sight to the immediate and the trivial. Values centred on material pleasure distance us from Life itself. It is easy for us to slip obliviously into a form of muddled mechanical existence, in which time flows by while we remain spiritually impoverished.

Bird watching might seem like just another hobby. But years of experience tells me that it provides a path leading to our communion with Nature. It lets us see the diversity and wonders of Life, and enables us to appreciate its beauty and serenity. All my bird watching friends are happy and optimistic. In both times good or bad, their mind stays calm and peaceful.

While promoting bird watching, I realize that beginners are generally keen on the identification of species and need a lot of help from photographs. A photographic guide that covers a good number of the bird species in Hong Kong is therefore badly needed. This book has materialized thanks to the generosity of members of the Hong Kong Bird Watching Society donating amazing photographs derived from many years of hard work in the field. The book also benefits from the text kindly written by members knowledgeable in the subject. Taken together, this guide is a most valuable work presenting to the general public the beauty of the birds in Hong Kong, in pictures and in words. We were all very pleased to discover that the book received an unexpectedly warm reception by the public on publication.

The photographic guide motivated many people to join the army of bird photographers. During the past decades, more and more superb photos comes up attracting more public's attention. This new edition has incorporated more bird species and photos after the 2009 edition which make this book much heavier.

All birdwatchers and photographers who have contributed towards the production of this book have done so as volunteers. They have done our community a great service indeed. I thank them most sincerely. We are most grateful for the support from Hong Kong Wetland Park of Agriculture, Fisheries and Conservation Department, which has enabled the early publication of the 2009 edition. We are also grateful for the support of an anonymous member of the Crested Bulbul Club of the Hong Kong Bird Watching Society make this new edition in 2020 come true.

I truly hope that this book will bring people closer to Nature and help them appreciate the joy of Life.

*CY Lam*
*Honorary President, Hong Kong Bird Watching Society*
*2020.05*

Shenzhen Bay
深圳灣
（后海灣）
Deep Bay

塱原
Long Valley

米埔
Mai Po

尖鼻咀
Tsim Bei Tsui

南生圍
Nam Sang Wai

香港
濕地公園
Hong Kong Wetland Park

大帽山
Tai Mo Shan

城門水塘
Shing Mun Reservoir

香港觀鳥熱點
**Hot Spots for Birdwatching in Hong Kong**

鴉洲
A Chau

沙螺洞
Sha Lo Tung

礐
Kau

籠公園
oon Park

蒲台
Po Toi

# 目錄 Contents

＊本書的鳥類分類及排序主要根據「國際鳥類學大會」(International Ornithological Congress) 的分類方法。

The classification and ranking of birds in this book are basically according to the classification method of "International Ornithological Congress".

# 香港觀鳥會

## 願景

人鳥和諧　自然長存

## 使命

香港觀鳥會致力教育、科研、生境管理與保育政策倡議，啟發及鼓勵公眾一起欣賞與保護野生雀鳥及其生境。

香港觀鳥會成立於1957年，一直以推廣欣賞及保育鳥類及其生境為宗旨。2002年獲認可為公共性質慈善機構。2013年成為國際鳥盟 (BirdLife International)的成員(Partner)，國際鳥盟是一個關注鳥類的國際性非政府組織聯盟機構，致力保育鳥類以及牠們的生境、全球的生物多樣性，促進自然資源的永續利用。

## 我們的工作

### 鳥類生態研究

掌握鳥類和有關生態環境的數據及資料，是制訂保育政策及實施相關措施的重要依據，因此，本會從不間斷地進行各項與鳥類有關的調查及研究工作。

### 環境監察及政策倡議

作為一個提供專業意見的綠色團體，本會持續地對各項自然保育政策、環境影響評估及大型發展的諮詢提供意見，亦同時監察各類型不同規模的發展項目對自然生態，特別是鳥類及其生境的影響。

### 自然教育

本會深信推動公眾欣賞野生鳥類，令市民親身體驗鳥類的美麗和珍貴，大家自然會由衷地保護牠們。因此，我們一直致力舉辦各項觀鳥及自然教育活動，把鳥類、大自然和市民連繫起來。

## 生境管理 —— 塱原與魚塘

### ▶「塱原自然保育管理計劃」

自2005年起，本會與長春社在上水塱原濕地開展香港首個農業式濕地的管理項目，保育塱原的生物多樣性、文化及景觀。計劃連結農友及土地擁有人，實踐一些有利鳥類和生物多樣性的濕地管理措施，共同建立有利鳥類的生活環境。

### ▶「香港魚塘生態保育計劃」

自2012年開始，本會於新界西北魚塘區開展「香港魚塘生態保育計劃」，約130多位漁民參與，覆蓋超過600公頃的魚塘。計劃主要目的是鼓勵漁民每年把魚塘的水位降低一次，讓水鳥可以在魚塘內捕食無經濟價值的雜魚及小蝦，改善及提高魚塘的生態價值。

## 跨地合作

本會與國際鳥盟於2005年合作展開中國項目。項目初期工作以能力建設為主，以鳥類監測、環境教育、觀鳥組織發展等主題舉辦培訓。其後專注鳥類的保育工作，包括受脅的中華鳳頭燕鷗、勺嘴鷸、靛冠噪鶥，以及華南地區非法捕鳥的問題。自1999年開始，本會成立「香港觀鳥會中國自然保育基金」，近年更擴展基金的範圍至亞洲各地，因此更名為「香港觀鳥會亞洲自然保育基金」。以上工作促進內地觀鳥機構的各方面發展及組織能力、增強各地鳥會之間的溝通和交流，推動鳥類保育工作。

# 支持香港觀鳥會

▶ 成為會員

▶ 捐款支持

▶ 訂閱香港觀鳥會電子報，緊貼鳥類和保育最新資訊！

---

電　話：(852) 2377 4387　　　　　電　郵：info@hkbws.org.hk
網　頁：www.hkbws.org.hk
地　址：香港九龍荔枝角青山道532號偉基大廈7樓C室

# The Hong Kong Bird Watching Society

## Vision

People and birds living in harmony as nature continues to thrive.

## Mission

HKBWS promotes appreciation and protection of birds and their habitats through education, research, habitat management and conservation advocacy.

The Hong Kong Bird Watching Society (HKBWS) was founded in 1957. The objective of the organization is to promote the appreciation and the conservation of birds as well as their habitats. It became a public charitable organization in 2002 and BirdLIfe International Partner in 2013. BirdLife International is a global partnership of conservation organizations that strives to conserve birds, their habitats and global biodiversity, working with people towards sustainably use of natural resources.

## Our Work

### Bird Ecology and Research

Data and knowledge of bird species and their habitat ecology are key factors in supporting the formulation and execution of conservation policy and measures. Researches and surveys on birds and their habitats have been conducting throughout the past six decades.

### Environmental Monitoring and Policy Advocacy

HKBWS acts as a green group providing professional views.  We keep providing comments on government conservation policies, environmental impact assessments and large development projects. We also monitor different kinds and size of development projects to avoid any adverse impact towards the natural environment, especially to birds and their living environment.

### Nature Education

We believe once the general public can directly experience the beauty and precious of birds, they will be more committed to the protection of birds. Therefore, we organize different kinds of bird-related educational programmes to link up among birds, nature and the people.

## Habitat Management

▶ **"Nature Conservation Management for Long Valley"**

Since 2005, HKBWS and The Conservancy Association began an agricultural wetland conservation project in Long Valley in order to conserve the biodiversity, culture and landscape of Long Valley. With the engagement of local farmers and landowners, wetland management measures favouring birds and other wetland-dependent wildlife were implemented

▶ **"Hong Kong Fishpond Conservation Scheme"**

In 2012, a conservation scheme named "Hong Kong Fishpond Conservation Scheme" has been launched in the Northwest New Territories. This scheme covered more than 600 hectares of fishponds and more than 130 fishermen engaged. The purpose of the scheme is to encourage fishermen lowering the water level to allow waterbirds feeding on the trash fishes and shrimps left in the ponds. As a whole, the conservation value of fishpond could be maintained and even strengthened.

## Transboundary Cooperation Programme

HKBWS and BirdLife International jointly established the China Programme in 2005. Initial work of the China programme focused on capacity building, which includes holding training workshops on bird monitoring, environmental education and organization development. Later, the Programme works more on bird conservation, including threated species like Chinese Crested Tern, Spoon-billed Sandpiper, Blue-crowned Laughingthrush, as well as problems of illegal hunting in South China. In 1999, HKBWS set up the "HKBWS China Conservation Fund", and later changed as "HKBWS Asia Conservation Fund" to extend the cooperation to the Asian countries. All these works enhance the capacity and development of bird watching societies in mainland China, encourage communication and exchange of information, as well as support and promote bird conservation work in different area.

# Support HKBWS

▶ Be a Member

▶ Donate to HKBWS

▶ Subscribe E-news - Stick with the latest news of birds and our work!

---

Tel:  (852) 2377 4387          E-mail:  info@hkbws.org.hk

Website:  www.hkbws.org.hk

Address:  7C, 532 Castle Peak Road, Lai Chi Kok, Kowloon, Hong Kong

# 觀鳥在香港
# Birdwatching in Hong Kong

　　香港的面積約一千一百平方公里，面積雖小，卻「五臟俱全」。香港不僅是國際大都會，更是世界知名的觀鳥勝地，香港現時的野鳥記錄已達五百五十種，種數佔全中國約三分之一、全球的二十分之一。無論身處何地，觀鳥都可以讓我們融入大自然，體會色彩斑斕、變化萬千的鳥類世界，就讓我們藉觀鳥來探索香港最美麗動人的一面吧！

Despite its small size (around 1,100km$^2$) and its image as a bustling international city, Hong Kong is teeming with bird life. Over 550 different species have been recorded here, which accounts for about one-third of all the species found in China, or one-twentieth of the global total. No matter where you live or work, birdwatching is a way to enjoy and be amazed by the wonders of the natural world that surrounds us all. Let us discover one of the most beautiful aspects of Hong Kong.

## 香港——觀鳥天堂
## Hong Kong – A Paradise for Birdwatching

　　為甚麼香港有這麼多野生雀鳥？

　　（1）物種按生物地理區分佈，由於區與區之間有各種天然屏障阻隔，例如海洋、高山、沙漠，因此不同生物地理區之間的物種會大異其趣。香港位於兩個生物地理區（古北界和東洋界）之間，因此我們可以同時見到兩個地理區的不同鳥種。

　　（2）香港位於其中一條主要的水鳥遷徙路線上 ──「東亞—澳大利西亞遷飛路線」，每年九月至翌年五月，大量水鳥在西伯利亞與澳洲、紐西蘭之間遷徙，很多水鳥便會在春、秋時份途經香港或在香港度過冬天。

　　（3）香港的天然環境得天獨厚，擁有各式各樣的生態環境，郊野公園約佔全港陸地面積四成多，更擁有不少重要生境。香港的生態環境大致可分為濕地（鹹淡水和淡水）、開闊原野、溪流、海洋、海岸和海島、林地和市區等，此外，颱風有時也會將海鳥帶到本港水域。

Why are there so many bird species in Hong Kong?

(1) Species are distributed in different bio-geographical areas separated by natural barriers such as oceans, mountains and deserts, each of which has its own characteristics. As Hong Kong lies between two major bio-geographical areas (Palaearctic and Oriental), we have the chance to see birds from both regions.

(2) Hong Kong is located on the East Asian – Australasian Flyway, one of the main waterbird migration routes. Between September and May, many birds pass through Hong Kong on the way between northern Asia and wintering grounds in the Southern hemisphere, including Australia and New Zealand. Some birds also winter in Hong Kong.

(3) Hong Kong is unique because it features a wide diversity of habitats within its borders, especially in the country parks, which cover some 40% of the land. These habitats include wetlands (brackish and freshwater), open country, streams, coastal area and islands, woodland and urban area. In addition, typhoons occasionally bring truly oceanic species into Hong Kong waters.

# 鳥類與生境
# Birds and Habitats

　　香港的野鳥分佈在不同的生境。香港有高近約一千米的高山，也有濕地、林地、開闊原野、海岸環境以至城市和鄉村，適合各種習性及要求不同的野鳥。

Birds live in a wide variety of habitats in Hong Kong. There are mountains up to nearly 1,000 meters in height, wetlands, woodlands, open country, coastal areas, and of course, cities and villages.

## 濕地 Wetlands

　　香港的濕地大致可分為淡水和鹹淡水濕地兩種。

Both freshwater and brackish water wetlands can be found in Hong Kong.

　　鹹淡水濕地在河水流入海時形成，河口的淡水混入海水，鹽分逐漸增加。同時，在河口的流速減弱，沉積物易於聚集，形成可供紅樹、蘆葦生長的底層。最著名的鹹淡水濕地位於米埔后海灣一帶，其中約15平方公里於1995年9月根據「拉姆薩爾公約」被列為國際重要濕地，冬季或春、秋時份可找到數以萬計的水鳥、涉禽和猛禽。

Brackish water wetlands formed when streams flow into the sea, creating an area of high salinity around estuaries as water flow becomes slower. This encourages the accumulation of sediment which provides a rich substrate for the growth of mangroves and reeds. The Pearl River Delta lies to the northwest of Hong Kong, and the largest brackish wetland is found here. About 15km$^2$ of the Mai Po Inner Deep Bay wetlands was designated a Ramsar Site in September 1995. During winter, spring and autumn, it hosts tens of thousands of waterbirds and waders, as well as the species that prey on them.

※ 呂德恆 Henry Lui

※ Sam Chang

淡水濕地包括魚塘和農耕土地。新界西北目前還有不少魚塘，可找到多種濕地鳥種。冬季時，漁民會將魚塘的水抽去，在塘中餘下的小水窪便會引來上百隻鷺鳥，捕食水中缺乏商業價值的魚。此外，魚塘濕潤的底層亦吸引不少水鳥到來覓食。

農耕土地亦是其中一種淡水濕地，主要集中在泛濫平原，靠農民從地下或河道引水灌溉。位於新界上水的塱原濕地是本港僅存最大的一片農耕濕地，以種植通菜和西洋菜的水耕農作物，被視為一個仍然保留傳統文化的地點。淡水濕地為多種鳥類提供棲息地，常見的有鷚、秧雞、鵪鶉、家燕、棕扇尾鶯、彩鷸、伯勞、卷尾、椋鳥等。

Freshwater wetlands, including fishponds and farmland, are located in the northwest New Territories. Fishponds are drained in winter as part of the fishpond management. This concentrates the 'trash fish' in the small pools that left behind, drawing dozens of egrets to feed. The muddy fishpond bottom also attracts good numbers of waders.

Some farmers pump water from underground to create a type of freshwater wetland, which is usually located in the flood plains. Long Valley, near Sheung Shui, also regarded as a cultural heritage site, is the largest remaining area of freshwater wetland where water spinach and watercress are cultivated. The place attracts many pipits, rails, wagtails, Barn Swallows Zitting Cisticolas, Greater Painted Snipes, shrikes, drongos and starlings.

※ 呂德恆 Henry Lui

## 溪流 Streams

香港沒有天然的大河流，城門河、錦田河等已因市區發展由溪流變成排水明渠。香港的天然溪流生境多位於郊野公園，較為人熟悉的有大埔滘、城門、梧桐寨、南涌河、大蠔河等。沿溪而下，有樹林、濕地等不同生境。溪流內及其兩旁是昆蟲生長的理想環境，因而吸引了愛在水邊捕食的雀鳥，例如鷺鳥、鶺鴒、燕尾、翠鳥、鶲、紅尾水鴝等鳥種。

There are no large natural rivers in Hong Kong. Shing Mun River and Kam Tin River are in fact concreted ditches which were built to channel water flow. Natural streams are found mainly in country parks. Well-known places with this kind of habitat include Tai Po Kau Nature Reserve, Shing Mun Reservoir, Ng Tung Chai, Nam Chung and Tai Ho River. Streams often pass through many different habitats, such as woodland and wetland. They provide habitats for insects, and at the same time good feeding habitats for egrets, herons, wagtails, forktails, kingfishers, flycatchers and Plumbeous Redstart.

※ 呂德恆 Henry Lui

※ 呂德恆 Henry Lui

※ 呂德恆 Henry Lui

## 海洋、沿岸和海島 Ocean, Coastal Area and Islands

香港有各種不同類型的海岸生境，例如石灘、沙灘、泥灘、岩壁和小島。這些地方除了可找到一些水鳥和岩鷺外，沿岸較偏僻的林地還有白腹海鵰繁殖。香港東面海域一些無人小島是燕鷗每年夏季繁殖的地點。在過境遷徙季節以至颱風期間，更不時會有少見的海鳥如鰹鳥、鸌、賊鷗等離岸生活的海鳥被吹到近岸的地方。

※ 王學思 Wong Hok Sze

Hong Kong has various types of coastal habitats, including rocky shores, sandy shores, mudflats, cliffs and small islands. In addition to certain waterbirds and Pacific Reef Egrets, White-bellied Sea Eagles also breed in these areas. Terns also breed in some small uninhabited islands in eastern waters during summer. During the migration season, and particularly if typhoons occur, there are good chances for seeing rare seabirds, such as boobies, petrels and jaegers, which can be blown inshore.

※ 呂德恆 Henry Lui

## 開闊原野 Open Country

開闊原野泛指一些開闊、有植被覆蓋但樹木不多、無人居住的
土地，例如灌叢、草坡和一些荒廢田野，有些更長期積水及長滿雜
草。這類環境在香港仍有不少，但現時亦面臨不少開發壓力，將其割裂或甚至
移平作城市發展用途。在林地外圍，可找到杜鵑、鴉鵑、鵐；入夜後，還可以
找到或聽到夜鷹和貓頭鷹的叫聲。在尖鼻咀、新田、落馬洲、錦田、洞梓、鹿
頸、榕樹澳等地，都有這類生境。

Open country is an open area with vegetation, limited trees, and no human
settlements. These include shrubland, grassland and abandoned farmland. Some
lowland areas are also flooded and have become overgrown with grasses. Although
this type of habitat is still fairly common in Hong Kong, they are facing imminent
development pressure, leading to fragmentation of habitat or even land formation
for urban development. On the fringe of woodland areas these habitats harbour
cuckoos, coucals, and buntings. In the evening, it is worth listening for the calls of
nightjars and owls. Locations with extensive open country in Hong Kong include
Tsim Bei Tsui, San Tin, Lok Ma Chau, Kam Tin, Tung Tze, Luk Keng, and Yung Shue O.

※ 呂德恆 Henry Lui

## 林地 Woodland

在十九世紀，外國人普遍形容香港是「一塊光禿無樹的石頭」，當時的植被多為各村落後山的風水林。現時香港大部分林地都是二次大戰後重生的次生林，大部分都在郊野公園內，受到相當好的保護。由於較少人為干擾，加上林木漸趨成熟，逐漸吸引喜愛林地的鳥種，如擬鶲、山椒鳥、鵑鵙、鶇、鶯、山雀、太陽鳥、鶺及啄花鳥等。熱門的觀林鳥地點有大埔滘自然護理區、城門水塘、甲龍、龍虎山、大潭水塘等。

※ 許桓峰 Alvin Hui

※ 呂德恆 Henry Lui

In the early 19th century, Hong Kong Island was famously described as "a barren rock, without a tree upon it". At that time, most woodland was found in small *fung shui* patches behind villages. Nowadays, most of the vegetation cover is secondary woodland, which has regenerated since the end of the Second World War. Most woodland is located within country parks, which provide a high degree of protection against development. These areas, with lower human disturbance, have gradually become more mature and attract species which favour forest habitat including barbets, minivets, cuckoo-shrikes, bulbuls, warblers, tits, sunbirds, thrushes and flowerpeckers. Well-known woodlands good for birdwatching include Tai Po Kau Nature Reserve, Shing Mun Reservoir, Kap Lung, Lung Fu Shan, and the Tai Tam Reservoir area.

## 高地 High Ground

　　香港的高地和山坡較乾燥，長滿了灌叢和蕨，是尋找猛禽(如鵰、鷂和隼)的理想地方。在較高的山坡，可以見到鷓鴣、大草鶯和山鷚。大帽山和飛鵝山都有這種高地生境。

※ 呂德恆 Henry Lui

※ 呂德恆 Henry Lui

The arid slopes of some high hills retain little water. Here the only vegetation is often shrubs and ferns. These are the best places for raptors, including eagles, buzzards, and falcons. There is a good chance of spotting Chinese Francolin and at higher altitudes Chinese Grassbird and Upland Pipits. Examples of these areas are Tai Mo Shan and Fei Ngo Shan.

## 市區 Urban Areas

香港的市區都集中在沿岸的平地，有些鳥類已適應了在市區生活，建築物可以保護牠們免受狂風暴雨和天敵的威脅。路旁及屋苑的樹木為鳥類提供晚上停棲的地方，屋簷成為雨燕和家燕營巢的上佳選擇，公園更是鳥類覓食和棲息的地方。樹麻雀、八哥、喜鵲、斑鳩、鵯等已能在都市繁衍下一代。

The urban areas of Hong Kong are concentrated on flat land along the coast. A number of bird species have adapted to the urban environment, which provides them with protection from natural threats such as thunderstorms, strong winds and predators. Trees and shrubs along roadsides and in housing estates are used as roosting areas by various bird species. Overhanging building features are used as nesting areas by swallows and swifts. Parks are also good feeding and roosting areas for a number of species such as Tree Sparrows, Crested Mynas, Magpies, Doves, Bulbuls, etc. which choose to breed in urban areas.

※ 呂德恆 Henry Lui

# 鳥類遷徙
# Bird Migration

香港錄得的野生雀鳥種類已超過550種，大部分屬於冬候鳥或過境遷徙鳥，其次是留鳥，夏候鳥及其他種類比較少，所以最佳觀鳥月份在9月至5月之間。

| | | |
|---|---|---|
| 春、秋季過境遷徙鳥 | （46%） | 遷徙途中在香港短暫休息後，再繼續南遷或北返 |
| 冬候鳥 | （25%） | 秋季飛來香港越冬，春季離開 |
| 留鳥 | （11%） | 香港全年可見 |
| 夏候鳥 | （4%） | 春季飛來香港，秋季才離開 |
| 其他（迷鳥、偶見鳥） | （14%） | 偶然在香港出現、迷途或情況不明 |

遷徙是動物（包括鳥類）對生存環境、氣候、種群數量及密度的反應。牠們可以在遷徙的過程中發掘不同或更適合的棲息地，增加生存機會，以及增強在非繁殖地的育幼能力。

※ 王學思 Wong Hok Sze

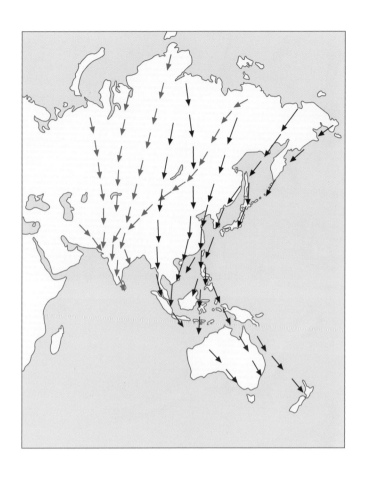

　在北半球緯度較高的地區，春、夏季可供應雀鳥的食物資源豐富。不過到了秋季，天氣轉冷，昆蟲和漿果等食物減少，令數以百萬計在西伯利亞及中國東北部地區繁殖的雀鳥南遷至較溫暖的地方如中國南部、東南亞、澳洲及紐西蘭等地越冬，待翌年3月至5月才重返故地繁殖。這種每年隨季節變化而遷徙的雀鳥，稱為「候鳥」或「遷徙鳥」。

　鳥類的繁殖地、繁殖時間、遷徙時間，以及飛行路線都有不同，形成不同的遷徙模式。候鳥通常都能憑着本能，聯群結隊越過自然障礙，在繁殖地和一個或多個非繁殖地之間往返。其中以水鳥的飛行路線最長，有些候鳥每年遷徙路程更長達二萬九千公里。在遷徙途中，牠們需要在「中途站」歇息，香港正是大量候鳥的「中途站」或「終點站」。

　候鳥遷徙需要大量能量，儲存在皮下和包圍體內組織的脂肪正是能量的來源，「中途站」能讓候鳥安全地補充身體消耗的能量。遷徙成功與否，關鍵在於旅途中有沒有理想的棲息與覓食條件（如食物、庇護所、水源），以及適合的環境因素（氣溫、降雨、捕獵和競爭者）。

※ 孔思義・黃亞萍 Jemi and John Holmes

※ 林文華 James Lam

※ 孔思義・黃亞萍 Jemi and John Holmes

Hong Kong has a list of over 550 recorded bird species. A high proportion of them are winter visitors and passage migrants, followed by resident birds; there are comparatively few summer visitors. The best time for watching birds is between September and May.

| | | |
|---|---|---|
| **Spring and Autumn migrants** | (46%) | These species pass through Hong Kong during their northward or southward migration. |
| **Winter visitors** | (25%) | Visitors which arrive in autumn and leave in spring |
| **Residents** | (11%) | These species spend the whole year in Hong Kong |
| **Summer visitors** | (4%) | Visitors which arrive in spring and leave in autumn |
| **Others (Vagrant and Occasional Visitors)** | (14%) | Species which visit Hong Kong occasionally, in some cases apparently as vagrants. |

Migration is a type of adaptation shown by animal species (including birds) to their living environment, weather, species population and density. It is a process which enables a species to search for alternative food supplies, increasing its chance of survival and its ability to take care of the next generation in a more suitable environment.

At higher latitudes in the Northern Hemisphere, food is more abundant during spring and summer. However food items such as insects and berries become increasingly difficult to find when the weather becomes cold in autumn, and extremely cold during winter. Millions of birds migrate from their northern breeding grounds in Siberia and northeast China to warmer in the southern areas (e.g. southern China, southeast Asia, Australia and New Zealand) where they spend the winter. They then return to their breeding grounds in early spring between March and May. Such birds are known as migrants.

Birds show different migration patterns: breeding grounds, breeding period and migration routes vary among different species. Acting out of instinct, birds are able to traverse immense natural barriers between breeding and wintering grounds. They frequently migrate in groups and make one or more stopovers en route to refuel. Some waterbirds and seabirds can achieve journeys of 29,000 km in one year! Hong Kong is an important stopover station or wintering ground for many migrants.

Migrating birds consume huge amounts of energy. Fat stored under the skin and internal organs is the main source of this energy. Successful migration depends on a number of factors, including ability to find suitable habitats, resource availability (food, water and shelter, in particular) and favourable environmental conditions (temperature, rainfall, predators and competitors, etc).

## 觀鳥的準備
## Preparing to Watch Bird

### 第一步：熟習辨認身邊的雀鳥
### Step 1: Become Familiar with Birds around You

開始觀鳥，最好就是在日常生活中練習辨認身邊出現的野鳥，例如在上學或上班途中，或偶然從窗外眺望。香港到處都有野鳥的蹤影，即使在繁忙的彌敦道上，隨時可以碰上五、六種雀鳥，例如八哥、原鴿、鵲鴝、紅耳鵯、樹麻雀、珠頸斑鳩等。

To start birdwatching, the basic step is to become familiar with birds that can be found around where you live, work or study. This can be practiced while on the way to work or school, or just by looking out of the window. Spotting birds is fun! Birds can be found everywhere in Hong Kong. Even on busy Nathan Road, five or six bird species including Crested Myna, Rock Dove, Magpie Robin, Red-whiskered Bulbul, Tree Sparrow, and Spotted Dove can be easily found within minutes.

### 第二步：配備工具
### Step 2: Get the Equipment

#### 光學儀器

一般來說，觀鳥主要的「工具」是我們的眼睛，但假如距離太遠，便需要借助光學儀器把雀鳥放大，以便清楚觀察。

1. 雙筒望遠鏡——用來觀察飛行中或近距離的雀鳥

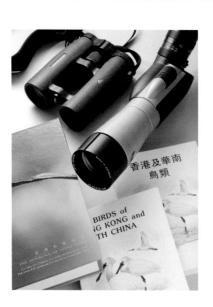

一般品牌都會在鏡上刻上一組數字，例如「10×40 7.3°」，「10×40」表示望遠鏡的放大倍數是10倍，物鏡直徑為40毫米；而「7.3°」則表示視場（鏡中可見的視野範圍）為7.3度。

選擇適當的雙筒望遠鏡需要考慮下列因素：

倍　　　數——觀鳥用的望遠鏡以7至10倍為佳。倍數太小難以看清楚細節，太大則無法穩定，影像亦較暗。物鏡直徑宜於30至50毫米之間，雖然物鏡愈大集光能力愈強，但太大則過於笨重，不利長時間使用。

相對亮度——相對亮度依（直徑 ÷ 倍數）$^2$公式計算，如10×40望遠鏡的相對亮度為16。相對亮度以9至25之間較理想。

鍍　　　膜——應選擇有透明鍍膜的望遠鏡。

視場角度——適宜在5.5度以上。

鏡身重量——觀鳥時望遠鏡掛於頸上，900克以下較為合適。

對　　　焦——應選擇手動調焦的望遠鏡，並且以中置對焦為佳。坊間有自動對焦或不用對焦的望遠鏡，不便用來觀察近處的雀鳥。最近對焦距離愈短愈好。

稜　　　鏡——傳統折角式稜鏡組合比較簡單，但是體積和重量都比較大。直筒式稜鏡構造緊密，稜鏡和鏡片不易移位，重量也較輕巧。不少直筒式稜鏡雙筒望遠鏡抗潮防塵能力較好，部分更經過充氮防水處理。

2. 單筒望遠鏡——用來觀察距離遠並且比較不活躍的雀鳥

單筒望遠鏡倍數較高,主要用來觀賞水鳥,因為距離通常較遠。物鏡直徑以60至80毫米為佳。如選用有變焦功能的目鏡,20至60倍變焦較為合適。

3. 三腳架

單筒望遠鏡必須架於三腳架上,三腳架需穩固,負重能力要高,以免在強風中抖動。可選擇有快速收放腳管的設計。

### 圖鑑

圖鑑幫助我們辨認雀鳥,以及提供鳥類的棲息地點、分佈範圍和行為習性等相關資料。一般來説,圖鑑分兩種,即攝影圖鑑和手繪圖鑑。除了本書之外,《香港及華南鳥類》(尹璉、費加倫、林超英著,政府新聞處,2006)(中文或英文版)也是觀鳥必然之選。

選擇鳥類圖鑑時,應考慮使用環境,如要拿到野外使用,可放在口袋的圖鑑會方便些,可能的話可以選購平裝版本,既實惠又輕便。

※ 呂德恆 Henry Lui

### 筆記簿、筆

▶ 筆記簿宜有硬皮、印有行線、袋裝大小、釘裝結實，另外可加一條橡皮
圈作書籤。應使用原子筆，避免用水筆，以免雨水令字跡變得模糊。

▶ 遇到未能辨認的鳥類，應立即做筆記，記下形態和特徵，然後向資深鳥
友請教，或者在香港觀鳥會網上討論區（www.hkbws.org.hk/BBS）留言討
論，交流經驗。

▶ 做筆記可以大大提升在野外辨識鳥類的能力。筆記內容愈詳細愈好，包
括日期、地點、天氣、鳥類特徵、形態、行為習性、叫聲、種群數量、
海拔高度等。

## Optical Instruments

Our eyes are the essential "equipment" for birdwatching. However we need optical instruments to watch birds that are too far away for the naked eye.

1. Binoculars – These are used for watching birds that are not too far away or birds in flight.

Binoculars carry a set of numbers such as $10 \times 40$  7.3°: $10 \times 40$ means the magnification is ten times, the diameter of the objective lens is 40mm, 7.3° means the binoculars have a field of view of 7.3°.

The following specifications are important in selecting a suitable pair of binoculars:

*Magnification -*    Binoculars for birdwatching normally have 7x to 10x magnification. Lower magnifications cannot show enough detail, while higher magnifications can create problems with vibration and a dimmer image. The objective lens should be 30mm to 50mm. A bigger lens means more weight and hence not suitable for prolonged use in the field.

*Relative brightness -* Relative brightness can be estimated from the formula (diameter of the objective lens $\div$ magnification)$^2$. A pair of 10 x 40 binoculars has relative brightness of 16. Values between 9 and 25 are most suitable for birdwatching.

*Coating -*    Binoculars with transparent coatings are preferred.

*Field of view -*    It is better to have an angle of view greater than 5.5°.

*Weight -*    Binoculars will usually be carried around the neck, normally the weight should be below 900g.

*Focus -*    Centre-focusing binoculars that focus both barrels at the same time are preferred. Auto-focus binoculars are not suitable for birdwatching. As for the shortest focusing distance, the closer is the better.

*Prism -*    Porro prism designs are simpler. However inherent from their design, porro prism binoculars are heavier. Roof prism binoculars are compact, light and comfortable to hold. Some roof prism binoculars are nitrogen-filled and are entirely dust-free and waterproof.

2. Telescopes – Telescopes have much higher magnification than binoculars but a smaller field of view. They are used for watching birds that are relatively inactive from great distances. Objective lenses between 60mm and 80mm are preferred. Where used, zoom eyepieces should be from 20x to 60x.

3. Tripod

A telescope should be placed on a sturdy tripod especially to reduce vibration caused by strong winds. Quick release features on some designs can facilitate deployment.

## Field Guides

A good field guide aids field identification, and also gives information on general characteristics such as habitat, distribution and behaviour. They come in two types: photographic guides and drawing guides. In addition, serious birdwatchers in Hong Kong should get a copy of "Birds of Hong Kong and South China" by C.Viney, K. Phillipps and C.Y. Lam, available in both Chinese and English editions.

Size matters - pocket size is preferred. Perfect-bind is also much better than hard cover, being cheaper and lighter to carry around.

### Notebook and Pen

▶ It is best to choose a small pocket-sized hardcover notebook with lines. A small rubber band can keep it open at the right place, while a ball-pen reduces the risk of ink running when the pages get wet.

▶ Use the notebook to take field notes on unfamiliar birds. Then seek help from experienced bird watchers, perhaps by posting your query on the forum of the Hong Kong Birdwatching Society website (www.hkbws.org.hk/BBS).

▶ Field notes should include the following information: date, place, weather, features, behaviour, calls, number, and altitude.

## 第三步：出發時的準備
## Step 3: Preparing for the Trip

避免穿着顏色太鮮艷的服飾，宜選擇綠、啡、藍等配合自然環境的顏色。

不同野鳥有不同的觀察時間，因此應在出發前了解目的地和路線，以便安排行程。觀察林鳥應在清晨時份，觀察海岸附近的濕地水鳥則要注意潮汐時間，宜於大潮前或後到泥灘附近守候。米埔泥灘的理想潮水高度約2.1米左右，尖鼻咀約為1.6米。觀賞猛禽可選擇中午時份到開闊原野，猛禽會利用從地面上升的熱空氣在空中盤旋。觀察農地或城市鳥類，宜於清晨或黃昏時份，因為

※ 呂德恆 Henry Lui

※ 呂德恆 Henry Lui

雀鳥在中午時不太活躍。海鳥可於夏季時到離岸小島附近海面遠距離觀察，千萬不要登島干擾雀鳥。

　　鳥種方面，出發前應搜集資料，了解當地的生態環境，配合當時的季節，在圖鑑上查閱可能會遇見的鳥種、辨識要點、常見程度等。準備愈充分，收穫愈豐富。

It is always the best to match the natural environment, that is wearing suitable colours, especially green, blue and brown. Avoid bright colours.

The best time for birdwatching is different for different habitats. If you are going to watch woodland birds, choose early morning when the birds are more active. Viewing wetland birds depends on the tide. Arrive at the birdwatching site before or after high tide. The best tidal level for waterbirds at Mai Po boardwalk is about 2.1m, while that for Tsim Bei Tsui is about 1.6m. If you are looking for raptors, it is best to go to more open locations around noon, when these birds soar using convection currents. For farmland or urban birds, early morning and late afternoon are the best time. Birds are generally not too active around noon. For seabirds, we need to go to remote islands during summer. Good views can be obtained from a boat. It is not necessary to land on the islands, and doing so disturbs the birds.

To prepare for a successful outing, it is good to collect information on habitat and season, and to search the field guide for information on possible species, their identification and abundance. Good preparation leads to productive trips.

## 觀鳥小貼士
## Tips of watching birds

發現鳥蹤時，立即保持靜止，原地舉起望遠鏡觀察，動作不要過大。如距離太遠，可輕步走近目標觀察，但切記點到即止，不要干擾雀鳥。

使用雙筒望遠鏡的正確方法，是先用眼睛尋找鳥的位置並盯緊，然後舉起望遠鏡瞄準和對焦。舉鏡前要同時留意鳥的位置及周圍的物件，如樹枝等，以便在鏡中定位。舉鏡後可能要作窄幅度上下掃瞄找尋目標，多加練習便可以很快上手。

When you find a bird, avoid unnecessary movement. Look at the bird immediately using your binoculars. If it is too far away, try to approach it indirectly and look again. Always avoid unnecessary disturbance to birds .

To use binoculars: keep looking at the bird, then hold the binoculars to the eyes, adjust the position and focus at the same time. Speed in locating birds with binoculars can also be increased by a quick scan to establish the location of large branches or objects in the vicinity, helping confirm where exactly the bird is.

## 提交觀鳥記錄
## Submit your Records

　　香港觀鳥會的紀錄委員會自1957年起收集香港的鳥類記錄，覆核鳥類狀況、反映環境變化等資料，這些資料對鳥類的保護工作及自然保育有莫大幫助。

　　本會鼓勵任何人士每次觀鳥後都整理觀鳥記錄及轉交本會，記錄的鳥種不一定是罕見雀鳥，事實上我們正需要很多普通鳥類的記錄，以便掌握本地鳥種數量、遷徙分佈和趨勢，以及展開相關的調查工作。

　　你可以在本會網頁(www.hkbws.org.hk)下載記錄表格，這個檔案亦包括最新的香港鳥類名錄，另外亦同時上載罕有雀鳥記錄表格，用作遞交罕見記錄。表格可以電郵(recorder@hkbws.org.hk)呈交或郵寄給本會，地址：香港九龍荔枝角青山道532號偉基大廈7樓C室。

Since 1957, the Records Committee of the Hong Kong Bird Watching Society has collected records of birds of Hong Kong, and reviewed their status. All this information is very important for the conservation of birds and the protection of their habitats.

We encourage readers to submit records to us after each birding trip. We welcome records of both common and rare species, that help us analyze migration patterns and population trends, as well as provide useful background information for us to initiate relevant studies.

You can download the record sheet from our website (www.hkbws.org.hk). The record sheet also includes an updated checklist of the birds of Hong Kong. A record form for rare species is also available there. Records can be submitted by e-mail (recorder@hkbws.org.hk) or by mail to: 7C, V Ga Building, 532 Castle Peak Road, Lai Chi Kok, Kowloon, Hong Kong.

## 觀鳥比賽
## Bird Watching Competition

　　香港歷史最悠久的觀鳥比賽，要算是世界自然基金會香港分會在1984年首次舉辦的「香港觀鳥大賽」(The Big Bird Race)，目的是為米埔自然保護區籌款，當時只有兩隊參加比賽，分別是香港觀鳥會隊及米埔隊。自此以後，該比賽成為一年一度的觀鳥界盛事，至今仍有舉辦，還不時加插不同的盃賽，讓不同界別或程度的觀鳥者參與。

　　香港觀鳥會自2002年起亦舉辦觀鳥比賽，主要作為會員及鳥友間的活動。比賽錄得的鳥種亦作為當年全港鳥類速查的一個重要記錄，協助我們了解香港的鳥況。香港觀鳥會的觀鳥比賽名稱亦曾多番改動，最早期名「猜尋呈」香港觀鳥記錄比賽，及後改名「香港觀鳥大賽」，近年則稱為「觀鳥馬拉松」，並希望藉此籌募經費。

　　觀鳥比賽的形式會因應參賽者的不同界別或程度而略有調整，不過大致沿用早年的「觀鳥大賽」規則演變出來。「經典」的觀鳥比賽規則是由大約四人組成一隊，亦可以額外邀請朋友任司機或記錄員，但不能參與比賽，比賽時間由大會指定的開始時間連續進行24小時，並自行計劃到全港任何地方記錄見到或聽到的鳥種。比賽中各隊員要同行，鳥種記錄需要三位或以上組員同時見到或聽到才有效。比賽完成，全隊要同時在指定地點報到，遲到則扣減分數。之後則有數十分鐘整理觀鳥記錄冊(log book)，再遞交大會評審。

　　比賽完結後多會於當日安排聚餐，頒獎禮也會在聚餐中進行。比賽結果以記錄的鳥種數目排序，設冠、亞、季軍。由於是籌款活動，也設有最高籌款獎

項。此外，為增加趣味性，每隊需要選出其中一種雀鳥作為比賽的最佳鳥種，以及一個最遺憾記錄不到的鳥種。評判會各選出一隊，分別成為最佳雀鳥大獎（Bird of the Day）及糊塗蟲大獎（Dip of the Day）的得主。

頒獎典禮中，各隊會分享比賽的過程及心得，有時更有海外觀鳥者參與，是觀鳥者互相認識和切磋的好時機。聚餐暨頒獎禮通常作為一個年度的大型聯誼活動，友誼第一，比賽第二，大家都非常享受相聚的時刻。

或許有不少人感到疑惑，觀鳥隊伍自行決定到任何地方觀鳥，裁判沒有可能24小時跟隨每隊，豈不是很容易作弊？怎樣知道觀鳥記錄是否正確？首先要明白，甚麼時間、地方會有甚麼雀鳥是有一定模式和歷史記錄可稽查及參考，而參賽隊伍到訪的觀鳥熱點來來去去不外是那幾個，裁判可從不同隊伍的記錄參考對照，只有一隊記錄到某一鳥種的機會不太多，裁判若有疑問，也會聯絡參賽隊伍問個究竟。而參賽隊伍多是有經驗的觀鳥者，大家的觀鳥水平互相也略有認識，過分奇怪的記錄不難被評判或其他隊伍識破。當然要作弊也並非不可能，不過要有一定經驗的觀鳥者，才可偽造有板有眼的記錄，可是經驗豐富和能力高超的觀鳥者又何需作弊呢？畢竟，觀鳥比賽是一個以「誠信」為最高守則的「牙骹戰」，多於一個為獎品以戰的比賽，獎品也只是一些紀念品，比賽本身是一個籌款活動，觀鳥者亦多會視作一個高高興興的聯誼活動，誰會去作弊破壞氣氛呢？

The longest bird watching competition history in Hong Kong should give credit to the first held Hong Kong Bird Watching Competition, The Big Bird Race, by World Wide Fund for Nature Hong Kong in 1984 with a target to raise fund for Mai Po Nature Reserve. There are only two racing teams participating the competition at the time. They are Hong Kong Bird Watching Society Team and the Mai Po Team respectively. Since then, the competition turns to be an annual apex of bird watching activity. It is still holding currently, different categories of competitions are added from time to time to attract participants from different society sectors or those with various degree of skills.

Hong Kong Bird Watching Society also holds bird watching competition itself starting from 2002 mainly providing activity for members and bird watching fans. Bird species recorded in the competition at the same time becomes an important reference to understand the bird situation in Hong Kong. The name of the Bird Watching Competition held by Hong Kong Bird Watching Society has changed for several times. The latest name changed as Bird Watching Marathon, with a task to raise fund for the Society.

Bird watching competition format will adjust according to sectors or skill levels of participants, nevertheless the regulations are mainly evolved those from The Bird Watching Race held earlier years. The classic bird watching regulations will

have a team formed with around four members, extra friends as driver or record keeper can be invited as supporting. The competition duration is set by the host, from the defined commencing time and proceeds continuously for 24 hours, with self-planning route anywhere in Hong Kong to record what are the bird species seen or heard. Team members are required to do the outing altogether during the competition, valid bird species record should be seen or heard at least with 3 or more participants at the same time. Full team members are to report to the designated place when the competition finishes, mark will be deducted for late arrival. After that, teams will have at least 30 minutes to tidy up the bird watching log book before submitting for host review.

Feast gathering usually will be held on the same day after the competition finishes, award ceremony is also arranged in the midst. Competition result bases on the chronological order per the number records of bird species, awards are champion, the first runner-up and the second runner-up. Owning to being a fund raising activity, there is highest fund raising prize too. Besides, to enhance amusement feature, each team needs to choose the best bird species and name the other bird species make you feel most regret unable to record. The judge will pick one team each as the winner, namely Bird of the Day as well the Dip of the Day respectively.

During the award ceremony, each team will share their competition processes and experiences. Sometimes we even have overseas bird watching participants, it is an excellent chance for birdwatchers knowing and communicating with each other. Feast gathering along with award presentation is usually regarded as a large scale of friend-making activity on an annual basis. Friendship comes first and competition second, all of us extremely enjoy this moment of gathering.

The arrangement may confuse quite a few people as bird watching teams make own decision to go anywhere to track birds. It is impossible for the judges to follow each single team for 24 hours continuously, otherwise cheating may happen! How come do we know if bird watching records are correct? Firstly, we need to understand that what time and where any type of birds appear are having their typical modes and historical records that is available for cross-check and reference. Actually there are only several hot spots that are visited by birdwatchers, judges can make reference to records of different teams, the chance is remote if any specific bird species is recorded by only one team. Should judges have doubt, they will contact and ask participating team for details. Since participants are mostly experienced and have some preliminary understanding of the bird watching skill among themselves. Extreme strange record is easily caught by judges and other teams. Nevertheless cheating may possible, it of course requires bird watcher having a certain level of experience to make up reasonable-look record. Whereas there is no need to cheat for those having experience and with great capability! All in all, bird watching competition is a gentleman competition working on individual integrity as the supreme principle, rather than a competition fighting for a prize. And the prizes are just small souvenirs. Competition being a fund-raising activity, bird watching participants would tend to regard it as an enjoyable friend-making activity. Who would care to cheat and spoil this friendly atmosphere?

# 觀鳥及鳥類攝影守則
# Code of Conduct for Birdwatching and Bird Photography

為了減少觀鳥活動或鳥類攝影對雀鳥的干擾，香港觀鳥會制訂了一套守則供市民參考，希望可以作為上述活動一套良好行為的模範。

## 1. 以鳥為先

無論是進行觀鳥活動或鳥類攝影，應盡量以不影響鳥類的正常活動為原則，以免造成干擾。

a. 如果發現雀鳥顯得不安，有規避或其他異常反應，便要馬上停止；
b. 如果觀看或拍攝的人太多，更應特別注意；
c. 不要嘗試影響雀鳥的行為，例如驚嚇、驅趕或使用誘餌；
d. 少用閃光燈；
e. 不要破壞自然環境。

## 2. 保護敏感地點

雀鳥的營巢地點、海鳥繁殖的小島、稀有鳥種停棲的地點等都特別容易受到干擾，要加倍留意。

a. 保持適當距離，避免令雀鳥受到脅逼；
b. 不要登上有海鳥繁殖的小島；
c. 不要干擾鳥巢或周圍的植被，以免親鳥棄巢或招來天敵襲擊；
d. 不要隨便公開或透露敏感地點的位置，向不認識守則的人清楚解釋，以免帶來干擾；
e. 留意自己的行為，以防招惹好奇的人干擾。

## 3. 舉報干擾

如果發現有人干擾或傷害雀鳥，在安全情況下宜向他們解釋和勸止。如果未能阻止，請拍照記錄，並盡快向漁農自然護理署或致電1823舉報。

## 4. 尊重他人

a. 避免干擾其他在場觀鳥和拍攝的人，讓大家都可以享受其中的樂趣；
b. 小心不要破壞當地的設施或農作物。

Since birds are sensitive to disturbance, special care is required to avoid bringing disturbance to their lives. In order to provide a model for good practices in birdwatching and bird photography, HKBWS has drawn up the following code as a reference.

## 1. The Welfare of Birds Comes First

Birdwatching and bird photography should be carried out with minimum interference to the birds. Disturbance must be avoided as far as possible.

a. Stop if the birds appear disturbed, begin to move away or exhibit other abnormal reaction;
b. Exercise additional precautions when the activity is undertaken with a large group of people;
c. Do not attempt to influence the behaviour of birds, e.g. by flushing, chasing or baiting;
d. Use flash only sparingly;
e. Do not damage the natural environment.

## 2. Protect Sensitive Sites

Sites such as nests, seabird colonies and the roosts of rarities are particularly vulnerable. Take extra care to minimize disturbance.

a. Keep a suitable distance to avoid stressing the birds;
b. Do not land on islands with breeding colonies;
c. Do not disturb nests and their surrounding vegetation, or the nest could be abandoned or become exposed to predators;
d. Share information about the sites with discretion and do not reveal it casually in public. Explain clearly to those who may not understand the Code, to avoid bringing disturbance to the site;
e. Beware that your actions may attract unwanted attention and hence disturbance to the site.

## 3. Report Disturbances

If you find people disturbing or causing harm to birds, advise against the act when it is safe to do so. If disturbance cannot be stopped, take photos and report to the Agriculture, Fisheries and Conservation Department or report by giving phone call to 1823 as soon as possible.

## 4. Respect Others

a. Share the fun – avoid disturbing other birdwatchers or photographers on site;
b. Take care not to damage facilities or crops at the site.

# 觀鳥地點
# Places to Visit

| 觀鳥地點<br>Birdwatching spots | 交通<br>Transportation | 目標鳥種<br>Target species | 預計觀鳥時間<br>Time required at the site |
|---|---|---|---|
| **🏛🏠 濕地 Wetlands** | | | |
| 尖鼻咀<br>Tsim Bei Tsui | 在元朗泰豐街乘35號專線小巴於尖鼻咀下車。<br>Take the green minibus no. 35 from Tai Fung Street, Yuen Long. Get off at terminal at Tsim Bei Tsui. | 多種猛禽和濕地鳥類如黑臉琵鷺；傍晚時份有機會看到鵰鴞。<br>A wide variety of raptors and wetland birds, e.g. Black-faced Spoonbill. Possibility of Eurasian Eagle-Owl in the evening. | 4小時<br>4 hours |
| 南生圍<br>Nam Sang Wai | 元朗乘往上水的76K巴士或17號紅色小巴，在南生圍路口下車。<br>Take the bus no. 76K or red minibus no. 17 from Yuen Long to Sheung Shui. Get off at the main road into Nam Sang Wai, then walk. | 濕地鳥類<br>Wetland birds | 2小時<br>2 hours |
| 米埔自然護理區<br>Mai Po Marshes Nature Reserve | 在元朗或上水乘76K巴士或17號小巴(上水新發街或元朗水車館街)，在米埔村下車，沿担竿洲路步行20分鐘便可到達(需持有由漁農自然護理署發出的「進入米埔沼澤區許可證」才可進入保護區)。<br>Take the bus no. 76K in Sheung Shui or Yuen Long, or red minibus no. 17 from Sheung Shui (San Fat Street) or Yuen Long (Shui Che Kun Street). Get off at Mai Po Village. Follow Tam Kon Chau Road and walk for about 20 minutes. (It is a legal requirement for all visitor to have a "Mai Po Entry Permit" issued by the AFCD.) | 多種濕地鳥類，如黃嘴白鷺、黑臉琵鷺、黑嘴鷗、勺嘴鷸、小青腳鷸等；以及猛禽，如白肩鵰和烏鵰。<br>A wide variety of wetland birds, such as Chinese Egret, Black-faced Spoonbill, Saunders's Gull, Spoon-billed Sandpiper and Nordmann's Greenshank, as well as raptors such as Eastern Imperial Eagle and Greater Spotted Eagle. | 5小時<br>5 hours |
| 米埔新村/担竿洲<br>Mai Po San Tsuen / Tam Kon Chau | 米埔附近的魚塘，交通同上。在米埔村下車，然後沿担捍洲路步行，沿途觀鳥。<br>Fishponds around Mai Po. Transport as above, then explore the fishponds along the access road. | 濕地鳥類及多種鵐<br>Wetland birds and buntings | 2小時<br>2 hours |

| 觀鳥地點<br>Birdwatching spots | 交通<br>Transportation | 目標鳥種<br>Target species | 預計觀鳥時間<br>Time required at the site |
|---|---|---|---|
| 塱原<br>Long Valley | 上水乘76K巴士；或於上水港鐵站鄰近的小巴站乘搭綠色專線小巴50K、51K或55K；在燕崗村下車(進入青山公路後第一條行人天橋)，沿右面小路進入。<br>Take the bus no. 76K or green minibus no. 50K, 51K or 55K from Sheung Shui MTR Station. Get off at the first footbridge at Yin Kong Tsuen, not long after the bus turns right to Castle Peak Road. Walk along the path at the right hand side. | 彩鷸、扇尾沙錐、棕扇尾鶯、田鵯、藍喉歌鴝和絲光椋鳥<br>Greater Painted-snipe, Common Snipe, Zitting Cisticola, Richard's Pipit, Bluethroat and Red-billed Starling. | 2-3小時<br>2-3 hours |
| 鹿頸、南涌<br>Luk Keng, Nam Chung | 粉嶺港鐵站乘56K往鹿頸的專線小巴，總站下車。夏季時，沿鹿頸路途中可眺望鴉洲鷺林。<br>Take the green minibus no. 56K from Fanling MTR station to Luk Keng. A Chau (with an egretry in summer) is visible along Luk Keng. | 普通翠鳥、白胸翡翠、藍翡翠、斑魚狗、黑冠鵑隼和多種鷺鳥<br>Common Kingfisher, White-throated Kingfisher, Black-capped Kingfisher, Pied Kingfisher, Black Baza, egrets and herons | 2小時<br>2 hours |
| 荔枝窩<br>Lai Chi Wo | 大埔墟港鐵站乘綠色專線小巴往烏蛟騰，然後步行經上苗田、下苗田、三椏涌往荔枝窩村。<br>Take the green minibus from Tai Po Market MTR station to Wu Kau Tang. Walk though Sheung Miu Tin, Ha Miu Tin, Sham A Chung to Lai Chi Wo. A long hike. | 冠魚狗、蛇鵰和多種鷺鳥<br>Crested Kingfisher, Crested Serpent Eagle, egrets and herons | 8小時<br>8 hours |
| 香港濕地公園<br>Hong Kong Wetland Park | 天水圍鐵路站轉乘輕鐵705或706號，於濕地公園站下車。<br>Take the Light Rail no. 705 or 706. Get off at Wetland Park Station. | 多種水鳥如黑臉琵鷺、水雉及彩鷸<br>A wide variety of wetland species such as Black-faced Spoonbill, Pheasant-tailed Jacana and Greater Painted-snipe | 2小時<br>2 hours |

| 觀鳥地點<br>Birdwatching spots | 交通<br>Transportation | 目標鳥種<br>Target species | 預計觀鳥時間<br>Time required at the site |
|---|---|---|---|
| 船灣、洞梓、汀角<br>Shuen Wan, Tung Tsz, Ting Kok | 大埔港鐵站乘75K往大美督巴士在三門仔、洞梓或汀角下車。<br>Bus no. 75K from Tai Po Market MTR Station to Sam Mun Tsai, Tung Tsz or Ting Kok | 濕地鳥類<br>Wetland birds | 3小時<br>3 hours |

## 🏔 高地 Uplands

| 觀鳥地點<br>Birdwatching spots | 交通<br>Transportation | 目標鳥種<br>Target species | 預計觀鳥時間<br>Time required at the site |
|---|---|---|---|
| 飛鵝山<br>Fei Ngo Shan (Kowloon Peak) | 彩虹港鐵站乘計程車經飛鵝山道往基維爾營，在往基維爾營的標誌下車，然後步行下山回彩虹或沿衛奕信徑往西貢蠔涌方向步行下山。<br>Best to take a taxi from Choi Hung MTR Station to Gilwell Campsite. Watch birds around the hillside, walk along the road back to Choi Hung or down the other side to Ho Chung. | 山鷚和中華鷓鴣<br>Upland Pipit and Chinese Francolin | 2小時<br>2 hours |
| 大帽山<br>Tai Mo Shan | 荃灣港鐵站乘51號巴士，在大帽山郊野公園路口下車，步行上山；或乘計程車前往大帽山近山頂的閘口，再往山上觀鳥或步行下山。<br>Take the bus no. 51 from Tsuen Wan MTR Station. Get off at the sign at the entrance to Tai Mo Shan Country Park. Alternatively, take a taxi to the last barrier of the Park, then walk along the concrete road to the summit. | 山鷚、鷓鴣、大草鶯和棕頭鴉雀<br>Upland Pipit, Chinese Francolin, Chinese Grassbird and Vinous-throated Parrotbill | 2小時<br>2 hours |

## 🌳 開闊原野 Open Country

| 觀鳥地點<br>Birdwatching spots | 交通<br>Transportation | 目標鳥種<br>Target species | 預計觀鳥時間<br>Time required at the site |
|---|---|---|---|
| 榕樹澳<br>Yung Shue O | 沙田市中心乘299號往西貢的巴士，在水浪窩下車，沿往榕樹澳的車路步行約30分鐘。<br>Take the bus no. 299 from Sha Tin Town Centre to Sai Kung. Get off at Shui Long Wo, there is a road leading to Yung Shue O. It might take half an hour to walk there. | 中華鷓鴣、松鴉、小鴉鵑、褐魚鴞、普通夜鷹和領角鴞<br>Chinese Francolin, Eurasian Jay, Lesser Coucal, Brown Fish Owl, Grey Nightjar and Collared Scops Owl | 2小時<br>2 hours |

| 觀鳥地點<br>Birdwatching spots | 交通<br>Transportation | 目標鳥種<br>Target species | 預計觀鳥時間<br>Time required at the site |
|---|---|---|---|
| 沙螺洞<br>Sha Lo Tung | 大埔墟港鐵站乘巴士74K，在鳳園下車，沿路上山；或可在大埔墟乘計程車直接前往沙螺洞。山上可通往鶴藪水塘。<br>Take the bus no. 74K from Tai Po Market MTR Station. Get off at Fung Yuen and walk up the hill. Alternatively, take a taxi to Sha Lo Tung directly. A trail from there leads to Hok Tau Reservoir. | 金頭扇尾鶯和多種鵐<br>Golden-headed Cisticola and buntings | 3小時<br>3 hours |
| 洲頭<br>Chau Tau | 元朗或上水乘76K巴士，或上水新發街或元朗水車館街乘17號小巴，在洲頭下車，沿路上山。<br>Take the bus no. 76K or red minibus no. 17 from Sheung Shui (San Fat Street) or Yuen Long (Shui Che Kun Street). Get off at Chau Tau and walk towards the hills. | 白腹隼鵰、中華鷓鴣和小鴉鵑 (小毛雞)；傍晚時份有機會看到林夜鷹和鵰鴞<br>Bonelli's Eagle, Chinese Francolin and Lesser Coucal. Possibility of Savanna Nightjar and Eurasian Eagle-owl in the evening. | 2小時<br>2 hours |

※ 郭子祈 Kwok Tsz Ki

| 觀鳥地點<br>Birdwatching spots | 交通<br>Transportation | 目標鳥種<br>Target species | 預計觀鳥時間<br>Time required at the site |
|---|---|---|---|
| ≋ 🌳 林地 、溪流 Woodlands and Streams | | | |
| 香港仔水塘<br>Aberdeen Reservoir | 香港仔市中心沿石排灣道步行上山，另一方便的方法是乘坐計程車直接前往香港仔郊野公園或任何往石排灣邨的交通工具。<br>From the Central Aberdeen, walk uphill along Shek Pai Wan Road to reach the entrance of Aberdeen Country Park. Alternatively, take a taxi to the Country Park directly or any available transport to Shek Pai Wan Estate. | 大嘴烏鴉、夜鷺和黑鳶 (麻鷹)<br>Large-billed Crow, Black-crowned Night Heron and Black Kite | 2 小時<br>2 hours |
| 摩星嶺<br>Mount Davis | 往堅尼地城乘 5B 或 47 號巴士到摩星嶺總站，從山腳上山；或乘計程車往山頂青年旅舍，步行下山。<br>Take the bus no. 5B or 47 to Kennedy Town. Get off at the terminal, and walk up Mount Davis Road. Alternatively, take a taxi to the Youth Hostel near the summit of of Mount Davis and walk down. | 髮冠卷尾、紅嘴藍鵲以及多種過境遷徙鳥<br>Hair-crested Drongo, Red-billed Blue Magpie and various passage migrants | 2 小時<br>2 hours |
| 龍虎山和柯士甸山<br>Lung Fu Shan and Mount Austin | 穿過香港大學校園，沿克頓道往山頂，或到達龍虎亭後，沿小山路下山。<br>From the University of Hong Kong, walk up Hatton Road towards the Peak, or from Mount Austin walk downhill and follow the narrow tracks to the left after reaching Lung Fu Pavilion. | 鳳頭鷹、黑喉噪鶥、紅嘴藍鵲、烏鶇、烏灰鶇和灰背鶇<br>Crested Goshawk, Black-throated Laughingthrush, Red-billed Blue Magpie, Chinese Blackbird, Japanese Thrush and Grey-backed Thrush | 2 小時<br>2 hours |
| 馬己仙峽<br>Magazine Gap | 中環交易廣場乘 15 號往山頂巴士在舊山頂道和僑福道交界的車站下車。<br>Take the bus no. 15 from Exchange Square, Central. Get off at the stop near the junction of Old Peak Road and Guildford Road. | 傍晚時份觀賞晚棲黑鳶 (麻鷹)<br>Roosting Black Kites at sunset | 1 小時<br>1 hour |

| 觀鳥地點<br>Birdwatching spots | 交通<br>Transportation | 目標鳥種<br>Target species | 預計觀鳥時間<br>Time required at the site |
|---|---|---|---|
| 大埔滘自然護理區<br>Tai Po Kau Nature Reserve | 大埔墟港鐵站乘72、72A、73A或74A巴士或乘計程車前往。<br>Take the bus no. 72, 72A, 73A or 74A or taxi from Tai Po Market MTR Station. | 多種林區鳥類，如海南藍仙鶲、白眉鵐、黑領噪鶥、黃頰山雀、山鶺鴒、橙腹葉鵯、白腹鳳鶥、絨額鳾、紫綬帶、灰喉山椒鳥、赤紅山椒鳥、藍翅希鶥和叉尾太陽鳥<br>Many woodland species such as Hainan Blue Flycatcher, Tristram's Bunting, Greater Necklaced Laughingthrush, Yellow-cheeked Tit, Forest Wagtail, Orange-bellied Leafbird, White-bellied Erpornis, Velvet-fronted Nuthatch, Japanese Paradise Flycatcher, Grey-chinned Minivet, Scarlet Minivet, Blue-winged Minla and Fork-tailed Sunbird | 4小時<br>4 hours |
| 城門水塘<br>Shing Mun Reservoir | 荃灣港鐵站附近兆和街乘82號專線小巴前往城門郊野公園。<br>Take the green minibus no. 82 from Siu Wo Street near Tsuen Wan MTR Station terminating at Shing Mun Country Park. | 多種林區鳥類，如海南藍仙鶲、紅頭穗鶥和棕頸鈎嘴鶥<br>Many woodland species such as Hainan Blue Flycatcher, Rufous-capped Babbler and Streak-breasted Scimitar Babbler | 3小時<br>3 hours |
| 梧桐寨<br>Ng Tung Chai | 太和港鐵站乘64K往元朗巴士，在梧桐寨站下車，沿路入梧桐寨村，經萬德寺上山，沿路經過多個瀑布。沿路可步行上大帽山頂。<br>Take the bus no. 64K from Tai Wo MTR Station. Get off at Ng Tung Chai bus stop. Follow path across farmland uphill to Ng Tung Chai Village, from where a road leads to a temple and further to the waterfalls and eventually to Tai Mo Shan. | 灰鶺鴒、灰樹鵲、日本歌鴝和棕腹大仙鶲<br>Grey Wagtail, Grey Treepie, Japanese Robin and Fujian Niltava | 3小時<br>3 hours |

| 觀鳥地點<br>Birdwatching spots | 交通<br>Transportation | 目標鳥種<br>Target species | 預計觀鳥時間<br>Time required at the site |
|---|---|---|---|
| 碗窰<br>Wun Yiu | 大埔墟港鐵站步行，穿過運頭塘村，到達運路，再沿路步行至碗窰。<br>Walk through Wan Tau Tong Estate from Tai Po Market MTR Station. On reaching Tat Wan Road, walk uphill to Wun Yiu. | 紅尾水鴝、銅藍鶲、三寶鳥、蛇鵰、赤紅山椒鳥和灰喉山椒鳥<br>Plumbeous Water Redstart, Verditer Flycatcher, Oriental Dollarbird, Crested Serpent Eagle, Scarlet Minivet and Grey-chinned Minivet | 1小時<br>1 hour |
| 大蠔河<br>Tai Ho River | 大嶼山東涌港鐵站乘計程車到白芒村，沿路經白芒學校、牛牯塱村到達大蠔灣。沿路上山步行1.5小時可到達梅窩，乘坐小輪返市區。<br>Take a taxi from Tung Chung MTR Station to Pak Mong Village. Walk along village trail to Tai Ho Wan passing Pak Mong School and Ngau Kwu Long. Then walk for about 1.5 hours uphill and down the other side to Mui Wo. There is a regular ferry back to Central. | 三寶鳥、岩鷺及多種冬候鳥為主的林區鳥類<br>Oriental Dollarbird, Pacific Reef Heron and various woodland species in particular, winter visitors | 4小時<br>4 hours |

※ 郭子祈 Kwok Tsz Ki

| 觀鳥地點<br>Birdwatching spots | 交通<br>Transportation | 目標鳥種<br>Target species | 預計觀鳥時間<br>Time required at the site |
|---|---|---|---|
|  海洋、沿岸和海島 Ocean, Coastal Areas and Islands | | | |
| 蒲台<br>Po Toi | 可在香港仔碼頭或赤柱乘船前往。可預先查詢最新班次。<br>Take a ferry at Aberdeen Pier or Stanley. You may check the updated ferry schedule. | 多種海鳥，包括黑枕燕鷗、褐翅燕鷗和紅頸瓣蹼鷸，還有罕見的過境遷徙鳥，例如黑鳽、黑枕黃鸝、黃眉姬鶲、灰山椒鳥等<br>Seabirds such as Black-naped Tern, Bridled Tern and Red-necked Phalarope, as well as rare passage migrants, such as Black Bittern, Black-naped Oriole, Narcissus Flycatcher, Ashy Minivet | 3小時<br>3 hours |
| 塔門<br>Tap Mun | 大學港鐵站步行15分鐘至馬料水碼頭，乘小輪前往（最早開8:30）。或由西貢黃石碼頭乘搭小輪前往。可預先查詢最新班次。<br>Take the earliest ferry from Ma Liu Shui Pier (15-minute walk from University MTR Station) at 8:30am daily. There is also a regular ferry from Wong Shek Pier in Sai Kung East Country Park. You may check the updated ferry schedule. | 於船上觀察白腹海鵰、岩鷺，以及海鳥和燕鷗等鳥類<br>White-bellied Sea Eagle, Pacific Reef Heron and terns from the ferry | 3小時<br>3 hours |
| 東平洲<br>Tung Ping Chau | 大學港鐵站步行15分鐘至馬料水碼頭，乘小輪前往，早上9:00開船，只於週六、日及公眾假期開出，當日有一班於下午5:15開出的回程船。可預先查詢最新班次。<br>Take a ferry leaving from Ma Liu Shui Pier (15-minute walk from University MTR Station). Ferries are only available on Saturdays, Sundays and public holidays, leaving at 9:00am. There is only one retrun boat at 5:15pm on the same day. | 燕鷗、白腹海鵰和春秋過境遷徙鳥，如海鳥和棕尾褐鶲<br>Terns, White-bellied Sea Eagle and passage migrants such as seabirds and Ferruginous Flycatcher | 8小時<br>8 hours |

| 觀鳥地點<br>Birdwatching spots | 交通<br>Transportation | 目標鳥種<br>Target species | 預計觀鳥時間<br>Time required at the site |
|---|---|---|---|

### 🏢🌳 市區 Urban Area

| 觀鳥地點<br>Birdwatching spots | 交通<br>Transportation | 目標鳥種<br>Target species | 預計觀鳥時間<br>Time required at the site |
|---|---|---|---|
| 香港公園<br>Hong Kong Park | 金鐘港鐵站太古廣場後面。<br>Easily accessed from Admiralty MTR Station, or by other means; next to Pacific Place shopping centre. | 小葵花鳳頭鸚鵡、紅領綠鸚鵡、黃眉柳鶯、叉尾太陽鳥和紅嘴藍鵲<br>Yellow-crested Cockatoo, Rose-ringed Parakeet, Yellow-browed Warbler, Fork-tailed Sunbird and Red-billed Blue Magpie | 1小時<br>1 hour |
| 香港動植物公園<br>Hong Kong Zoological and Botanical Gardens | 中環港鐵站步行到花園道。公園入口位於花園道上山方向的右邊。<br>From Central MTR Station, walk up Garden Road. The Gardens are on the right. | 小葵花鳳頭鸚鵡、黃眉柳鶯、叉尾太陽鳥和紅嘴藍鵲<br>Yellow-crested Cockatoo, Yellow-browed Warbler, Fork-tailed Sunbird and Red-billed Blue Magpie | 1小時<br>1 hour |
| 九龍公園<br>Kowloon Park | 入口位於尖沙咀港鐵站A1出口。<br>Enter from Tsim Sha Tsui MTR Station Exit A1. | 夜鷺、白胸翡翠，亞歷山大鸚鵡、黃眉柳鶯、紫綬帶、叉尾太陽鳥、灰背鶇、紅嘴藍鵲和黑領椋鳥<br>Black-crowned Night Heron, White-throated Kingfisher, Alexandrine Parakeet, Yellow-browed Warbler, Japanese Paradise Flycatcher, Fork-tailed Sunbird, Grey-backed Thrush, Red-billed Blue Magpie and Black-collared Starling | 2小時<br>2 hours |
| 彭福公園<br>Penfold Park | 沙田火炭港鐵站附近。<br>Near Fo Tan MTR Station. | 常見市區鳥類及多種鷺鳥<br>Many common urban bird species, and also egrets and herons | 2小時<br>2 hours |

※ 郭子祈 Kwok Tsz Ki

# 鳥種分類
# Family Description

## 鷓鴣、鵪鶉和雉雞
## Francolins, Quails and Pheasants

### 雉科 Phasianidae（雞形目 Galliformes）

外形笨重，頭部細小，大部分時間在陸地行走。不易觀察，但叫聲明顯易認。

Heavy-looking birds with small heads. Spend a lot of time on the ground. Difficult to see but easy to identify by their call.

※ 陳佳瑋 Chan Kai Wai

## 鴨、雁和天鵝 Ducks, Geese and Swans

### 鴨科 Anatidae（雁形目 Anseriformes）

嘴寬而扁平，腳短，多位於體後部，前三趾間有蹼，喜歡漂浮於水面，善泳。兩性外貌同色或異色，異色者雄鴨較雌鴨大，色彩亦較鮮艷。翅上多有翼鏡，呈金屬光澤。食物種類多樣，大部分為雜食性，食物有雜草種子、水生植物、昆蟲、貝類、魚類和兩棲類等。飛行時拍翼急速。

Aquatic birds characterized by broad flat bills. Their legs are short with webs between the first 3 toes. Good dabblers or divers. Sexes alike or differ. For those sexes differ, male ducks generally are brighter in colours. Most species have glossy speculum on wings. Mostly omnivorous, feed on grass seeds, aquatic plants, insects, shellfish and amphibians. Fly with rapid wing beats.

※ 江敏兒、黃理沛 Michelle and Peter Wong

## 夜鷹 Nightjars

### 夜鷹科 Caprimulgidae（夜鷹目 Caprimulgiformes）

與鴞形目一樣屬夜行性鳥類，但嘴部較弱，基部寬闊，捕食飛行中的昆蟲。嘴鬚長而發達。飛羽較長而闊，翅尖亦長。喜棲息於空曠草地，能靜止於地上或樹枝上，保護色極佳。

Nocturnal. Wide, weak bill with long bristles for catching insects in flight. Long, broad and pointed wings. Favour open grassland. Well camouflaged on ground or in trees.

※ 江敏兒、黃理沛 Michelle and Peter Wong

＊此鳥種分類是按兩冊介紹的雀鳥而定。
　Family description is according to the birds mentioned in the guide book.

## 雨燕 Swifts

### 雨燕科 Apodidae（雨燕目 Apodiformes）

中至小型空中覓食雀鳥，身體短小，頸短。翼長而尖。尾短，呈方形或長叉尾狀。嘴短而闊，腿十分短而有毛，腳小而強壯，腳趾直而尖。

Small to medium-sized aerial insect feeders. Bodies compact, necks short. Wings long and pointed. Tails vary from short and square to long and forked. Bills short and broad, with wide gape. Legs extremely short and feathered; feet small and strong, with sharp, pointed toes.

※ 深藍 Owen Chiang

## 杜鵑和鴉鵑 Cuckoos and Coucals

### 杜鵑科 Cuculidae（鵑形目 Cuculiformes）

體型和鴿子相近但較修長。嘴尖稍向下彎，尾長而闊成楔形。羽色多為灰色和褐色，常躲在茂密的樹冠中，要靠獨特的叫聲辨認。杜鵑科鳥類自己不築巢，多在其他鳥類巢中產卵寄生（巢寄生）。

Size comparable to doves but slimmer. Bill slightly decurved. Long and broad tail forming wedge shape. Plumage mostly grey or brown, Cuckoos are often hidden in dense canopy. Identified by distinctive calls. They lay eggs in other birds' nests (Brood parasitism).

※ 江敏兒 · 黃理沛 Michelle and Peter Wong

## 鳩和鴿 Doves and Pigeons

### 鳩鴿科 Columbidae（鴿形目 Columbiformes）

體型較圓碩，腳短而強壯，適合在地上行走。羽色多樣，有些鳥種顏色鮮艷。棲息地廣泛分佈，包括樹木、灌叢、以至市區公園。主要以種子及果實為食，築巢於樹、灌叢、岩崖或建築物上，以樹枝組成一個盤狀巢。

Stout bodies and short, strong legs, adapted to walking on the ground. Often colourful. They favour a wide range of habitats, including woodland, scrubland and urban parks. Feed mainly on seeds and fruits. Build platform-shaped nests of twigs on trees, scrubland, rock cliffs or buildings.

※ 何志剛 Pippen Ho

## 秧雞 Rails, Crakes and Coots

### 秧雞科 Rallidae（鶴形目 Gruiformes）

體型較小的雞，頭小而頸長，嘴短而強，四趾及跗蹠部較長。不善飛行，逃避敵人時會拔足狂奔。喜歡棲息於沼澤及密集的灌叢中，以小魚、水生昆蟲和植物嫩草為食。

Small head and long neck, short and strong bill. Toes and tarsus are longer. Poor flyers, runing to escape predators. Favour marshes and dense shrubland. Diet of small fish, aquatic insects and seedlings.

※ 郭匯昌 Andy Kwok

## 鶴 Cranes

### 鶴科 Gruidae（Gruiformes 鶴形目）

儀態優雅的大型涉禽，嘴、頸和腳都很長。雌雄同色，飛行時頸部伸直，組成人字形行列。

Large but elegant wading birds. Long bills, necks and legs. Sexes alike. Fly in V-formation with outstretched necks.

※ 孔思義，黃亞萍 Jemi and John Holmes

## 鷿鷈 Grebes

### 鷿鷈科 Podicipedidae（鷿鷈目 Podicipediformes）

體型似鴨但較扁平，頸部長而細，嘴直而尖，趾間有瓣狀蹼，善於潛泳。起飛時會在水面奔跑，留下一列浪花。在蘆葦間的水面築巢，剛孵出的幼鳥喜停在成鳥的背上。主要棲息於濕地，以水生植物為食。

Diving birds resembling ducks but look much flatter, with small long necks and pointed sharp bill. Toes, with webs around toes for diving. Run on water surface for a short distance on take-off. Build nests on water surface between reeds. When hatched, juveniles often sit on the back of adults. Favour wetlands and feed on aquatic plants.

※ 江敏兒，黃理沛 Michelle and Peter Wong

## 三趾鶉 Buttonquails

### 三趾鶉科 Turnicidae（鴴形目 Charadriiformes）

體型似鵪鶉，但較細小。大多只有三趾，沒有後趾。雌雄異色，雌鳥較大而顏色鮮艷。生活在開闊的沼澤或耕地，非常怕人，受驚時會突然飛出。

Resemble Japanese Quail but smaller. Three-toed, no hind claw. Sexes differ, females are more colourful. Inhabit open country including marshes and farmland. They stay well-hidden in vegetation, bursting into flight when flushed.

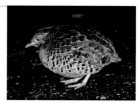

※ 孔思義，黃亞萍 John and Jemi Holmes

## 石鴴 Stone-curlews

### 石鴴科 Burhinidae（鴴形目 Charadriiformes）

體型較大的涉禽，腿長，色彩暗淡，棲於開闊多石地帶或海灘。

Larger waders with long legs and cryptic coloration. Live in open, stony land or on beaches.

※ 江敏兒，黃理沛 Michelle and Peter Wong

## 蠣鷸 Oystercatchers

### 蠣鷸科 Haematopodidae（鴴形目 Charadriiformes）

軀體粗壯的涉禽，腳短，嘴粗長。

Bulky shorebirds with short legs and stout bill.

※ 陳志雄 Allen Chan

## 長腳鷸和反嘴鷸 Stilts and Avocets

### 反嘴鷸科 Recurvirostridae（鴴形目 Charadriiformes）

體型較大的涉禽，嘴細長而上翹。脛裸露而跗蹠較長，尾短而方。常棲息於淺水區如泥灘。

Larger waders. Fine upcurved bills. Long, bare tarus and tibia. Tails short and square. Favour shallow water including mudflats.

※ 孔思義、黃亞萍 Jemi and John Holmes

## 鴴和麥雞 Plovers and Lapwings

### 鴴科 Charadriidae（鴴形目 Charadriiformes）

小、中型涉禽，嘴短而直，尖端較寬。翼形尖短，尾羽短小。主要羽色為棕色、黑色和白色。棲息於水邊，以無脊椎動物為食。

Small to medium-sized waders. Their bills are short and straight but thicker at tip. Wings are short, usually pointed. Most of them have brown, black and white colour pattern. Usually found at water edge and feed on small invertebrates.

※ 深藍 Owen Chiang

## 彩鷸 Painted-snipes

### 彩鷸科 Rostratulidae（鴴形目 Charadriiformes）

嘴直而長，尾羽較短。雌鳥比雄鳥大，且色澤鮮艷。常棲於低窪沼澤，躲藏於矮叢中，生性怕人。清晨或黃昏到耕地上覓食，以昆蟲和無脊椎動物為食。

Bills straight and long, short tails. Females are larger and more colourful. Inhabit lowland marshes. Secretive. Feed on amphibians and invertebrates in farmland at dawn and dusk.

※ 林文華 James Lam

## 水雉 Jacanas

### 水雉科 Jacanidae（鴴形目 Charadriiformes）

本科尾羽一般較短，唯水雉特長。四趾特長，有堅硬的爪。喜棲息於淡水沼澤，喜吃昆蟲、兩棲類和無脊椎動物。

Long toes and strong claws facilitate walking on floating vegetation. Habitat freshwater marshes, feeding on insects, amphibians and invertebrates.

※ 江敏兒、黃理沛 Michelle and Peter Wong

## 鷸、沙錐、瓣蹼鷸等
## Sandpipers, Snipes, Phalaropes and allies

### 鷸科 Scolopacidae（鴴形目 Charadriiformes）

本科涉禽體型大小不一，由細小的濱鷸到大型的白腰杓鷸。嘴的形狀多樣，有的向上翹，有的向下彎，有的甚至呈勺形。腳和頸都較長。常成群於泥灘濕地覓食，以軟體動物、甲殼類為主要食物。

Waders with a wide range of sizes, from small Great Knot to large Eurasian Curlew. They have different sizes and bill shapes, including upcurved, decurved and some spoonbill shapes. Legs and neck are long. Feed on molluscs and shellfish on mudflat.

※ 夏敦天 Martin Hale

## 燕鴴 Pratincoles

### 燕鴴科 Glareolidae（鴴形目 Charadriiformes）

嘴短闊而尖端稍曲，初級飛羽長度超過尾端，尾呈叉狀，飛行時似燕子。棲於海岸沼澤濕地，以昆蟲為主食，是唯一可在飛行中捕食的涉禽。

Bill short and broad. Primaries are longer than tail tip. Forked-tailed, resembling Swallow in flight. Inhabit coastal wetlands, feed on insects. It is the only waders that feed on insects in flight.

※ 林文華 James Lam

## 鷗和燕鷗 Gulls and Terns

### 鷗科 Laridae（鴴形目 Charadriiformes）

體型小至中等，羽色較單調，以黑、白、灰和褐色為主。翅長而尖，尾長而呈開叉或扇形。本科多屬海洋性鳥類，但也會在濕地泥灘等地方出現。

Medium to large-sized. Simple colour combination, mainly black, white, grey and brown. Wings long and pointed. Tails may be fork or fan-shaped. At sea or along mudflat and shoreline habitats.

※ 黃卓研 Cherry Wong

## 賊鷗 Jaegers

### 賊鷗科 Stercorariidae（鴴形目 Charadriiformes）

大小如鷗類，靠搶掠維生，追逐飛行中的鷗和燕鷗。賊鷗在接近極圈緯度繁殖，長途跋涉遷徙。雌雄同色，但毛色變化多端。

Jaeger are gull-sized, predatory seabirds that harass gulls and terns in flight to force them to jettison their catch. They breed in sub-polar latitudes and most undertake long migrations. Sexes alike, but plumage patterns are variable.

※ 江敏兒‧黃理沛 Michelle and Peter Wong

## 海雀 Murrelets

### 海雀科 Alcidae（鴴形目 Charadriiformes）

主要生活在海洋中。身體長、頸短、翼短和窄。飛行呈直線方式，快速而有力地拍翼。

Highly aquatic and exclusively marine. Bodies elongate; necks short, wings short and narrow. Flight direct and strong with rapid whirring wing-beats.

※ Geoff Welch

## 鸏 Tropicbirds

### 鸏科 Phaethontidae（鸏形目 Phaethontiformes）

熱帶海洋上中等大小和非常善於飛行的海鳥，飛行姿態優雅似鳩鴿，盤旋升至空中。俯衝入海覓食，經常在海面浮游，並把長尾豎起。身體主要白色，成鳥以嘴和長尾的顏色配搭來辨別鳥種。

Medium-sized highly aerial seabirds of tropical seas. Graceful pigeon-like flight soaring high into the air. Plunge-dive and often sit on the sea with their long tails cocked. Mostly white and adults specially identified by a combination of bill and tail-streamer colour.

※ 宋亦希 Sung Yik Hei

## 潛鳥 Loons

### 潛鳥科 Gaviidae（潛鳥目 Gaviiformes）

體型大，吃魚為主。用腳推動游泳及潛水，只在開闊水面出現，冬天一般在海上。雌雄同色，辨別鳥種不易，香港非常罕見。

Large, mainly fish-eating, foot-propelled swimming/diving birds. Only found on open water, usually at sea in winter. Sexes alike. Identification is difficult. Very rare in Hong Kong.

※ 江敏兒．黃理沛 Michelle and Peter Wong

## 鸌 Petrels and Shearwaters

### 鸌科 Procellariidae（鸌形目 Procellariiformes）

長途遷徙的遠洋海鳥，只在築巢或受風暴影響時才靠近陸地。翼尖長，不常拍動，常見於中國南海。

Oceanic seabirds which undertake long migrations. Only seen close to land when nesting or storm-driven. Long narrow rigidly held wings. Often seen in the South China Sea.

※ 阿忠剛 Pippen Ho

## 鸛 Storks

### 鸛科 Ciconiidae（鸛形目 Ciconiiformes）

大型水鳥，體形粗壯，嘴鈍而長，腿甚長，脛部半裸露。翅長而寬，飛行時頸部伸直，尾甚短。

Very large waterbirds. Stout body, long and thick bill. Long legs, tibia partly feathered. Long and broad wings, fly with neck extended, tail relatively short.

※ 譚耀良 Tam Yiu Leung

## 軍艦鳥 Frigatebirds

### 軍艦鳥科 Fregatidae（鰹鳥目 Suliformes）

大型海鳥。嘴長成鉤形。翼形長而窄，末端尖，展翅可長達二米。尾長開叉。羽色以黑和深褐為主。常靠海面的氣流滑翔。喜歡搶奪其他海鳥的漁獲，亦是捕魚能手。

Large-sized oceanic birds. Long and strongly hooked bill. Wings very long, narrow and pointed, with wingspan over 2 metres. Long and distinctive forked tail. Black and dark brown feather. Superb gliders, soaring effortlessly on thermals. Renowned for their piratical behaviour on other seabirds catches. They are also skillful fishers.

※ 黃卓研 Cherry Wong

## 鰹鳥 Boobies

### 鰹鳥科 Sulidae（鰹鳥目 Suliformes）

大型海鳥，軀體雪茄形，翼尖長，尾楔狀，嘴粗長而尖。振翅有力，沿直線前進，在岩山島嶼上集體築巢。

Large seabirds with cigar-shaped bodies. Long pointed wings, wedge-shaped tails and stout pointed bills. Strong direct flight. Nest colonially on rocky islands.

※ 譚耀良 Tam Yiu Leung

## 鸕鷀 Cormorants

### 鸕鷀科 Phalacrocoracidae（鰹鳥目 Suliformes）

體羽以黑色為主，嘴長而尖端有鈎，潛水能力極高，是捕魚能手。由於體羽缺防水油脂，常展翅待乾。喜歡成群聚集於樹枝上，排泄物會將全株植物染成白色。

Plumage mainly black in colour. Bills long, pointed and hooked at tip. Good at diving and fishing. Often seen with wings extended for drying due to lack of oily waterproof secretion. Favour gathering in groups on tree branches, which are often stained white by their droppings.

※ 夏敖天 Martin Hale

## 琵鷺和䴉 Spoonbills and Ibises

### 䴉科 Threskiornithidae（鵜形目 Pelecaniformes）

䴉的嘴長，嘴尖向下彎；琵鷺的嘴尖端扁平呈琵琶狀。飛行時，嘴和頸部伸直。

Ibises have long and decurved bills. Spoonbills have bills with a spoon-shaped end. Bill and neck are outstretched in flight.

※ 盧嘉孟 Lo Kar Man

## 鷺鳥和鳽 Egrets, Herons and Bitterns

### 鷺科 Ardeidae（鵜形目 Pelecaniformes）

體型多樣，由較小的鳽到較大的蒼鷺都有。嘴長而直，腿和頸也很長，脛下部常裸露。善於在淺水區覓食，是濕地如魚塘、溪流、海岸及沼澤等常見的水鳥。飛行時頸向後縮，腳向後伸直。繁殖期常大群在樹林中營巢，聚集成鷺鳥林。食物主要為魚類、兩棲動物和昆蟲等。

Sizes range from small bitterns to large grey herons. Bills long and straight, legs and neck are very long. Naked tibia adapted for feeding in shallow water. Often found at fishponds, coast and wetland areas. In flight, neck pulled back and legs stretched behind body. Breed in large groups in woodland, forming egretries. Feed mainly on fish, amphibians and insects.

※ 李雅婷 Anita Lee

## 鵜鶘 Pelicans

### 鵜鶘科 Pelecanidae（鵜形目 Pelecaniformes）

體型巨大的水鳥，可長達1.8米。嘴長而扁平，下嘴有大型喉囊。善於游泳和飛翔，常結小群於空中翱翔。俗名「塘鵝」。

Large waterbirds with wingspan up to 1.8 metres. Characterised by a long bill with conspicuous pouch underneath. Pelicans are good at hunting fish. Often flies and fishes in small groups.

※ 夏敖天 Martin Hale

# 魚鷹 Ospreys

## 鶚科 Pandionidae（鷹形目 Accipitriformes）

廣泛分佈。魚鷹的外趾能反轉，趾底多刺突，常衝入水中捕捉魚類。

Worldwide distribution. Reversible outer toe and sharp spicules on the underside of the foot used to catch and hold their prey, which is caught by diving feet-first into water.

※ 黃卓研 Cherry Wong

# 鵰和鷹 Eagles and Hawks

## 鷹科 Accipitridae（鷹形目 Accipitriformes）

本科的鳥種由細小的雀鷹到大型的禿鷲，體型各異，雌鳥比雄鳥大。嘴尖、向下彎曲成鉤狀，翅大，尾羽形狀不一，善於飛行。具強而有力的腳和趾，尖銳的鉤爪有助獵食。食性複雜，由小型的兩棲類、爬蟲類至較大的哺乳類；海岸濕地的猛禽則以魚類、海蛇為食糧。棲息於不同生境，通常在空中飛行時見到。

Large and varied family, from small Besra to large Black Vulture. Females are generally larger than males. Broad wings and tails with various shapes support powerful flight. Strong legs, toes and sharp talons facilitate hunting. Feed on a wide variety of food, from small reptiles, amphibians to larger mammals. Species that depend on wetlands feed on fish and water snakes. Found in a wide variety of habitats. Many are easy to see in flight.

※ 黃卓研 Cherry Wong

# 鴞（貓頭鷹）Owls

## 鴟鴞科 Strigidae（鴞形目 Strigiformes）

體型大小不一，頭形寬大，眼圓大，眼周圍的羽毛組成面盤，部分鳥種頭兩側有角羽。嘴強而有力，下彎成鉤狀。為夜行性猛禽，視覺和聽覺敏銳，飛行時無聲。以昆蟲和蜥蜴為食。

Wide range of sizes. Broad heads with round and big forward-looking eyes, with feathers around the eyes forming a facial disk. Some species have long eartufts. Strong hook-shaped bill. Nocturnal predators with good eye-sight and hearing, silent flight. Prey includes insects and lizards.

※ 江敏兒，黃理沛 Michelle and Peter Wong

# 佛法僧 Rollers

## 佛法僧科 Coraciidae（佛法僧目 Coraciiformes）

體型較粗壯，嘴大而嘴基闊，飛行時翼長而闊，棲息於開闊環境，以昆蟲、爬蟲及兩棲動物為食。

Stout body. Large bill with wide bill-base. In flight they show long and broad wings, favouring open country. Feed on insects, reptiles and amphibians.

※ 郭匯昌 Andy Kwok

## 戴勝 Hoopoes

### 戴勝科 Upupidae（犀鳥目 Bucerotiformes）

冠羽明顯突出，嘴細長而微向下彎。波浪形飛行。棲息於開闊生境，喜在地上找尋昆蟲和小型無脊椎動物為食。

Unmistakable crown feathers. Bills slender and slightly decurved. Undulating flight. Prefer open country. Feed on insects and small invertebrates.

※ 李佩玲 Eling Lee

## 翠鳥 Kingfishers

### 翠鳥科 Alcedinidae（佛法僧目 Coraciiformes）

體型小，頭大，嘴長而粗直。羽色多變，常具金屬光澤。棲於林地的翠鳥以昆蟲或小型動物為主食，棲於濕地的翠鳥以魚、蝦等為食。多在河岸或土崖挖洞營巢。

Birds with small body and big head. Bill long, thick and straight. Colourful plumage, often with metallic gloss. Open country-dwelling kingtishers favour insects and small animals. Wetland species favour fish, shrimps, etc. In breeding season, they build nests at river banks.

※ 陳家強 Isaac Chan

## 蜂虎 Bee-eaters

### 蜂虎科 Meropidae（佛法僧目 Coraciiformes）

體型小，嘴細長而尖，由基部開始稍微向下彎。羽色以藍、褐為主，鮮艷美麗。棲息於開闊環境，常停在電線上，在空中捕捉昆蟲為食。

Small size. Long and sharp, decurved bill. Beautiful blue and brown plumage. Favour open country, often perching on electric wires. Feed on insects in flight.

※ 崔汝棠 Francis Chu

## 亞洲擬鴷 Asian Barbets

### 鬚鴷科 Megalaimidae（鴷形目 Piciformes）

體型粗壯，嘴大而粗壯，上嘴略長於下嘴。棲於成熟森林中，以種子、果實及昆蟲為食。

Stout body, bill strong and large. Upper bill slightly longer than lower bill. Favour mature woodland. Feed on seeds, fruits and insects.

※ 黃卓研 Cherry Wong

## 啄木鳥 Woodpeckers

### 啄木鳥科 Picidae（鴷形目 Piciformes）

中至小型雀鳥，趾爪有力和尾短，特別適應爬樹，大部分有粗壯的嘴，飛行呈波浪形，叫聲粗糙。

Medium-small size birds. Powerful feet and short stiff tails are specially adapted for climbing trees. Most have strong bills. Undulating flight. Harsh and screaming call notes.

※ 黃才安 Wong Choi On

## 隼 Falcons

### 隼科 Falconidae（隼形目 Falconiformes）

體型較小的猛禽，雌鳥比雄鳥大。飛行時翅膀長而尖，尾短，頭短。能在空中停留，以高速飛行捕獵食物。生性兇猛，常以較小的雀鳥為食。

Small raptors. Females larger than males. In flight they show long and pointed wings, short tail and neck. They hover, and dive at high speed to catch prey, Which can include small birds.

※ 江敏兒‧黃理沛 Michelle and Peter Wong

## 鳳頭鸚鵡 Cockatoos

### 鳳頭鸚鵡科 Cacatuidae（鸚形目 Psittaciformes）

體型多變，具強而有力的嘴，尖端明顯鉤曲。翼形強而闊大，尾長。羽色多變，鮮艷奪目。頭頂具明顯冠羽。叫聲多變響亮，常模仿其他鳥類聲音。原居地為熱帶雨林，以硬殼果實為主要食糧。

Varied in sizes. Strong and powerful hooked bills. Wings strong and broad with long tail. Colourful plumage and conspicuous crested feather. Very loud voice and often mimic other sounds. Their native habitat is tropical rainforest where they feed mainly on hard shell nuts.

※ 陳志雄 Allen Chan

## 鸚鵡 Parakeets

### 鸚鵡科 Psittacidae（鸚形目 Psittaciformes）

嘴鉤曲如猛禽，能打開硬殼、咬破果實和樹葉。腳短而強壯，善爬樹。羽色較艷麗，多為綠色和紅色。喜群居，叫聲響亮。

Hook-shaped bill resembles that of raptors. Able to open nut shells and demolish fruit and foliage. Legs short and strong, good at climbing trees. Colourful plumage, mostly red and green. Often in noisy groups.

※ 黃卓研 Cherry Wong

## 八色鶇 Pittas

### 八色鶇科 Pittidae（雀形目 Passeriformes）

羽色鮮艷但隱蔽難見的林鳥，隱藏於濃密的叢林下層。叫聲響亮，飛行時拍翼呼呼作響。

Enigmatic and elusive, brightly coloured forest birds that keep to thick undergrowth. Loud calls and whirring flight.

※ Geoff Welch

## 山椒鳥和鵑鵙 Minivets and Cuckooshrikes

### 山椒鳥科 Campephagidae（雀形目 Passeriformes）

體型大小不一，身體瘦長，腳弱小。嘴短而壯，基部略闊，嘴端向下彎。山椒鳥多成群活動，並不時鳴叫。鵑鵙則單獨或數隻出沒。棲於茂密林中，以昆蟲、果實為食。

Sizes vary. Slim body, small legs. Short and strong bills with wide base and decurved tips. Minivets often fly and call in groups. Cuckooshrikes usually appear as singles, and sometimes in small groups. Favour dense woodland. Feed on insects and fruits.

※ 江敏兒‧黃理沛 Michelle and Peter Wong

## 伯勞 Shrikes

### 伯勞科 Laniidae（雀形目 Passeriformes）

身體中至小型，嘴強壯，有利鉤。尾長，腳強健，有鉤爪，大多有黑色過眼紋。雌雄羽色相似，幼鳥下體有橫斑。棲息於開闊平原至林區邊緣，有很強的領域行為。捕食昆蟲、蛙、蜥蜴、小鳥或鼠類，會將獵物插在樹枝上儲存。

Small to medium-sized. Bill strong and tipped with sharp hook. Long tail, strong legs, hooked claws. Dark eye-stripe on most species. Sexes alike, juveniles have barred underparts. Habitats include open country to edge of woodland. Strong territorial behaviour. Prey includes insects, frog, lizards, small birds and ratsk. Prey "stored" by impaling on spikes, hence they are named "butcher bird".

※ 江敏兒、黃理沛 Michelle and Peter Wong

## 白腹鳳鶥 White-bellied Erpornis

### 鶯雀科 Vireonidae（雀形目 Passeriformes）

此科大部分鳥種見於美洲，只有兩個屬於亞洲出現。中小型雀鳥，以林鳥為主，喜於林區中上層活動。食性廣泛，由細小的脊椎動物、昆蟲以至果實均是牠們的食物。

This family of birds mainly occur in the Americans and only two genera occur in Asia. Medium small-sized birds mainly inhabit mid-upper level in woodland. They have a wide-range of diet including small vertebrates, insects and fruits.

※ 江敏兒、黃理沛 Michelle and Peter Wong

## 黃鸝 Orioles

### 黃鸝科 Oriolidae（雀形目 Passeriformes）

體型中等，嘴尖略向下彎，翅尖長，尾甚短呈扇狀。雄鳥羽色較雌鳥鮮艷。棲於樹林間，喜在樹冠活動。以昆蟲和果實為食。

Medium-sized. Bill tip slightly decurved. Wings are pointed and long, tails short and slightly fan-shaped. Male is more colourful than females. Active in woodland and tree canopies. Feed on insects and fruits.

※ 呂德恆 Henry Lui

## 卷尾 Drongos

### 卷尾科 Dicruridae（雀形目 Passeriformes）

體型中等，嘴稍向下彎，嘴尖有鉤。尾長呈叉狀，有些種類尾羽向上卷曲，體羽多為灰或黑色，雌雄相似。常停在樹梢上，以昆蟲包括蜻蜓為食。

Medium-sized. Bill tip slightly decurved and hooked. Long forked tails and black plumage, sexes alike. Often perch on exposed branches. Prey on insects including dragonflies.

※ 江敏兒、黃理沛 Michelle and Peter Wong

# 王鶲和綬帶
# Monarchs and Paradise Flycatchers

## 王鶲科 Monarchinae（雀形目 Passeriformes）

毛色鮮艷、外型似鶲的大型鳥類，嘴強壯。

Large, brightly plumaged flycatcher-like birds with strong bills.

※ 郭匯昌 Andy Kwok

# 烏鴉、喜鵲等 Crows, Magpies and allies

## 鴉科 Corvidae（雀形目 Passeriformes）

大型鳥，嘴和腳都很粗壯，喜在開闊地上行走。大多為雜食性的鳥類，吃種子、果實、小動物和屍體。

Large perching birds. Bills and legs thick and strong. Often active on tho ground. Omnivorous diet of seeds, fruits and small animals

※ 江敏兒、黃理沛 Michelle and Peter Wong

# 太平鳥 Waxwings

## 太平鳥科 Bombycillidae（雀形目 Passeriformes）

高遷徙性群居鳥，數目時有間歇性爆發。身體渾圓，頸短，嘴短而嘴基寬闊。翼尖長，尾短呈方形。羽毛柔軟、濃密和呈絲質光澤，有濃密短毛的冠羽。

Migratory and nomadic, with periodic irruptions. Body plump-looking; neck short. Bill short, thick and broad at base. Wings long and pointed. Tail short and square. Plumage soft, dense and silky. Crest with short dense feathers.

※ 霍棟豪 Stanley Fok

# 方尾鶲 Grey-headed Canary-flycatcher

## 玉鶲科 Stenostiridae（雀形目 Passeriformes）

此科鳥種全都是活躍的食蟲鳥種；會於樹枝上下飛躍捕食飛行中的昆蟲，或於樹枝上跳躍捕捉獵物，亦習慣經常拍動及伸展翅膀。

Birds in this family are all active insectivores. They hunt for flying insects by leaping from tree branches or hopping among branches for other preys. Often seen beating or stretching their wings.

※ 李啟康 Lee Kai Hong

# 山雀 Tits

## 山雀科Paridae（雀形目 Passeriformes）

體型大小不一，外形健碩。翅近圓形，尾圓而微叉。喜於樹林及高大灌叢間出沒，捕捉昆蟲為食。

Variable sizes. Stout-shaped, round wings, round and slightly forked tail. Favour woodland and shrubland with large trees. Feed on insects.

※ 郭匯昌 Andy Kwok

## 攀雀 Penduline Tits

### 攀雀科 Remizidae（雀形目 Passeriformes）

小型雀鳥，嘴錐狀，翅短圓而尖，尾方形。常見於蘆葦叢中，善於攀緣，有時會腹部朝天倒懸。以昆蟲為食，有時也吃種子。

Small-sized. Cone-shaped bills, round and pointed wings, square tails. Favour reedbed, good at climbing and sometimes hang themselves upside-down. Feed on insects and sometimes seeds.

※ 黃卓研 Cherry Wong

## 百靈 Larks

### 百靈科 Alaudidae（雀形目 Passeriformes）

體型細小如麻雀，頭有冠羽，嘴較細小而呈圓錐形，後爪長而直。常見於平坦開闊環境，如草地、耕地、沼澤地等。主要以雜草種子、嫩芽為食，也吃昆蟲。常在空中邊飛邊唱，叫聲悅耳。

As small as Tree Sparrows, with small crown and small cone-shaped bill. Hind claws long and straight. Often found in flat open country, e.g. grassland, farmland and marshes. Favour grass seeds, buds and insects. Often call sweetly in flight.

※ 陳家強 Isaac Chan

## 鵯 Bulbuls

### 鵯科 Pycnonotidae（雀形目 Passeriformes）

中小型鳥類，嘴形多變，如雀嘴鵯屬的嘴短厚而向下彎，鵯屬的則細尖。大多成群活動，生境由市區公園、灌叢林地以至濕地紅樹林等。喜食果實和種子，也吃昆蟲。

Small to medium size. Bills vary, for example Collared Finchbill has thick decurved bill, Bulbuls have small and pointed bills. Most of them are gregarious. Favours urban parks, shrubland or mangroves. Feeds on seeds and fruits, sometimes insects.

※ 羅錦文 Law Kam Man

## 燕和沙燕 Swallows and Martins

### 燕科 Hirundinidae（雀形目 Passeriformes）

體型細小而輕巧，嘴形短闊，略呈三角形。翅狹長而尾羽呈叉狀。常成群在開闊地方出現，亦喜站在電線上。飛行速度快，在空中捕食昆蟲。巢以泥土、雜草築成半碗形，置於屋簷之下或峭壁懸岩之間。

Small size. Bills are broad and triangular. Wings are long and narrow, tail forked. Often appear in flocks in open country, perching on electric wires. In flight, they fly at high speed and feed on flying insects. Their nests are half-bowl shaped, made of mud and grass and built beneath overhangs, rocks and cliffs.

※ 江敏兒．黃理沛 Michelle and Peter Wong

## 鷦鶥 Wren-babblers

### 鱗胸鷦鶥科 Pnoepygidae（雀形目 Passeriformes）

此科鳥種為亞洲東南部特有的小型鳥種，尾巴極短至近乎無尾。活躍於林區下層，生性隱蔽，會發出獨特易認的尖銳叫聲，要於林中尋找此科雀鳥建議先尋找其叫聲。

This family of small birds are endemic to South-east part of Asia. Their extremely short tail almost appeared missing. Active but secretive at the low level of woodland. Often give a distinctive sharp call, which is the key to find this bird in the undergrowth.

※ 何國海 Danny Ho

## 樹鶯 Bush Warblers

### 樹鶯科 Cettiidae（雀形目 Passeriformes）

此科鳥種大部分雀鳥都是行蹤隱蔽的褐色鳥種，於密集的樹叢中活動。大部分鳥種於地面活動，警戒時亦不會驚飛，會像老鼠般匆忙逃走；另有部分鳥種亦會於植被的中上層活動。

Most species in this family are secretive brown birds inhabiting dense bushes. Rather than fly away, most of them remain on the ground of undergrowth and escape like a mouse when frightened. Some species inhabit the mid-upper level of vegetations.

※ 關朗曦 Matthew Kwan

## 長尾山雀 Long-tailed Tits

### 長尾山雀科 Aegithalidae（雀形目 Passeriformes）

體型細小，嘴短而粗厚，尾較長，翅短而圓，體羽蓬鬆，雌雄羽色相似。

Small size. Short and thick bills with long tails. Short and broad wing with fluffy plumage. Sexes alike.

※ 黃卓研 Cherry Wong

## 柳鶯 Leaf Warblers

### 柳鶯科 Phylloscopidae（雀形目 Passeriformes）

體型細小的鳥種，於植被的中上層活動，以昆蟲為食。成員以綠色及褐色為主色調，部分有眉紋及翼帶。部分鳥種的外型差異極少，難以分辨；但是各種柳鶯繁殖時都有其獨特的歌聲，足以分辨出不同鳥種，可惜香港並非柳鶯繁殖地故牠們極少鳴唱。

Small-sized birds inhabit mid-upper level of vegetation. Mainly feed on insects. Most member of this family are green or brown, some of them have supercilia and wing bars. Several of them are indistinguishable due to their highly similar appearances. Each of them have a distinctive call during breeding season and these calls are diagnostic for species identification. Unfortunately, Hong Kong is not their breeding range and they are rarely heard singing.

※ 關朗曦 Matthew Kwan

## 葦鶯 Reed Warblers

### 葦鶯科 Acrocephalidae（雀形目 Passeriformes）

小至中型鳥種，於多種生境均有分佈，以褐色及橄欖色為主色調，不少成員背上都有條紋。此科很多鳥種都是太平洋小島上的特有種。於香港出沒的成員常見於濕地生境，並以蘆葦床為甚。

Small to medium-sized birds occurring in various habitats. Mainly brown or olive-green. Some of them have streaks on their back. Many species in this family are Pacific Islands endemics. In Hong Kong, most species of this family occur in wetland, especially reedbed.

※ 深藍 Owen Chiang

## 短翅鶯和蝗鶯
## Bush Warblers and Grasshopper Warblers

### 蝗鶯科 Locustellidae（雀形目 Passeriformes）

此科包含很多行蹤隱蔽、於濕地及灌叢出沒的鳥種。牠們善於在濃密的植被中行走穿梭，較少飛行，相對上較喜歡像老鼠般流竄。

This family includes many secretive species which occur in wetland and bushes. They are good at running among dense vegetations like mice, not often fly.

※ 文權溢 Bill Man

## 扇尾鶯、鷦鶯等
## Cisticolas, Prinias and allies

### 扇尾鶯科 Cisticolidae（雀形目 Passeriformes）

與鶯科非常接近的草地鳥類。

Grassland birds closely related to warblers.

※ 林文華 James Lam

## 鈎嘴鶥和穗鶥
## Scimitar Babblers and Babblers

### 林鶥科 Timaliidae（雀形目 Passeriformes）

中等大小陸棲鳥，嘴強直而稍向下彎，腳強壯，善於奔走和攀爬樹枝，不善飛行。大多於灌叢中或樹林的下層活動，以昆蟲、果實和其他植物為食。歌聲嘹亮悦耳。

Medium-sized, terrestrial or arboreal. Bills vary in this family. Strong legs adapted for walking on ground (Scimitar-babbler and Laughingthrush) or hopping from branch to branch (Minlas and Yuhinas). Favour shrubland or woodland, feeding on insects, fruits and other plants. Calls variable.

※ 郭匯昌 Andy Kwok

## 雀鶥和草鶯 Fulvettas and Grassbirds

### 幽鶥科 Pellorneidae（雀形目 Passeriformes）

此科鳥種多於林區低層或地面活動，經常鳴叫，觀鳥時往往先聞其聲後才見其蹤影。林區其他雀鳥漸趨寂靜時仍會聽到此科鳥種的叫聲。

Species in this family inhabit low or bottom level of woodland. Their call are often heard before the birds are located. They keep calling even when the woodland gone quiet without noise from other birds.

※ 陳志雄 Allen Chan

## 噪鶥、希鶥等
## Laughingthrushes, Minlas and allies

### 噪鶥科 Leiothrichidae（雀形目 Passeriformes）

多為小至中等體型的鳥種，大部分都有較長的尾巴及強壯的身體。此科的亞洲鳥種多於林區植被棲息，經常小群出沒。

Mostly small to medium-sized species with relatively long tails and strong bodies. Asian species of this family often occur in groups in woodland.

※ 黃卓研 Cherry Wong

## 林鶯和鴉雀等 Sylviid Warblers, Parrotbills and allies

### 鶯鶥科 Sylviidae（雀形目 Passeriformes）

此科大部分成員分布於歐亞非大陸，只有一種居於美洲大陸。此科鳥種分布於多種生境，包括濕地、溫帶森林、熱帶雨林及竹林。

Most members of this family occur in Afro-Eurasia, only one species occur in Americans. They inhabit various habitats including wetland, temperate forest, tropical rain forest and bamboo forest.

※ 呂德恆 Henry Lui

## 繡眼鳥 White-eyes

### 繡眼鳥科 Zosteropidae（雀形目 Passeriformes）

體型細小，全身大致綠色，眼周有白圈，嘴小而尖。常成群活動，穿梭於樹林及其外圍，以昆蟲和果實為食。

Small and green. Conspicuous white eyerings, small and sharp bill. Often in flocks in woodland and forest edge. Feed on insects and fruits.

※ 羅錦文 Law Kam Man

## 鳾 Nuthatches

### 鳾科 Sittidae（雀形目 Passeriformes）

小型雀鳥，嘴強而直，腳短但很發達和強壯。常於樹幹上快速攀爬穿梭，找尋昆蟲為食。

Small size. Strong and straight bill, short and strong feet with long toes adapted to climbing on tree trunks to look for insects.

※ 李佩玲 Eling Lee

## 椋鳥和八哥 Starlings and Mynas

### 椋鳥科 Sturnidae（雀形目 Passeriformes）

體型大小適中，嘴直而尖，尾短呈方形。腿長而強健，多為陸棲，常於樹上活動。性合群，叫聲嘈雜。以果實、漿果為食，也吃昆蟲。

Medium-sized. Bills straight and pointed. Short and square tails. Legs long and strong. Mostly terrestrial and favour perching on trees. Gregarious and noisy. Feed on fruits, berries and sometimes insects.

※ 呂德恆 Henry Lui

## 鶇 Thrushes

### 鶇科 Turdidae（雀形目 Passeriformes）

中型雀鳥，嘴較短，翅長而平，尾外形不一。大多以昆蟲、幼蟲、漿果為食。棲息地多樣化，有林地、河流、灌叢、草地、沼澤、岩石等。

Medium sized. Short bill, long and flat wing, tails varied. Feed on insects, grubs and berries. Habitats include woodland, river, shrubland, grasslands, marshland and cliffs.

※ 夏敖天 Martin Hale

## 鶲和鴝 Flycatchers and Chats

### 鶲科 Muscicapidae（雀形目 Passeriformes）

體型略小，嘴扁平而基部較闊，嘴鬚發達，腳較弱。體羽大都是褐、灰、藍色，雌雄異色，雄鳥較鮮艷。常見於茂盛的樹林中，通常會在空中飛捕昆蟲。

Small-sized. Bills flat with broad base and bristles. Weak legs. Plumage is mainly brown, grey or blue. Sexes differ, males are more colourful. Often found in dense woodland. Feed on flying insects.

※ 郭匯昌 Andy Kwok

## 葉鶇 Leafbirds

### 葉鶇科 Chloropseidae（雀形目 Passeriformes）

體型較鵯大，嘴與頭長度相若，形狀細尖。雌雄羽色相異，但大多以綠色為主，雄性較鮮艷。常混在其他鳥群之間，在樹冠頂以花粉、果實、昆蟲為食。

Larger than bulbuls. Bill small and pointed, with similar length as the head. Sexes differ, both are mainly green but males are more colourful. Often flock with other birds. Feed on fruits and insects in woodland canopy.

※ 陳志雄 Allen Chan

## 啄花鳥 Flowerpeckers

### 啄花鳥科 Dicaeidae（雀形目 Passeriformes）

體型細小，嘴小翼小，尾短，常於樹頂槲寄生類植物間活動。雌雄異色，雄性色彩較艷麗。

Small size. Small bill, small wings and short tail. Often active between parasitic mistletoes growths on top of trees. Males are more colourful.

※ 李君哲 Jasper Lee

## 太陽鳥 Sunbirds

### 花蜜鳥科 Nectariniidae（雀形目 Passeriformes）

體型纖細，嘴細長而稍向下彎。雄性羽色華麗，有金屬光澤，雌鳥較暗淡。飛行時常發出尖銳的叫聲，活躍於市區公園或林間的樹冠及花叢間。

Small and slender-bodied, long and slightly decurved bill. Males have colourful and glossy plumage, while females are mainly green. Call at high pitch in flight, often active on top of tree canopy, or at flowering plants in urban parks or woodland areas.

※ 陳志雄 Allen Chan

## 麻雀 Sparrows

### 雀科 Passeridae（雀形目 Passeriformes）

體型細小，短尾而矮胖鳥類，具備用來吃植物種子的厚嘴，性好合群。

Small-sized, plump greyish brown birds with short tails. Powerful bills for eating seeds. Gregarious.

※ 黃卓研 Cherry Wong

## 文鳥 Munias

### 梅花雀科 Estrildidae（雀形目 Passeriformes）

體型細小。嘴粗厚和圓錐形。翼短闊，翼尖圓和鈍。喜歡在開闊原野的草被生活，有時亦見於樹林和蘆葦。群居性。

Small size. Bill thick and conical-shaped. Wings short and broad with rounded or bluntly pointed tip. Typically birds of open country grasses, but also found in woodland and reedbeds. Gregarious.

※ 何志剛 Pippen Ho

## 織雀（織布鳥）Weavers

### 織雀科 Ploceidae（雀形目 Passeriformes）

體型中等。嘴粗厚呈圓錐形。多群居生活，在樹洞內或屋簷下築巢。以種子為食，繁殖時也吃昆蟲。

Medium-sized. Bill short, thick and conical. Mostly live in flocks. Nests in tree holes and under overhangs of buildings. Feed on seeds, and sometimes insects when breeding.

※ 關朗曦、關子凱 Matthew and TH Kwan

## 鶺鴒和鷚 Wagtails and Pipits

### 鶺鴒科 Motacillidae（雀形目 Passeriformes）

體型纖小，嘴形細長，翅尖長，尾羽外側近乎白色。多棲息於草地、河或沼澤邊。善於在地上走動，停留時常上下或左右擺尾。飛行時成波浪形起伏，邊飛邊叫。多以昆蟲為食，也食植物種子。

Small. Bill slender and long, wings pointed. Outer tail feathers whitish. Inhabit grassland, edge of rivers or marshes. Often flick tail at rest. Undulating flight with calls. Feed on insects and sometimes seeds.

※ 江敏兒・黃理沛 Michelle and Peter Wong

## 燕雀和金翅雀類
## Brambling and Cardueline Finches

### 燕雀科 Fringillidae（雀形目 Passeriformes）

體型細小，嘴強厚，上下嘴喙互相緊接。生活於樹林、矮叢及蘆葦叢中，以果實和種子為食。

Small birds with powerful bills. They live in woodland, shrubland and sometimes reedbeds, feeding on fruits and seeds.

※ 李啟康 Lee Kai Hong

## 鵐 Buntings

### 鵐科 Emberizidae（雀形目 Passeriformes）

嘴圓錐形而頗尖，嘴峰直，上下喙不完全切合而有縫隙。食物主要是種子，也吃昆蟲。

Small birds with straight, sharp conical bills. Upper and lower bills are not well fitted. Feed on seeds and sometimes insects.

※ 江敏兒・黃理沛 Michelle and Peter Wong

# 香港有記錄的全球瀕危鳥類
# Globally Threatened Birds in Hong Kong

| 中文鳥名<br>Chinese Name | 英文鳥名<br>English Name | 學名<br>Scientific Name | 全球保育狀況<br>Global Conservation Status |
|---|---|---|---|
| 青頭潛鴨 | Baer's Pochard | *Aythya baeri* | 極度瀕危<br>Critical Endangered (CR) |
| 白鶴 | Siberian Crane | *Leucogeranus leucogeranus* | |
| 勺嘴鷸 | Spoon-billed Sandpiper | *Calidris pygmaea* | |
| 白腹軍艦鳥 | Christmas Frigatebird | *Fregata andrewsi* | |
| 小葵花<br>鳳頭鸚鵡 | Yellow-crested Cockatoo | *Cacatua sulphurea* | |
| 黃胸鵐 | Yellow-breasted Bunting | *Emberiza aureola* | |
| 紅腰杓鷸 | Far Eastern Curlew | *Numenius madagascariensis* | 瀕危<br>Endangered (EN) |
| 大濱鷸 | Great Knot | *Calidris tenuirostris* | |
| 小青腳鷸 | Nordmann's Greenshank | *Tringa guttifer* | |
| 東方白鸛 | Oriental Stork | *Ciconia boyciana* | |
| 黑臉琵鷺 | Black-faced Spoonbill | *Platalea minor* | |
| 栗鳽 | Japanese Night Heron | *Gorsachius goisagi* | |
| 草原鵰 | Steppe Eagle | *Aquila nipalensis* | |
| 鵲色鸝 | Silver Oriole | *Oriolus mellianus* | |
| 小白額雁 | Lesser White-fronted Goose | *Anser erythropus* | 易危<br>Vulnerable (VU) |
| 棕頸鴨 | Philippine Duck | *Anas luzonica* | |
| 紅頭潛鴨 | Common Pochard | *Aythya ferina* | |
| 角鸊鷉 | Horned Grebe | *Podiceps auritus* | |
| 三趾鷗 | Black-legged Kittiwake | *Rissa tridactyla* | |
| 黑嘴鷗 | Saunders's Gull | *Chroicocephalus saundersi* | |
| 遺鷗 | Relict Gull | *Ichthyaetus relictus* | |
| 白腰燕鷗 | Aleutian Tern | *Onychoprion aleuticus* | |
| 冠海雀 | Japanese Murrelet | *Synthliboramphus wumizusume* | |
| 黃嘴白鷺 | Chinese Egret | *Egretta eulophotes* | |
| 卷羽鵜鶘 | Dalmatian Pelican | *Pelecanus crispus* | |
| 烏鵰 | Greater Spotted Eagle | *Clanga clanga* | |
| 白肩鵰 | Eastern Imperial Eagle | *Aquila heliaca* | |
| 仙八色鶇 | Fairy Pitta | *Pitta nympha* | |
| 白頸鴉 | Collared Crow | *Corvus torquatus* | |
| 飯島柳鶯 | Ijima's Leaf Warbler | *Phylloscopus ijimae* | |

| 中文鳥名<br>Chinese Name | 英文鳥名<br>English Name | 學名<br>Scientific Name | 全球保育狀況<br>Global Conservation Status |
|---|---|---|---|
| 遠東葦鶯 | Manchurian Reed Warbler | *Acrocephalus tangorum* | 易危<br>Vulnerable (VU) |
| 史氏蝗鶯 | Styan's Grasshopper Warbler | *Helopsaltes pleskei* | |
| 大草鶯 | Chinese Grassbird | *Graminicola striatus* | |
| 白喉林鶲 | Brown-chested Jungle Flycatcher | *Cyornis brunneatus* | |
| 田鵐 | Rustic Bunting | *Emberiza rustica* | |
| 硫磺鵐 | Japanese Yellow Bunting | *Emberiza sulphurata* | |
| 鵪鶉 | Japanese Quail | *Coturnix japonica* | 近危<br>Near Threatened (NT) |
| 羅紋鴨 | Falcated Duck | *Mareca falcata* | |
| 白眼潛鴨 | Ferruginous Duck | *Aythya nyroca* | |
| 黑海番鴨 | Black Scoter | *Melanitta americana* | |
| 紅頂綠鳩 | Whistling Green Pigeon | *Treron formosae* | |
| 斑脇田雞 | Band-bellied Crake | *Porzana paykullii* | |
| 大石鴴 | Great Stone-Curlew | *Esacus recurvirostris* | |
| 蠣鷸 | Eurasian Oystercatcher | *Haematopus ostralegus* | |
| 鳳頭麥雞 | Northern Lapwing | *Vanellus vanellus* | |
| 白腰杓鷸 | Eurasian Curlew | *Numenius arquata* | |
| 斑尾塍鷸 | Bar-tailed Godwit | *Limosa lapponica* | |
| 黑尾塍鷸 | Black-tailed Godwit | *Limosa limosa* | |
| 紅腹濱鷸 | Red Knot | *Calidris canutus* | |
| 彎嘴濱鷸 | Curlew Sandpiper | *Calidris ferruginea* | |
| 紅頸濱鷸 | Red-necked Stint | *Calidris ruficollis* | |
| 飾胸鷸 | Buff-breasted Sandpiper | *Calidris subruficollis* | |
| 半蹼鷸 | Asian Dowitcher | *Limnodromus semipalmatus* | |
| 灰尾漂鷸 | Grey-tailed Tattler | *Tringa brevipes* | |
| 黃嘴潛鳥 | Yellow-billed Loon | *Gavia adamsii* | |
| 白額鸌 | Streaked Shearwater | *Calonectris leucomelas* | |
| 黑頭白鹮 | Black-headed Ibis | *Threskiornis melanocephalus* | |
| 禿鷲 | Cinereous Vulture | *Aegypius monachus* | |
| 亞歷山大鸚鵡 | Alexandrine Parakeet | *Psittacula eupatria* | |
| 紫綬帶 | Japanese Paradise Flycatcher | *Terpsiphone atrocaudata* | |
| 小太平鳥 | Japanese Waxwing | *Bombycilla japonica* | |
| 斑背大尾鶯 | Marsh Grassbird | *Helopsaltes pryeri* | |
| 琉璃藍鶲 | Zappey's Flycatcher | *Cyanoptila cumatilis* | |
| 紅頸葦鵐 | Japanese Reed Bunting | *Emberiza yessoensis* | |

＊ 取自國際自然保護聯盟瀕危物種紅皮書 Retrieved from The International Unoin for Conservation of Nature (IUCN) Red List of Threatened Species

# 鳥類身體辨識
# Illustrated Glossary

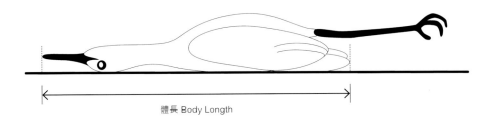

體長 Body Length

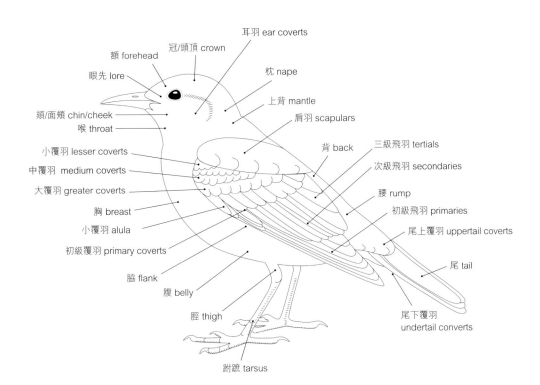

耳羽 ear coverts
冠/頭頂 crown
額 forehead
枕 nape
眼先 lore
上背 mantle
頦/面頰 chin/cheek
肩羽 scapulars
喉 throat
背 back
三級飛羽 tertials
小覆羽 lesser coverts
次級飛羽 secondaries
中覆羽 medium coverts
大覆羽 greater coverts
腰 rump
胸 breast
初級飛羽 primaries
小覆羽 alula
尾上覆羽 uppertail coverts
初級覆羽 primary coverts
脇 flank
尾 tail
腹 belly
脛 thigh
尾下覆羽 undertail converts
跗蹠 tarsus

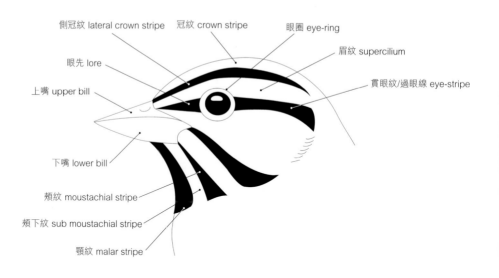

側冠紋 lateral crown stripe　冠紋 crown stripe　眼圈 eye-ring

眉紋 supercilium

眼先 lore

貫眼紋/過眼線 eye-stripe

上嘴 upper bill

下嘴 lower bill

頰紋 moustachial stripe

頰下紋 sub moustachial stripe

顎紋 malar stripe

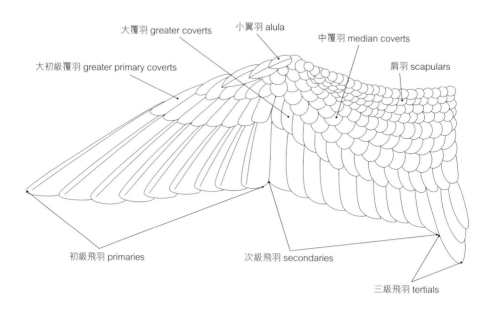

大覆羽 greater coverts　小翼羽 alula　中覆羽 median coverts

大初級覆羽 greater primary coverts

肩羽 scapulars

初級飛羽 primaries

次級飛羽 secondaries

三級飛羽 tertials

# 鳥類結構及功能
# The Structure and Function of Birds

鳥類最獨特的身體特徵，是皮膚表面長有羽毛。羽毛主要分三類：正羽、絨羽、纖羽，而當中正羽亦再細分為廓羽、飛羽、尾羽等。廓羽在身軀表面，塑造身體的輪廓，挺直的飛羽和尾羽幫助鳥類飛行；絨羽生長在正羽下，有保溫作用；細小的纖羽緊貼在皮膚表面，估計與觸覺有關。

鳥類屬脊椎動物，擁有內骨骼。若與其他脊椎動物如哺乳類、兩棲類比較，鳥類的前肢特化成翅膀，有大片胸骨附着發達的胸肌控制前肢活動，頸椎和喙較長。因着飛行的緣故，骨內有很多空間，亦有不少骨癒合成較簡單的組合，以減輕重量。

由於前肢特化成翅膀作飛行用途，一些其他脊椎動物前肢進行的動作，便由喙部或後肢代替。這些動作包括捕殺獵物、撿拾食物或巢枝、處理食物、疏理羽毛等等，游禽亦會以趾間長有蹼的後肢划水游泳。

因着飛行的需要，雀鳥體內不少器官亦與其他脊椎動物有異。飛行需要很大能量，鳥類呼吸系統在肺部以外有多個氣囊，以便在吸氣和呼氣時，均有氧氣供應血液細胞氧化食物，產生能量。鳥類會取食高能量食物，食物進入食道後，會分別經過嗉囊、前胃和砂囊進行消化，砂囊內多有砂子或小石，以幫助磨碎較硬的食物。會飛行的鳥類，腸道較其他脊椎動物短，減少宿便留在體內，減輕重量。

鳥類的視覺十分發達，眼睛跟身軀比例較其他脊椎動物大，眼內視覺細胞密集。大部分鳥類雙眼長在頭的兩側，擁有廣闊視野，有利察看捕獵者來襲；部分捕獵性鳥類如貓頭鷹，兩眼則靠近在一起，視覺便有立體感，有助準確判斷獵物的距離。

不同的鳥種因着生活需要，嗅覺能力不一，部分鳥種亦有靈敏的嗅覺。相比人類，鳥類舌頭味蕾數量不多，估計沒有複雜的味覺。部分鳥類喙部擁有觸覺，有利於覓食。由於鳥類擁有極強的導航能力，除了依靠視覺和嗅覺，不少人認為鳥類擁有感應磁場的能力，那就有賴研究人員解開這個謎團了。

※ 郭子祈 Kwok Tsz Ki

The most unique physical characteristic of birds is the feathers that grow on the surface of their skin. There are three main types of feather: pennaceous feathers, down feathers and filoplumes. Pennaceous feathers are further divided into contour feathers, flight feathers, tail feathers, etc. Contour feathers grow on the surface of the body and form the body shape; rigid flight feathers and tail feathers assist flight; down feathers lie under the pennaceous feathers, forming a layer of insulation. Tiny filoplumes are found next to the skin and are thought to provide sensory information.

Birds are a class of vertebrates with internal skeletons. If you compare birds with other classes of vertebrates, such as mammals and amphibians, you will notice that a bird's forelimbs have evolved into wings, which are controlled by massive chest muscles attached to a keeled breastbone. Birds also have longer cervical vertebrae and beaks. To facilitate flying, birds have porous bones, with some of their bones fused into a relatively simple structure to reduce weight.

Since a bird's forelimbs have evolved into wings for flight, it uses its beak or hindlimbs for certain tasks which other vertebrates perform with their forelimbs, such as hunting, picking up food or nesting materials, processing food, preening, etc.. Aquatic birds also use their webbed hindlimbs as paddles when swimming.

Due to the need to fly, birds have many internal organs which are different from those of other vertebrates. To meet the high energy demand of flying, the respiratory system of birds include a number of air sacs in addition to the lungs, so as to provide a constant supply of oxygen for blood cells to oxidize the food and produce energy during both inhalation and exhalation. Birds will eat high-energy food. After entering the esophagus, food will pass through the crop, the proventriculus and the gizzard. The gizzard typically contains small pieces of grit or stone to aid in the grinding of hard food. Birds have shorter intestines than other vertebrates in order to reduce weight and facilitate flying by retaining less waste in their bodies.

Birds have superior vision. Compared to other vertebrates, the eyes of birds are larger in proportion to their body size, containing dense visual cells. In many birds, the eyes are positioned on the sides of their head, giving them a wide field of view, which can be useful for detecting predators and evading attacks. Certain birds of prey, such as owls, have forward facing eyes, giving them good binocular vision and hence the ability to judge the distance of the prey precisely.

※ 郭子祈 Kwok Tsz Ki

Depending on their way of living, some bird species have better-developed senses of smell than other species. Birds have much fewer taste buds than humans, so their sense of taste is probably not well refined. Some birds have touch receptors in their beaks, which allows them to feed easily. Birds have excellent navigation skills. Besides relying on their vision and sense of smell, some people believe birds can also use Earth's magnetic field for navigation, which remains a mystery to be solved by researchers.

# 辨認雀鳥的技巧
# Bird Identification Skill

　　觀鳥初學者一般最關注是辨認鳥種的技能，認為分辨雀鳥屬觀鳥首要步驟。香港現時有五百多種鳥類，不少更屬罕見；而一些類別如鷗類、鶯類，只有少數資深觀鳥者才能掌握辨認的技巧，初學者假如能辨認常見的數十至百多個鳥種已很不錯！

　　硬生生死記數十種雀鳥外表特徵當然很困難，我們建議可先分辨屬於哪個類別，再進一步細分。鳥類的外形是最基本的辨認依據，有些類別雀鳥外形獨特，例如鷺鳥、鴨、翠鳥、貓頭鷹等等，我們都有很多機會看過，對牠們的外形也有一定的印象。

　　辨認雀鳥通常需要查閱鳥類圖鑑，香港有歷史悠久且種類齊全的鳥類圖鑑，本書亦是很好的參考。觀鳥者可按圖索驥，比對鳥兒的外貌。除了圖片，圖鑑一般亦有顯示雀鳥的大小、顏色、特徵、叫聲、出沒季節、分布等等，部分鳥類亦有獨特的行為習性，所以觀察時不只要記下鳥兒的外表，也要留意行為習性、叫聲，以及時間和環境等資料。

　　由於與雀鳥有一定距離，初學者對雀鳥的大小尺寸並不容易判斷，再加上望遠鏡的倍率，更不易掌握。建議初學者以一些常見鳥作比對，例如把以下從小到大排列的雀鳥記熟，即暗綠繡眼鳥、樹麻雀、白頭鵯、珠頸斑鳩、喜鵲、黑鳶（麻鷹），再比較需要辨認的鳥種與這幾種雀鳥的大小，會較易入手。

　　雀鳥的顏色和特徵是判別的要點，但要留意有機會受不同因素影響。有些顏色在受光和不受光的情況下容易弄錯，有些羽毛有金屬光澤，不同受光角度會呈現不同色彩，有些雌雄異色，有些幼鳥或未成年鳥擁有不同樣貌，有些在繁殖期和非繁殖期更會有截然不同的羽色。而一些特徵如斑點、斑紋，有機會因羽毛磨損或個體差異而變得不顯眼。

　　用文字表達叫聲，有時會令人摸不着頭腦。現時網上有豐富的鳥聲資源，可以免費下載聆聽，但要留意有些鳥種在不同地域，可能帶有少許不同的「口音」。

　　不少鳥種會有特定的出沒季節和生境，這些都是重要的線索。香港有長期的觀鳥紀錄，可以參考一些鳥類報告和鳥類名錄，查閱個別鳥種的相關資料。

　　多到野外練習是提升觀鳥技巧的不二法門，有別於只觀看圖片或影片般的「紙上談兵」，親身觀鳥更能感覺得更全面。而熟習一些常見鳥、累積經驗之後，

※ 郭子祈 Kwok Tsz Ki

憑「感覺」便可以一眼辨認出常見鳥種，亦能輕易察覺區別出一些少見的鳥種加以留意，再進一步查證。

假如觀鳥時亦同時拍攝雀鳥，可以將辨認不到的照片放在香港觀鳥會或相關的網上討論區，請教鳥友如何辨別，很多友善的鳥友樂意提供幫助。不過若想更有效率提升觀鳥技巧，建議以觀察為主，攝鳥為輔，緊記拍攝只是輔助性質。由於攝影機拍攝的畫面細緻度不及人眼，要拍攝得好，鳥兒在遠處供觀賞也不足夠，往往要守候更近的距離、更佳的角度，因此若太在意拍攝，便很容易錯失很多觀察的體驗，不利提升觀鳥技巧。

Beginners to birdwatching normally emphasize greatly the skill of bird identification. They believe that the most significant first step to birdwatching is to identify the bird species. Currently, there are more than five hundred bird species in Hong Kong, of which many are rare. For some families of birds such as gulls and warblers, only a few seasoned birdwatchers have mastered the skill for their identification. It is considerable achievement already if beginners can identify a few tens to more than a hundred species.

To try to mechanically memorized the characteristic appearances of tens of bird species is definitely very difficult. We suggest that one starts with the families before further differentiation into finer subdivisions. The external features of birds are the basis of identification. Some family of birds such as egrets and herons, ducks, kingfishers, and owls are commonly encountered and their distinctive appearances are familiar to many.

The identification of birds often requires making reference to field guides. There are guides of longstanding reputation and comprehensive coverage in Hong Kong. This book is also a good reference. Birdwatchers can compare the appearance of the birds  with the pictures in the guidebooks. Apart from pictures and photographs, description of the size, colour, special features, call, sighting seasons and distribution of different species is also given. Some species have characteristic behavior and habits. During birdwatching, it is not only necessary to record the appearance of the birds but also to take note of the behaviour, call, time and environment of the sighting.

Since birds are often observed at a distance, new birders may have difficulties in judging their size. Lack of familiarity with the visual effect of the magnification of binoculars further hampers judgement. It is recommended that birders try to use commonly seen birds as references. For example, one can memorize the order in increasing size of the following birds: Swinhoe's White-eye, Eurasian Tree Sparrow, Chinese bulbul, Spotted Dove, Oriental Magpie and Black Kite. It will then be easy to judge by comparing the size of the bird to be identified with those on the list.

Although colour and features are important information for identification, it is necessary to remember that they may be influenced by different factors. Colour can be misjudged when seen in the light or in the shade. Some feathers have metallic sheen and show different colour when illuminated at different angle. Some males and females are different in colour. Juveniles and immature birds can have appearances quite different from the adults. Some birds have very distinct colouration in the breeding and non-breeding periods. Characteristics such as spots and stripes may become indistinct due to individual variation or worn feathers.

Bird calls described in words are often difficult to comprehend. Today, there are extensive online resources of recordings of bird calls that can be freely downloaded. However, one need to be aware that some bird species can have slightly different regional "accents".

Many birds have different sighting season and habitat. These are important clues for identification. There are longstanding birdwatching records in Hong Kong. One can refer to birdwatching reports and directories for information on the relevant species.

Frequent participation in field studies is the only way to improve one's bird identification skill. Unlike the "on-paper" exercises of examining photographs and watching videos, first-hand observation provides much fuller experience. After familiarization with common birds and accumulation of experiences, it is possible to identify common birds at first-sight. Also, rarer birds can readily be recognized and targeted for further investigation.

If photographs are taken during birdwatching, pictures of unidentified birds can be uploaded to the discussion forum of the Hong Kong Bird Watching Society or other related online forums for assistance. Many friendly birdwatchers are happy to help. However, in order to enhance one's birdwatching skill, observation should always be the first priority, with photography taking only a subsidiary role. The camera is much inferior to the human eyes in terms of resolution. To take good pictures, it is not sufficient to watch from a distance. Rather, one needs to wait for the birds to approach closer or to move to location with better shooting angle. If too much attention is given to photography, a lot of opportunities for observation will be sacrificed. This can hinder the progress of one's bird watching skill.

鏡頭下的 **267** 種

# 香港鳥類

**267 Kinds of Bird
under the Lens**

# 如何使用這本書
# How to Use this Book

① | 中、英文名稱
Chinese and English common names

② | 學名
Scientific name

214 | 反嘴鷸科
**Recurvirostridae**

① **反嘴鷸** ⓐ fǎn zuǐ yù ⑫
ⓑ 鷸：音核

③ 體長 length : 42-45cm

④

① Pied Avocet | *Recurvirostra avosetta* ②

⑬ ⑭

其他名稱 Other names：反嘴鴴, Avocet

⑤

⑨

⑩

黑白分明。嘴黑色細長，末端向上彎，腳灰色。頭頂至後頸、翼角和翼尖，以及翼上覆羽皆為黑色，其他部分大致白色。 ⑥

Unmistakable black-and-white bird, with long, narrow up-curved bill. Grey legs. Black from crown to hindneck, on wing tips and carpal joints, and on upperwing coverts. Other parts are white.

⑧

[1] Mai Po 米埔：Samson So 蘇毅雄
[2] adult 成鳥：Nam Sang Wai 南生圍，8-Dec-07, 07 年 12 月 8 日：Doris Chu 朱詠兒
[3] 1st winter 第一年冬天：Nam Sang Wai 南生圍，Nov-03, 03 年 11 月：Jemi and John Holmes 孔思義、黃亞萍
[4] adult in courtship 成鳥，正在交配：6-May-07, 07 年 5 月 6 日：Michelle and Peter Wong 江敏兒、黃理沛
[5] Mai Po 米埔：Marcus Ho 何萬邦
[6] Long Valley 塱原：Dec-06, 06 年 12 月：Matthew and TH Kwan 關朗曦、關子凱
[7] Mai Po 米埔：Mar-07, 07 年 3 月：Bill Man 文權溢
[8] 1st winter 第一年冬天：Mai Po 米埔：Nov-04, 04 年 11 月：Pippen Ho 何志剛

⑦

| | 春季過境遷徙鳥<br>Spring Passage Migrant | | | 夏候鳥<br>Summer Visitor | | 秋季過境遷徙鳥<br>Autumn Passage Migrant | | | 冬候鳥<br>Winter Visitor | | |
|---|---|---|---|---|---|---|---|---|---|---|---|
| ⑪ 常見月份 | 1 | 2 | 3 | 4 | 5 | 6 | 7 | 8 | 9 | 10 | 11 | 12 |
| | 留鳥<br>Resident | | | | 迷鳥<br>Vagrant | | | | 偶見鳥<br>Occasional Visitor | | | |

③ | 體長
Body length

⑥ | 描述鳥種外形特徵和特別行為
Description of characteristics and behaviour

④ | 體長和輪廓與本書的對比
Size compared to size of this book

⑦ | 外形特徵
Characteristics

⑤ | 本地拍攝的照片
Photos taken in Hong Kong

⑧ | 圖片說明
Caption

圖片說明的術語解釋：

| 幼鳥<br>juvenile | 離巢後至第一次換羽之間的鳥。<br>The first immature plumage after the nestling stage. |
|---|---|
| 成鳥<br>adult | 鳥類的羽毛變化到了最後一個階段，即以後的羽色及模式不會再有變化。<br>The bird acquires its final or definitive plumage that is then repeated for life. |
| 未成年鳥<br>immature | 除了成鳥之外的所有階段。<br>Denotes all plumages phases except adult. |
| 繁殖羽<br>breeding | 鳥類在繁殖季節呈現異常鮮艷的顏色。<br>Plumage that turn more attractive during breeding season. |
| 非繁殖羽雄鳥<br>male eclipse | 部分雄性鳥類在繁殖期過後身上出現類似雌鳥的毛色，主要見於鴨類和太陽鳥。<br>Post-nuptical, female-like plumage that occurs in males of some groups of birds - notably ducks and sunbirds. |
| 第一年冬天<br>1st winter | 鳥類在幼鳥階段之後所披的羽色。<br>Plumage adopted by many species after juvenile plumage. |
| 第一年夏天<br>1st summer | 鳥類在第一年冬天之後轉變成的羽色。<br>The plumage following the first winter plumage in birds. |

⑨ | 地圖 Map

| 指出該鳥較常出現的地區、或曾經出現的地區，而並非是該鳥在全港的分佈圖。<br>Areas that the birds are usually found, or once found. It is not a distribution map of the bird in Hong Kong. |
|---|
| 注：全粉紅色的地圖指在全港廣泛地區出現<br>Note: Pink coloured map indicates the bird is widely distributed in Hong Kong |

⑩ | 生態環境 Habitats

| | |
|---|---|
| | 濕地（淡水—魚塘、濕農地）<br>Wetland (freshwater - fishponds, wet agricultural land) |
| | 濕地（鹹淡水—蘆葦、紅樹林、基圍、泥灘）<br>Wetland (brackish - reedbed, mangrove, gei wai, mudflat) |
| | 溪流<br>Streams |
| | 海洋、沿岸和海島<br>Ocean, Coastal Waters and Islands |
| | 開闊原野（灌木叢、草地、仍有耕作或棄耕的農地）<br>Open Country (shrubland, grassland, active and abandoned farmland) |
| | 林地<br>Woodland |
| | 高地<br>Upland / High Ground |
| | 市區<br>Urban Area |

⑪ | 居留狀況
Status icon

| 常見<br>Common | 不常見<br>Uncommon | 稀少<br>Scarce | 罕有<br>Rare | 無記錄<br>No record |
|---|---|---|---|---|

| 常見月份 Icon for abundance throughout the year

⑫ | 普通話 ㊦ 拼音和廣東話 ㊪ 難字讀音
Putonghua ㊦ and Cantonese ㊪ pronunciation

| 注：這是拼音，並不是鳥種名字<br>Note: This is the pronunciation, not the name of the bird |
|---|

⑬ | 名稱讀音及鳥鳴聲 The pronunciation of the bird name and the sound of bird

⑭ | 鳥種片段 The video of the bird

雉科
**Phasianidae**

# 中華鷓鴣 <sup>普</sup> zhè gū

體長 length：31-34cm

## Chinese Francolin | *Francolinus pintadeanus*

其他名稱 Other names：鷓鴣

全身顏色斑駁。雄鳥頭頂、貫眼紋和頰紋黑色，眉紋棕色，眼先至耳羽及喉部白色。體羽黑色有濃密白點，肩羽棕色。雌鳥顏色較淺，喉部至下體有黑色及淺棕色細橫紋。春夏期間，在開闊的草坡或疏落的叢林間鳴叫，叫聲為沙啞的「ke，ke，ke，ka-ka-」，很遠都可聽到。受驚時會跑進草叢中。

Overall plumage mottled. Male has black crown, eye-stripes and moustachial stripe. Brown supercilium. Obvious white colour on the throat and the area between lore and ear coverts. Dark body covered with dense white spots, and a brown wing coverts. Female is paler in colour, with narrow light brown streaks between the throat and underparts. During spring and summer, often calls "ke, ke, ke, ka-ka-" in open grassland and sparse shrubland, easily heard at distance. Escapes into shrubs nearby when alarmed.

1 adult male 雄成鳥；Lamma Island 南丫島；May-08, 08 年 5 月；Chan Kai Wai 陳佳瑋
2 adult male 雄成鳥；Lamma Island 南丫島；15-May-07, 07 年 5 月 15 日；Owen Chiang 深藍
3 male 雄鳥；Ma Tso Lung 馬草壟；21-Apr-00, 00 年 4 月 21 日；Lo Kar Man 盧嘉孟
4 adult female 雌成鳥；Ping Che 坪輋；Jun-09, 09 年 6 月；Martin Hale 夏敖天
5 adult 成鳥；Sheung Shui 上水；May-18, 18 年 5 月；Kwok Tsz Ki 郭子祈
6 adult male 雄成鳥；Lamma Island 南丫島；May-07, 07 年 5 月；Harry Li 李炳偉

| | 春季過境遷徙鳥<br>Spring Passage Migrant | | 夏候鳥<br>Summer Visitor | | 秋季過境遷徙鳥<br>Autumn Passage Migrant | | 冬候鳥<br>Winter Visitor | |
|---|---|---|---|---|---|---|---|---|
| 常見月份 | 1 | 2 | 3 | 4 | 5 | 6 | 7 | 8 | 9 | 10 | 11 | 12 |

| 留鳥<br>Resident | 迷鳥<br>Vagrant | 偶見鳥<br>Occasional Visitor |
|---|---|---|

# 鵪鶉 <sup></sup>〔普〕ān chún

體長 length：17-19cm

## Japanese Quail | *Coturnix japonica*

其他名稱 Other names：日本鵪鶉

體型細小渾圓。雌雄羽色相同，上體深褐色及有白色縱紋，有明顯的淺色眼眉，腹部白色，脇部有黑色和褐色縱紋，尾羽短。受驚時會低飛一陣，再找地方躲藏。

Small round body. Sexes alike. Upperparts are dark brown with white streaks. Obvious pale eyebrows and white belly. Black and brown streaks on flanks. Short tail. When disturbed, flies away for short distance before dropping into cover.

1 female 雌鳥：Kowloon Tong 九龍塘；spring 2001, 01 年春天；Yam Wing Yiu 任永耀
2 adult female 雌成鳥：Long Valley 塱原：Oct-09, 09 年 10 月；Michelle and Peter Wong 江敏兒、黃理沛
3 Long Valley 塱原：Marcus Ho 何萬邦
4 adult 成鳥：Pui O 貝澳：Nov-13, 13 年 11 月；Ken Fung 馮漢城
5 adult 成鳥：Long Valley 塱原：Nov-19, 19 年 11 月；Roman Lo 羅文凱
6 adult 成鳥：Long Valley 塱原：Nov-19, 19 年 11 月；Roman Lo 羅文凱

| 春季過境遷徙鳥 Spring Passage Migrant | | | 夏候鳥 Summer Visitor | | | 秋季過境遷徙鳥 Autumn Passage Migrant | | | 冬候鳥 Winter Visitor | | |
|---|---|---|---|---|---|---|---|---|---|---|---|
| 1 | 2 | 3 | 4 | 5 | 6 | 7 | 8 | 9 | 10 | 11 | 12 |

常見月份

| 留鳥 Resident | 迷鳥 Vagrant | 偶見鳥 Occasional Visitor |
|---|---|---|

# 環頸雉

(普) huán jǐng zhì
(粵) 雉：音字

體長 length：53-89cm

## Common Pheasant | *Phasianus colchicus*

其他名稱 Other names：雉雞, Ring-necked Pheasant

雄鳥軀體顏色鮮艷，非常易認。錐形尾巴長40-45厘米，嘴形像雞嘴，寬大的眼周裸皮鮮紅色，有些亞種有白色頸圈。雌鳥型小(60厘米)淡褐色，上身有深色斑點。所有記錄被判斷為逸鳥。

Male has distinctive and colourful body with a 40-45cm long tapered tail. Rooster-like bill. Red face wattles. White neck ring in some sub-species. Female is smaller (60cm). Pale-brownish body. Upperparts are darkly spotted. All records are considered as birds escaped from captivity.

1 adult male 雄成鳥；Long Valley 塱原；May-05, 05 年 5 月；Henry Lui 呂德恆
2 adult male 雄成鳥；Long Valley 塱原；May-05, 05 年 5 月；Henry Lui 呂德恆
3 adult male 雄成鳥；Sai Kung Outdoor Recreation Centre 西貢戶外康樂中心；Feb-03, 03 年 2 月；Henry Lui 呂德恆
4 adult male 雄成鳥；Long Valley 塱原；May-05, 05 年 5 月；Henry Lui 呂德恆

| | 春季過境遷徙鳥<br>Spring Passage Migrant | | 夏候鳥<br>Summer Visitor | | 秋季過境遷徙鳥<br>Autumn Passage Migrant | | 冬候鳥<br>Winter Visitor | | |
|---|---|---|---|---|---|---|---|---|---|
| 常見月份 | 1 | 2 | 3 | 4 | 5 | 6 | 7 | 8 | 9 | 10 | 11 | 12 |

| 留鳥<br>Resident | 迷鳥<br>Vagrant | 偶見鳥<br>Occasional Visitor |
|---|---|---|

# 栗樹鴨 (普) lì shù yā

## Lesser Whistling Duck | *Dendrocygna javanica*

體長 length：38-42cm

中 型鴨，腳長，姿態像雁鵝。羽毛大致褐色，頭頂至上體深褐色，嘴和腳黑色，嘴端呈鈎狀。

Medium-sized duck. Long-legged duck with a goose-like posture. Generally brown, dark brown from cap to upperparts. Bill and legs are black, hooked tip on bill.

1 adult 成鳥：Nam Sang Wai 南生圍；30-Jul-06, 06 年 7 月 30 日：Michelle and Peter Wong 江敏兒、黃理沛
2 juvenile 幼鳥：Mai Po 米埔；Dec-06, 06 年 12 月：Kinni Ho 何建業
3 juvenile 幼鳥：Owen Chiang 深藍
4 juvenile 幼鳥：Mai Po 米埔；Dec-06, 06 年 12 月：Kinni Ho 何建業

| 春季過境遷徙鳥<br>Spring Passage Migrant | | | 夏候鳥<br>Summer Visitor | | | 秋季過境遷徙鳥<br>Autumn Passage Migrant | | | 冬候鳥<br>Winter Visitor | | |
|---|---|---|---|---|---|---|---|---|---|---|---|
| 1 | 2 | 3 | 4 | 5 | 6 | 7 | 8 | 9 | 10 | 11 | 12 |
| 留鳥<br>Resident | | | | 迷鳥<br>Vagrant | | | | 偶見鳥<br>Occasional Visitor | | | |

常見月份

# 灰雁 <sup>普</sup> huī yàn

體長 length：76-89cm

## Greylag Goose | *Anser anser*

羽色灰褐的大型雁，雌雄同色。嘴全粉紅色，頭至頸灰褐色，胸部至下體較淺色，上體灰色而羽緣幼白色邊，臀部白色，腳粉紅色。

Large greyish-brown goose. Male and female have similar appearance. Pink bill. Greyish-brown from head to neck. Paler breast and underparts. Grey upperparts with white fringes on feathers. White vent. Pink legs.

1 adult 成鳥；Mai Po 米埔；Dec-18, 18 年 12 月；Leo Sit 薛國華
2 adult 成鳥；Mai Po 米埔；Dec-18, 18 年 12 月；Leo Sit 薛國華
3 adult 成鳥；Mai Po 米埔；Aug-14, 14 年 8 月；Kinni Ho 何建業
4 adult 成鳥；Mai Po 米埔；Dec-18, 18 年 12 月；Leo Sit 薛國華

| 春季過境遷徙鳥 Spring Passage Migrant | | 夏候鳥 Summer Visitor | | 秋季過境遷徙鳥 Autumn Passage Migrant | | 冬候鳥 Winter Visitor | |
|---|---|---|---|---|---|---|---|

| 常見月份 | 1 | 2 | 3 | 4 | 5 | 6 | 7 | 8 | 9 | 10 | 11 | 12 |
|---|---|---|---|---|---|---|---|---|---|---|---|---|

| 留鳥 Resident | 迷鳥 Vagrant | 偶見鳥 Occasional Visitor |
|---|---|---|

# 寒林豆雁 (普) dòu yàn

體長 length：66-89cm

## Taiga Bean Goose | *Anser fabalis*

大型雁，身體大致深褐色，臀部白色。黑色帶橙色或黃色環帶的嘴短厚，腳橙色。

Large-sized goose. Overall dark brown with white vent. Short, thick black bill with orange or yellow ring at bill base. Orange legs.

1 adult 成鳥：Mai Po 米埔；1-Jan-09, 09 年 1 月 1 日；Geoffery Li 李振成
2 adult 成鳥：Mai Po 米埔；Nov-17, 17 年 11 月；Kwok Tsz Ki 郭子祈
3 adult 成鳥：Mai Po 米埔；Nov-17, 17 年 11 月；Kwok Tsz Ki 郭子祈
4 adult 成鳥：Mai Po 米埔；Nov-17, 17 年 11 月；Kwok Tsz Ki 郭子祈
5 adult 成鳥, race *middendorffii*, *middendorffii* 亞種：Mai Po 米埔；1-Jan-09, 09 年 1 月 1 日；Geoffery Li 李振成

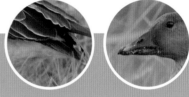

| 春季過境遷徙鳥 Spring Passage Migrant | | | 夏候鳥 Summer Visitor | | | 秋季過境遷徙鳥 Autumn Passage Migrant | | | 冬候鳥 Winter Visitor | | |
|---|---|---|---|---|---|---|---|---|---|---|---|
| 1 | 2 | 3 | 4 | 5 | 6 | 7 | 8 | 9 | 10 | 11 | 12 |

常見月份

| 留鳥 Resident | 迷鳥 Vagrant | 偶見鳥 Occasional Visitor |
|---|---|---|

# 小白額雁 <sup>普</sup> xiǎo bái é yàn

體長 length：53-66cm

## Lesser White-fronted Goose | *Anser erythropus*

**體**型相對較細的灰色雁。嘴短而淡粉紅，有白色斑塊圍繞嘴基並伸延至額部。頭至上體深灰，頸短，上背有淡色羽緣。下體深灰，腹部有小片深色斑塊，尾下覆羽白色。腳橙黃色。

Relatively small-sized goose. Short, pink bill with white band around bill base extending to forehead. Head to upperparts dark grey, short neck, upperparts has paler feather fringe. Underparts dark grey, with a small black patch on belly. White vent. Orange yellow legs.

1 juvenile 幼鳥；Lok Ma Chau 落馬洲；Nov-06, 06 年 11 月；Martin Hale 夏敖天
2 juvenile 幼鳥；Lok Ma Chau 落馬洲；Nov-06, 06 年 11 月；Martin Hale 夏敖天
3 juvenile 幼鳥；Lok Ma Chau 落馬洲；Nov-06, 06 年 11 月；Allen Chan 陳志雄
4 juvenile 幼鳥；Mai Po 米埔；Dec-06, 06 年 12 月；Matthew and TH Kwan 關朗曦．關子凱
5 juvenile 幼鳥；Lok Ma Chau 落馬洲；Nov-06, 06 年 11 月；Jemi and John Holmes 孔思義．黃亞萍
6 juvenile 幼鳥；Lok Ma Chau 落馬洲；Nov-06, 06 年 11 月；Allen Chan 陳志雄
7 juvenile 幼鳥；Lok Ma Chau 落馬洲；Nov-06, 06 年 11 月；Martin Hale 夏敖天

| 春季過境遷徙鳥 Spring Passage Migrant | | | 夏候鳥 Summer Visitor | | | 秋季過境遷徙鳥 Autumn Passage Migrant | | | 冬候鳥 Winter Visitor | | |
|---|---|---|---|---|---|---|---|---|---|---|---|
| 1 | 2 | 3 | 4 | 5 | 6 | 7 | 8 | 9 | 10 | 11 | 12 |
| 留鳥 Resident | | | | 迷鳥 Vagrant | | | | 偶見鳥 Occasional Visitor | | | |

常見月份

# 白額雁 <sup></sup>bái é yàn

體長 length：65-86cm

## Greater White-fronted Goose | *Anser albifrons*

大型灰色雁。嘴淡粉紅，有白色斑塊圍繞嘴基。頭至上體深灰，上背有淡色羽緣。下體深灰，腹部黑色，尾下覆羽白色。腳橙黃色。

Large-sized greyish goose. Pink bill with white band around bill base. Head and upperparts dark grey, with paler feather fringes. Underparts darker, black belly. White vent. Orange yellow legs.

1 juvenile 幼鳥；Lok Ma Chau 落馬洲；8-Nov-04, 04 年 11 月 8 日；Martin Hale 夏敖天
2 Immature 未成年鳥；Mai Po 米埔；25 Oct-16, 16 年 10 月 25 日；Beetle Cheng 鄭諾銘
3 Immature 未成年鳥；Mai Po 米埔；25 Oct-16, 16 年 10 月 25 日；Beetle Cheng 鄭諾銘
4 juvenile 幼鳥；Lok Ma Chau 落馬洲；Nov-04, 04 年 11 月；Jemi and John Holmes 孔思義、黃亞萍

| | 春季過境遷徙鳥<br>Spring Passage Migrant | | | 夏候鳥<br>Summer Visitor | | | 秋季過境遷徙鳥<br>Autumn Passage Migrant | | | 冬候鳥<br>Winter Visitor | | |
|---|---|---|---|---|---|---|---|---|---|---|---|---|
| 常見月份 | 1 | 2 | 3 | 4 | 5 | 6 | 7 | 8 | 9 | 10 | 11 | 12 |
| | 留鳥<br>Resident | | | | 迷鳥<br>Vagrant | | | | 偶見鳥<br>Occasional Visitor | | | |

# 翹鼻麻鴨

(普) qiào bí má yā

體長 length：58-67cm

## Common Shelduck | *Tadorna tadorna*

大型鴨，羽色對比鮮明，頭和翼墨綠色，胸部有一褐色橫帶，其餘羽毛大致白色。嘴紅色，繁殖時雄鳥嘴上有球狀突出物。常小群聚集，退潮後在泥灘上將嘴浸入水窪中覓食。

Large-sized duck with contrasting colours. Head and wings are blackish green, with a brown band on the breast. All the other feathers are mainly white. Bill is red. Breeding males have a knob above the bill. Usually in groups. Feeds on mudflats at low tide by dipping bill in puddles.

1 adult female 雌成鳥；Mai Po 米埔；Owen Chiang 深藍
2 adult female 雌成鳥；Mai Po 米埔；Feb-07, 07 年 2 月；Jemi and John Holmes 孔思義、黃亞萍
3 adult female 雌成鳥；Mai Po 米埔；Feb-07, 07 年 2 月；Jemi and John Holmes 孔思義、黃亞萍
4 non-breeding adult male 非繁殖羽雄成鳥；Mai Po 米埔；Karl Ng 伍耀成

| 春季過境遷徙鳥<br>Spring Passage Migrant | | | 夏候鳥<br>Summer Visitor | | | 秋季過境遷徙鳥<br>Autumn Passage Migrant | | | 冬候鳥<br>Winter Visitor | | |
|---|---|---|---|---|---|---|---|---|---|---|---|
| 1 | 2 | 3 | 4 | 5 | 6 | 7 | 8 | 9 | 10 | 11 | 12 |

常見月份

| 留鳥<br>Resident | 迷鳥<br>Vagrant | 偶見鳥<br>Occasional Visitor |
|---|---|---|

鴨科
Anatidae

# 赤麻鴨 普chì má yā

體長 length：61-71cm

Ruddy Shelduck | *Tadorna ferruginea*

大 型鴨，全身橙褐色，頸有一深色環，十分易認。飛行時翼底白色，翼尖深色。

Large duck with distinctive orange brown plumage overall and a dark ring on the neck. White underwing in flight with dark wing tips.

[1] adult female 雌成鳥；Mai Po 米埔；Apr-07, 07 年 4 月；Kinni Ho 何建業
[2] adult female 雌成鳥；Mai Po 米埔；Apr-07, 07 年 4 月；Kinni Ho 何建業
[3] adult female 雌成鳥；Mai Po 米埔；14-May-05, 05 年 5 月 14 日；Michelle and Peter Wong 江敏兒、黃理沛
[4] adult female 雌成鳥；Mai Po 米埔；Apr-07, 07 年 4 月；James Lam 林文華
[5] adult female 雌成鳥；Mai Po 米埔；May-05, 05 年 5 月；Henry Lui 呂德恆
[6] adult female 雌成鳥；Mai Po 米埔；14-May-05, 05 年 5 月 14 日；Michelle and Peter Wong 江敏兒、黃理沛

| 春季過境遷徙鳥<br>Spring Passage Migrant | | | | 夏候鳥<br>Summer Visitor | | 秋季過境遷徙鳥<br>Autumn Passage Migrant | | | 冬候鳥<br>Winter Visitor | | |
|---|---|---|---|---|---|---|---|---|---|---|---|
| 常見月份 1 | 2 | 3 | 4 | 5 | 6 | 7 | 8 | 9 | 10 | 11 | 12 |
| 留鳥<br>Resident | | | | 迷鳥<br>Vagrant | | | | 偶見鳥<br>Occasional Visitor | | | |

# 鴛鴦

(普)yuān yāng

## Mandarin Duck | *Aix galericulata*

體長 length：41-51cm

**體**型小，喜愛淡水環境的鴨，雄鳥有色彩奪目的繁殖羽，粗大的白色眉紋延伸至眼後。雌鳥上體羽毛是發亮灰色，配上白色的眼圈和過眼線、灰色的嘴，下體具有白色點。非繁殖期的雄鳥和雌鳥接近，但嘴是紅色。

Small-sized freshwater duck. Breeding male has elegant colourful plumage, unmistakable white eye stripe extending behind eyes. Female with bright grey upperparts and white eye-ring extending to supercilium, grey bill. Whitish spots covering the underparts. Non-breeding male resembles female, but with red bill.

1 adult male and female 雄成鳥和雌成鳥；Kam Tin 錦田；Nov-07, 07 年 11 月；Sonia and Kenneth Fung 馮啟文、蕭敏晶
2 adult male and female 雄成鳥和雌成鳥；Kam Tin 錦田；Jan-07, 07 年 1 月；Andy Kwok 郭匯昌
3 adult male and female 雄成鳥和雌成鳥；Kam Tin 錦田；Jan-07, 07 年 1 月；Owen Chiang 深藍
4 adult female 雌成鳥；Mai Po 米埔；Oct- 18, 18 年 10 月；Kinni Ho 何建業

| 春季過境遷徙鳥<br>Spring Passage Migrant | | | 夏候鳥<br>Summer Visitor | | | 秋季過境遷徙鳥<br>Autumn Passage Migrant | | | 冬候鳥<br>Winter Visitor | | |
|---|---|---|---|---|---|---|---|---|---|---|---|
| 1 | 2 | 3 | 4 | 5 | 6 | 7 | 8 | 9 | 10 | 11 | 12 |
| 留鳥<br>Resident | | | | 迷鳥<br>Vagrant | | | | 偶見鳥<br>Occasional Visitor | | | |

常見月份

# 棉鳧

(普) mián fú
(粵) 鳧：音符

體長 length：30-38cm

## Cotton Pygmy Goose | *Nettapus coromandelianus*

**體**型細小的鴨。嘴灰色。雄鳥虹膜紅色，頭、頸和胸白色，嘴至頭頂、胸帶、上體至尾深綠色，下體兩側具有深色幼紋。飛行時有明顯白色翼帶。雌鳥虹膜褐色，深褐色取代雄鳥的深綠色，頭頂至後頸深褐色，下體兩側有淡褐色斑，飛行時無翼斑。

Small-sized duck. Grey bill. Male bird has red iris. White head, neck to breast. Prominent dark green color band from bill to crown. Dark green breast band, and upperparts to tail. Underpart white with narrow stripes on both sides. In flight, it shows a thick white band on both wings. Female bird has brownish iris, with brownish color replacing the dark green in male bird. Crown to hindneck darkish brown. underparts and flanks have pale brownish patches. No wing bands in flight.

1 adult female 雌成鳥；Mai Po 米埔；29-Oct-06, 06 年 10 月 29 日；Or Ka Man 柯嘉敏
2 adult female 雌成鳥；Mai Po 米埔；Oct- 15, 15 年 10 月；Kinni Ho 何建業
3 adult male 雄成鳥；Lok Ma Chau 落馬洲；May-15, 15 年 5 月；Sam Chan 陳巨輝
4 adult female 雌成鳥；Mai Po 米埔；Francis Chu 崔汝棠

| | 春季過境遷徙鳥<br>Spring Passage Migrant | | | 夏候鳥<br>Summer Visitor | | 秋季過境遷徙鳥<br>Autumn Passage Migrant | | | 冬候鳥<br>Winter Visitor | | |
|---|---|---|---|---|---|---|---|---|---|---|---|
| 常見月份 | 1 | 2 | 3 | 4 | 5 | 6 | 7 | 8 | 9 | 10 | 11 | 12 |
| | 留鳥<br>Resident | | | 迷鳥<br>Vagrant | | | 偶見鳥<br>Occasional Visitor | | | | | |

# 花臉鴨
(普)huā liǎn yā

體長 length：39-43cm

Baikal Teal | *Sibirionetta formosa*

小型鴨。嘴灰色，胸部沾褐色。雄鳥面部花紋對比鮮明，十分易認。雌鳥像雌性白眉鴨，全身大致褐色，但嘴基有顯著白斑，面部隱約有雄鳥的花紋。

Small-sized duck. The bill is grey and the breast is tinted brown in both sexes. Male is unmistakable with contrasting facial pattern. Female resembles Garganey, mainly brown in colour, but with obvious white patch at the base of bill and a hint of the male facial pattern.

[1] adult female 雌成鳥：Sheung Shui 上水；Dec-08, 08 年 12 月；Jemi and John Holmes 孔思義，黃亞萍
[2] adult female 雌成鳥：Sheung Shui 上水；Dec-08, 08 年 12 月；Jemi and John Holmes 孔思義，黃亞萍
[3] male 雄鳥；Mai Po 米埔；Martin Hale 夏敖天
[4] adult male 雄成鳥；Jacky Chan 陳家華

| 春季過境遷徙鳥<br>Spring Passage Migrant | | | | 夏候鳥<br>Summer Visitor | | | 秋季過境遷徙鳥<br>Autumn Passage Migrant | | | 冬候鳥<br>Winter Visitor | | |
|---|---|---|---|---|---|---|---|---|---|---|---|---|
| 1 | 2 | 3 | 4 | 5 | 6 | 7 | 8 | 9 | 10 | 11 | 12 | 常見月份 |

| 留鳥<br>Resident | 迷鳥<br>Vagrant | 偶見鳥<br>Occasional Visitor |
|---|---|---|

# 白眉鴨

(普) bái méi yā

體長 length：37-41cm

## Garganey | *Spatula querquedula*

1

小 型鴨，嘴灰黑色，頭頂較扁平，和綠翅鴨相似，但游泳時軀體更貼近水面。雄鳥頭部有明顯白色眉紋，伸延至頸後側，飛行時翼上覆羽藍灰色。雌鳥有深色貫眼紋和淺色眉紋，喉白色。

Small-sized duck with greyish black bill and flat crown. Resembles a Common Teal but sits lower in the water when swimming. Male has sharp white head stripe extending to the nape. In flight, it shows bluish grey upperwing coverts. Female is brownish, with dark eyestripe, light supercilium and white throat.

1 adult male 雄成鳥；Mai Po 米埔；Apr-08, 08 年 4 月；Cherry Wong 黃卓研
2 adult female 雌成鳥；Nam Sang Wai 南生圍；Dec-07, 07 年 12 月；Andy Kwok 郭匯昌
3 adult male 雄成鳥；Mai Po 米埔；Mar-07, 07 年 3 月；Henry Lui 呂德恆
4 adult eclipse male 非繁殖羽雄成鳥；Mai Po 米埔；Jan-04, 04 年 1 月；Henry Lui 呂德恆
5 adult female 雌成鳥；Mai Po 米埔；Jan-07, 07 年 1 月；Cherry Wong 黃卓研
6 adult eclipse male 非繁殖羽雄成鳥；Nam Sang Wai 南生圍；Dec-07, 07 年 12 月；Andy Kwok 郭匯昌

| | 春季過境遷徙鳥<br>Spring Passage Migrant | | | 夏候鳥<br>Summer Visitor | | 秋季過境遷徙鳥<br>Autumn Passage Migrant | | 冬候鳥<br>Winter Visitor | | | |
|---|---|---|---|---|---|---|---|---|---|---|---|
| 常見月份 | 1 | 2 | 3 | 4 | 5 | 6 | 7 | 8 | 9 | 10 | 11 | 12 |
| | 留鳥<br>Resident | | | | 迷鳥<br>Vagrant | | | 偶見鳥<br>Occasional Visitor | | | |

# 琵嘴鴨 <sup>普</sup>pí zuǐ yā

體長length：43-56cm

## Northern Shoveler | *Spatula clypeata*

[1]

中型鴨，嘴黑色、闊大成匙狀。雄鳥頭部深綠色，眼黃色，胸部至脇部白色，腹部褐色。雌鳥全身大致褐色。

Medium-sized duck. Both sexes have black and broad spoon-like bills. Male has dark green head, yellow eyes, white breast and flanks, brown belly. Female is mainly brown in colour.

1 adult male 雄成鳥：Nam Sang Wai 南生圍；Jan-05, 05 年 1 月；Jemi and John Holmes 孔思義、黃亞萍
2 adult female 雌成鳥：Nam Sang Wai 南生圍；Dec-04, 04 年 12 月；Martin Hale 夏敖天
3 adult male and female 雄成鳥和雌成鳥：Nam Sang Wai 南生圍；Feb-06, 06 年 2 月；Henry Lui 呂德恆
4 adult eclipse male 非繁殖羽雄成鳥：Mai Po 米埔；20-Feb-07, 07 年 2 月 20 日；Michelle and Peter Wong 江敏兒、黃理沛
5 adult male 雄成鳥：Nam Sang Wai 南生圍；Feb-07, 07 年 2 月；Herman Ip 葉紀江
6 adult female 雌成鳥：Mai Po 米埔；Dec-06, 06 年 12 月；Cherry Wong 黃卓研
7 adult female 雌成鳥：Nam Sang Wai 南生圍；Jan-08, 08 年 1 月；Kami Hui 許淑君

| 春季過境遷徙鳥<br>Spring Passage Migrant | | | | 夏候鳥<br>Summer Visitor | | 秋季過境遷徙鳥<br>Autumn Passage Migrant | | | 冬候鳥<br>Winter Visitor | | |
|---|---|---|---|---|---|---|---|---|---|---|---|
| 1 | 2 | 3 | 4 | 5 | 6 | 7 | 8 | 9 | 10 | 11 | 12 |
| 留鳥<br>Resident | | | | 迷鳥<br>Vagrant | | | | 偶見鳥<br>Occasional Visitor | | | |

常見月份

# 赤膀鴨 <sup>普</sup>chì bǎng yā

體長 length：46-58cm

Gadwall | *Mareca strepera*

其他名稱 Other names：紫膀鴨

中型鴨，頭和翼背褐色，翼鏡有白斑。雄鳥胸部至下體大致灰色，臀部黑色。雌鳥和綠頭鴨雌鳥十分相似，全身大致褐色，但嘴邊橙色，偏愛淡水或鹹淡水。

Medium-sized duck. The head and back of wings are brown. White panel on speculum in both sexes. Male is generally grey from breast to underparts, and black undertail coverts. Female resembles Mallard, mainly brown in colour, but edge of bill is orange. Prefers fresh or brackish water.

[1] adult eclipse male and female 非繁殖羽雄鳥和雌鳥；Mai Po 米埔；25-Dec-06, 06 年 12 月 25 日；Michelle and Peter Wong 江敏兒、黃理沛
[2] adult female 雌成鳥；Mai Po 米埔；25-Dec-06, 06 年 12 月 25 日；Michelle and Peter Wong 江敏兒、黃理沛
[3] breeding male 繁殖羽雄鳥；Mai Po 米埔；Geoff Carey 賈知行

| 春季過境遷徙鳥<br>Spring Passage Migrant | | | 夏候鳥<br>Summer Visitor | | 秋季過境遷徙鳥<br>Autumn Passage Migrant | | | 冬候鳥<br>Winter Visitor | | |
|---|---|---|---|---|---|---|---|---|---|---|
| 1 | 2 | 3 | 4 | 5 | 6 | 7 | 8 | 9 | 10 | 11 | 12 |

常見月份

| 留鳥<br>Resident | 迷鳥<br>Vagrant | 偶見鳥<br>Occasional Visitor |
|---|---|---|

# 羅紋鴨 (普)luó wén yā

體長 length：46-54cm

## Falcated Duck | *Mareca falcata*

體型大的鴨類，雄鳥整體銀灰色，頭部具金屬褐色和綠色的毛色，頸部白色。雄鳥和雌鳥的嘴深色。停棲時，雄鳥有長而彎的黑白三級飛羽垂掛尾旁。雌鳥上體佈滿褐色和深色鱗狀斑，嘴部深色。

Large-sized duck. Male overall silvery grey, with brown and green glossy head and white neck. Both sexes are dark-billed. At rest, male's long black-and-white tertials drape over tail. Female's upperparts are of brown and dark scale-like feathers, and dark bill.

1 adult male 雄成鳥；Mai Po 米埔：Owen Chiang 深藍
2 adult male 雄成鳥；Mai Po 米埔：16-Apr-06, 06 年 4 月 16 日；Michelle and Peter Wong 江敏兒、黃理沛
3 adult male 雄成鳥；Mai Po 米埔：16-Apr-06, 06 年 4 月 16 日；Michelle and Peter Wong 江敏兒、黃理沛
4 adult female 雌成鳥；Mai Po 米埔：Feb-20, 20 年 2 月；Henry Lui 呂德恆

| 春季過境遷徙鳥 Spring Passage Migrant | | | 夏候鳥 Summer Visitor | | 秋季過境遷徙鳥 Autumn Passage Migrant | | 冬候鳥 Winter Visitor | | | |
|---|---|---|---|---|---|---|---|---|---|---|
| 1 | 2 | 3 | 4 | 5 | 6 | 7 | 8 | 9 | 10 | 11 | 12 |

常見月份

| 留鳥 Resident | | 迷鳥 Vagrant | | 偶見鳥 Occasional Visitor | |
|---|---|---|---|---|---|

# 赤頸鴨

(普) chì jǐng yā

體長 length：45-51cm

## Eurasian Wigeon | *Mareca penelope*

[1]

中型鴨，嘴灰色。雄鳥頭褐色，額奶黃色，胸淡紅褐色，靜止時脇部有白色細橫紋，翼上有明顯白斑。雌鳥全身大致深褐色。叫聲為嘹亮的「彪－彪－」聲，時常集體在池塘中游泳。

Medium-sized duck. Grey bill. Male has brown head, milky yellow forehead, light reddish brown breast, white narrow strips on flanks and a prominent white patch on wings at rest. Female is mainly dark brown. Voice is a loud "bill-bill-". Usually in groups swimming in ponds.

[1] adult male moulting into breeding plumage 雄成鳥轉換繁殖羽；Nam Sang Wai 南生圍；Feb-05, 05 年 2 月；Cherry Wong 黃卓研
[2] breeding male 繁殖羽雄鳥；Mai Po 米埔；Jan-04, 04 年 1 月；Henry Lui 呂德恆
[3] adult eclipse male 非繁殖羽雄成鳥；Mai Po 米埔；2-Dec-06, 06 年 12 月 2 日；Owen Chiang 深藍
[4] breeding male 繁殖羽雄鳥；Nam Sang Wai 南生圍；Jan-05, 05 年 1 月；Henry Lui 呂德恆
[5] adult male moulting into breeding plumage 雄成鳥轉換繁殖羽；Nam Sang Wai 南生圍；3-Dec-06, 06 年 12 月 3 日；Michelle and Peter Wong 江敏兒、黃理沛
[6] breeding male 繁殖羽雄鳥；Mai Po 米埔；25-Jan-09, 09 年 1 月 25 日；Owen Chiang 深藍

| | 春季過境遷徙鳥<br>Spring Passage Migrant | | | 夏候鳥<br>Summer Visitor | | 秋季過境遷徙鳥<br>Autumn Passage Migrant | | 冬候鳥<br>Winter Visitor | | | |
|---|---|---|---|---|---|---|---|---|---|---|---|
| 常見月份 | 1 | 2 | 3 | 4 | 5 | 6 | 7 | 8 | 9 | 10 | 11 | 12 |
| | 留鳥<br>Resident | | | | 迷鳥<br>Vagrant | | | | 偶見鳥<br>Occasional Visitor | | | |

# 綠眉鴨

(普) lù méi yā

體長 length：45-56cm

## American Wigeon | *Mareca americana*

貌似赤頸鴨但體型稍大的中型鴨。嘴灰色而前端黑色，雄鳥額部奶白色，繁殖期有一抹金屬綠色的粗眼線。臉至頸灰白色，胸腹棕褐色，腋羽全白（有別於赤頸鴨），飛羽黑色而羽緣幼白色邊，臀部白色，尾黑色。

Medium-sized duck. Resembles Eurasian Wigeon but slightly bigger. Grey bill with black tip. Male has milky white forehead. Thick metallic green eye-stripes in breeding plumage. Greyish-white from face to neck. Brown breast and belly. Distinguishes from Eurasian Wigeon by pure white axillaries. Black flight feathers with thin white fringes. White vent. Black tail.

1 adult male 雄成鳥：Mai Po 米埔：Jan- 16, 16 年 1 月：Kinni Ho 何建業
2 adult male 雄成鳥：Mai Po 米埔：Jan- 16, 16 年 1 月：Natalie Chan 陳佩霞
3 adult male 雄成鳥：Mai Po 米埔：Jan- 16, 16 年 1 月：Natalie Chan 陳佩霞
4 adult male 雄成鳥：Mai Po 米埔：Jan- 16, 16 年 1 月：Kinni Ho 何建業

| 春季過境遷徙鳥<br>Spring Passage Migrant | | | 夏候鳥<br>Summer Visitor | | | 秋季過境遷徙鳥<br>Autumn Passage Migrant | | | 冬候鳥<br>Winter Visitor | | |
|---|---|---|---|---|---|---|---|---|---|---|---|
| 1 | 2 | 3 | 4 | 5 | 6 | 7 | 8 | 9 | 10 | 11 | 12 |

常見月份

| 留鳥<br>Resident | 迷鳥<br>Vagrant | 偶見鳥<br>Occasional Visitor |
|---|---|---|

# 棕頸鴨

(普) zōng jǐng yā

體長 length：48-58cm

## Philippine Duck | *Anas luzonica*

體 型大的鴨，雌雄相似。嘴黑色，頭至上頸的紅棕色與黑色的頭頂和貫眼紋對比鮮明。身軀大致灰褐色，有深色鱗狀紋，翼鏡紫藍色，腳深灰色。

Large-sized duck. Male and female have similar appearance. Black bill. Reddish-brown head and upper neck contrast with black crown and eye-stripes. Body mainly greyish-brown with dark scaly pattern. Purplish-blue specula. Dark grey legs.

1 adult 成鳥；Mai Po 米埔；Mar-10, 10 年 3 月；Jac Lau 劉振鴻
2 adult 成鳥；Mai Po 米埔；Mar-10, 10 年 3 月；Jac Lau 劉振鴻
3 adult 成鳥；Mai Po 米埔；Mar-10, 10 年 3 月；Jac Lau 劉振鴻
4 adult 成鳥；Mai Po 米埔；Mar-10, 10 年 3 月；Jac Lau 劉振鴻

| 春季過境遷徙鳥<br>Spring Passage Migrant | | | 夏候鳥<br>Summer Visitor | | | 秋季過境遷徙鳥<br>Autumn Passage Migrant | | | 冬候鳥<br>Winter Visitor | | |
|---|---|---|---|---|---|---|---|---|---|---|---|
| 1 | 2 | 3 | 4 | 5 | 6 | 7 | 8 | 9 | 10 | 11 | 12 |

常見月份

| 留鳥<br>Resident | 迷鳥<br>Vagrant | 偶見鳥<br>Occasional Visitor |
|---|---|---|

# 印緬斑嘴鴨

（普）yìn miǎn bān zuǐ yā

體長 length：58-63cm

## Indian Spot-billed Duck | *Anas poecilorhyncha*

其他名稱 Other names：Spot-billed Duck

大型淺褐色淺水鴨，雌雄羽色大致相同。嘴黑色而末端黃色，與中華斑嘴鴨最明顯的分別是嘴後有明顯紅色的眼先和深綠色翼鏡，靜止時可見。身軀、背、頭頂和枕部顏色較深。

Large light brown dabbling duck with similar plumage in both sexes. Bill is black with yellow nib. Red lore and deep green speculum visible at rest can easily distinguish from Chinese Spot-billed Duck. The body, back, cap and nape appeared darker.

[1] juvenile 幼鳥：Mai Po 米埔：May-06, 06 年 5 月：Kinni Ho 何建業
[2] adult 成鳥：Mai Po 米埔：26-Jul-03, 03 年 7 月 26 日：Michelle and Peter Wong 江敏兒・黃理沛
[3] adult 成鳥：Mai Po 米埔：29-Jul-03, 03 年 7 月 29 日：Michelle and Peter Wong 江敏兒・黃理沛
[4] adult 成鳥：Mai Po 米埔：29-Jul-03, 03 年 7 月 29 日：Michelle and Peter Wong 江敏兒・黃理沛

| 春季過境遷徙鳥 Spring Passage Migrant | | | 夏候鳥 Summer Visitor | | | 秋季過境遷徙鳥 Autumn Passage Migrant | | | 冬候鳥 Winter Visitor | | |
|---|---|---|---|---|---|---|---|---|---|---|---|
| 1 | 2 | 3 | 4 | 5 | 6 | 7 | 8 | 9 | 10 | 11 | 12 |
| 留鳥 Resident | | | | 迷鳥 Vagrant | | | | 偶見鳥 Occasional Visitor | | | |

常見月份

# 中華斑嘴鴨

（普）zhōng huá bān zuǐ yā

體長 length：58-63cm

## Chinese Spot-billed Duck | *Anas zonorhyncha*

其他名稱 Other names：Chinese Spotbill

大型深褐色淺水鴨，雌雄羽色大致相同。嘴黑色而末端黃色，有明顯深色冠紋、眼線和面頰橫紋與淡色頭部成對比。靜止時可見紫色翼鏡。常見於米埔，並在該處繁殖。

Large buffy-brown dabbling duck with similar plumage in both sexes. Bill is black with yellow nib. Dark crown line, eyeline and cheek stripe stand out from paler sides of head. Purple speculum visible at rest. Common in Mai Po, where it breeds.

[1] adult male 雄成鳥：Mai Po 米埔；Jul-04, 04 年 7 月；Henry Lui 呂德恆
[2] adult female 雌成鳥：Mai Po 米埔；Dec-06, 06 年 12 月；Cherry Wong 黃卓研
[3] adult male on right 雄成鳥在右；Fung Lok Wai 豐樂圍；Nov-04, 04 年 11 月；Jemi and John Holmes 孔思義、黃亞萍

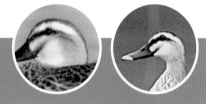

| 春季過境遷徙鳥 Spring Passage Migrant | | | 夏候鳥 Summer Visitor | | | 秋季過境遷徙鳥 Autumn Passage Migrant | | | 冬候鳥 Winter Visitor | | |
|---|---|---|---|---|---|---|---|---|---|---|---|
| 1 | 2 | 3 | 4 | 5 | 6 | 7 | 8 | 9 | 10 | 11 | 12 |

常見月份

| 留鳥 Resident | 迷鳥 Vagrant | 偶見鳥 Occasional Visitor |
|---|---|---|

# 綠頭鴨 <sub>普</sub>lù tóu yā

體長 length：55-70cm

## Mallard | *Anas platyrhynchos*

大型鴨。雄鳥對比鮮明，十分易認，嘴黃色，頭深綠色，胸口深褐色，間有白線相隔。尾近似黑色，軀體淡灰色。雌鳥全身羽毛大致褐色，嘴有橙色斑。

Large-sized duck. Male is unmistakable with high contrast of yellow bill, deep green head, brown breast with white line between head and breast. Blackish tail and light grey body. Female is brownish overall, and its bill is dark with orange colour.

1 male 雄鳥：Ho Chung 蠔涌；Michelle & Peter Wong 江敏兒，黃理沛
2 adult male 雄成鳥：Lamma Island 南丫島；11-May-02, 02 年 5 月 11 日：Michelle and Peter Wong 江敏兒，黃理沛
3 adult male 雄成鳥：Lamma Island 南丫島；11-May-02, 02 年 5 月 11 日：Michelle and Peter Wong 江敏兒，黃理沛
4 adult male 雄成鳥：Lamma Island 南丫島；11-May-02, 02 年 5 月 11 日：Michelle and Peter Wong 江敏兒，黃理沛

| 春季過境遷徙鳥<br>Spring Passage Migrant | | | 夏候鳥<br>Summer Visitor | | 秋季過境遷徙鳥<br>Autumn Passage Migrant | | | 冬候鳥<br>Winter Visitor | | |
|---|---|---|---|---|---|---|---|---|---|---|
| 1 | 2 | 3 | 4 | 5 | 6 | 7 | 8 | 9 | 10 | 11 | 12 |
| 留鳥<br>Resident | | | 迷鳥<br>Vagrant | | | 偶見鳥<br>Occasional Visitor | | |

常見月份

# 美洲綠翅鴨

(普) měi zhōu lù chì yā

體長 length：34-38cm

## Green-winged Teal | *Anas carolinensis*

外貌和綠翅鴨相似，雄鳥以胸側一條白色顯眼直紋，以及翼邊欠缺白橫紋判別。嘴黑色，頸短，有綠色翼鏡。雄鳥頭部深褐色，有深綠色的粗眼線，軀體灰色，臀部奶黃色帶有黑邊。雌鳥全身大致褐色。

Resembles Eurasian Teal. Male can be distinguished by a prominent white stripe on each side of breast and the lack of the white edge on wings. Black bill. Short neck. Green specula. Male has dark brown head with thick dark green eye-stripes. Grey body. Milky yellow vent with black edge. Female is mainly brown.

1 adult 成鳥：Mai Po 米埔：Apr-11, 11 年 4 月：Allen Chan 陳志雄
2 adult 成鳥：Mai Po 米埔：Apr-11, 11 年 4 月：Allen Chan 陳志雄
3 adult 成鳥：Mai Po 米埔：Jemi and John Holmes 孔思義、黃亞萍
4 adult 成鳥：Mai Po 米埔：Jemi and John Holmes 孔思義、黃亞萍

| 春季過境遷徙鳥<br>Spring Passage Migrant | | | 夏候鳥<br>Summer Visitor | | | 秋季過境遷徙鳥<br>Autumn Passage Migrant | | | 冬候鳥<br>Winter Visitor | | |
|---|---|---|---|---|---|---|---|---|---|---|---|
| 1 | 2 | 3 | 4 | 5 | 6 | 7 | 8 | 9 | 10 | 11 | 12 |
| 留鳥<br>Resident | | | | 迷鳥<br>Vagrant | | | | 偶見鳥<br>Occasional Visitor | | | |

常見月份

鴨科
**Anatidae**

# 針尾鴨
(普)zhēn wěi yā

體長 length：50-66cm

Northern Pintail | *Anas acuta*

中 至大型鴨。嘴灰色，頸細長，尾羽尖長。雄鳥頭和後頸深褐色，胸部白色向上延伸至頸兩側成細線，指向後枕。軀體大致灰色。雌鳥全身大致褐色。覓食時頭及上身沒入水中，屁股朝天。

Medium to large-sized duck. Grey bill, narrow long neck, and sharp long tails in both sexes. Male head and hindneck are deep brown, with distinctive white line running up from white breast on both sides of the neck, forming two narrow lines pointing to the nape. Body is generally grey. Female is mainly brown. Upends when feeding.

1 adult male and female 雄成鳥和雌成鳥；Mai Po 米埔；Mar-07, 07 年 3 月；James Lam 林文華
2 adult eclipse / 1st winter male 非繁殖羽成鳥 / 第一年冬天雄鳥；San Tin 新田；Dec-06, 06 年 12 月；James Lam 林文華
3 adult male 雄成鳥；Mai Po 米埔；10-Feb-07, 07 年 2 月 10 日；Michelle and Peter Wong 江敏兒、黃理沛
4 Mai Po 米埔；Jan-07, 07 年 1 月；Henry Lui 呂德恆
5 adult male 雄成鳥；Nam Sang Wai 南生圍；Feb-07, 07 年 2 月；Cherry Wong 黃卓研
6 adult female 雌成鳥；Nam Sang Wai 南生圍；Jan-08, 08 年 1 月；Cherry Wong 黃卓研
7 adult female 雌成鳥；Nam Sang Wai 南生圍；Dec-07, 07 年 12 月；Michelle and Peter Wong 江敏兒、黃理沛
8 adult male 雄成鳥；Mai Po 米埔；10-Feb-07, 07 年 2 月 10 日；Michelle and Peter Wong 江敏兒、黃理沛

| | 春季過境遷徙鳥<br>Spring Passage Migrant | | | 夏候鳥<br>Summer Visitor | | 秋季過境遷徙鳥<br>Autumn Passage Migrant | | 冬候鳥<br>Winter Visitor | | |
|---|---|---|---|---|---|---|---|---|---|---|
| 常見月份 | 1 | 2 | 3 | 4 | 5 | 6 | 7 | 8 | 9 | 10 | 11 | 12 |
| | 留鳥<br>Resident | | | 迷鳥<br>Vagrant | | | 偶見鳥<br>Occasional Visitor | | |

# 綠翅鴨 ⑮lù chì yā

## Eurasian Teal | *Anas crecca*

體長 length：34-38cm

香港最細小的鴨。嘴黑色，頸短，有綠色翼鏡。雄鳥頭部深褐色和深綠色，有細長黃線分隔，軀體灰色，臀部奶黃色帶有黑邊。雌鳥全身大致褐色。

Smallest duck in Hong Kong. Black bill, short neck, green speculum in both sexes. Head of male is dark brown and green, marked out by a thin yellow line. Body is grey and undertail covert is milky yellow with black edge. Female is mainly brown.

1 adult male 雄成鳥；Nam Sang Wai 南生圍；Feb-05, 05 年 2 月；Cherry Wong 黃卓研
2 adult male 雄成鳥；Nam Sang Wai 南生圍；Dec-07, 07 年 12 月；Michelle and Peter Wong 江敏兒、黃理沛
3 Mai Po 米埔；Feb-07, 07 年 2 月；Bill Man 文權溢
4 adult male 雄成鳥；Mai Po 米埔；Jan-04, 04 年 1 月；Henry Lui 呂德恒
5 adult male 雄成鳥；Nam Sang Wai 南生圍；Dec-05, 05 年 12 月；Jemi and John Holmes 孔思義、黃亞萍
6 adult female 雌成鳥；Nam Sang Wai 南生圍；Feb-07, 07 年 2 月；Herman Ip 葉紀江

| 春季過境遷徙鳥 Spring Passage Migrant | | | | 夏候鳥 Summer Visitor | | 秋季過境遷徙鳥 Autumn Passage Migrant | | 冬候鳥 Winter Visitor | | | |
|---|---|---|---|---|---|---|---|---|---|---|---|
| 1 | 2 | 3 | 4 | 5 | 6 | 7 | 8 | 9 | 10 | 11 | 12 |
| 留鳥 Resident | | | | 迷鳥 Vagrant | | | | 偶見鳥 Occasional Visitor | | | |

常見月份

鴨科
Anatidae

# 紅頭潛鴨 ㊪hóng tóu qián yā

體長 length：42-49cm

## Common Pochard | *Aythya ferina*

1

中型鴨，嘴和腳灰色。雄鳥頭紅褐色，胸黑色，和近白色的軀體對比鮮明。雌鳥全身大致褐色，臉上有特別的淡褐色花紋。

Medium-sized duck. Bill and legs are grey in both sexes. Male has reddish brown head and black breast, contrasting with the pale greyish body. Female is mainly brown but has distinctive pale brownish facial pattern.

① adult male 雄成鳥；Nam Sang Wai 南生圍；Dec-07, 07 年 12 月；Michelle and Peter Wong 江敏兒、黃理沛
② adult female 雌成鳥；Nam Sang Wai 南生圍；Dec-07, 07 年 12 月；Michelle and Peter Wong 江敏兒、黃理沛
③ adult male 雄成鳥；Nam Sang Wai 南生圍；Dec-07, 07 年 12 月；Michelle and Peter Wong 江敏兒、黃理沛
④ adult female 雌成鳥；Nam Sang Wai 南生圍；Dec-07, 07 年 12 月；Michelle and Peter Wong 江敏兒、黃理沛
⑤ adult male and female 雄成鳥和雌成鳥；Nam Sang Wai 南生圍；Dec-07, 07 年 12 月；Michelle and Peter Wong 江敏兒、黃理沛
⑥ adult male 雄成鳥；Nam Sang Wai 南生圍；Dec-07, 07 年 12 月；Michelle and Peter Wong 江敏兒、黃理沛
⑦ adult female 雌成鳥；Nam Sang Wai 南生圍；Dec-07, 07 年 12 月；Michelle and Peter Wong 江敏兒、黃理沛

| 春季過境遷徙鳥<br>Spring Passage Migrant | | 夏候鳥<br>Summer Visitor | | 秋季過境遷徙鳥<br>Autumn Passage Migrant | | 冬候鳥<br>Winter Visitor | |
|---|---|---|---|---|---|---|---|

| 常見月份 | 1 | 2 | 3 | 4 | 5 | 6 | 7 | 8 | 9 | 10 | 11 | 12 |
|---|---|---|---|---|---|---|---|---|---|---|---|---|

| 留鳥<br>Resident | 迷鳥<br>Vagrant | 偶見鳥<br>Occasional Visitor |
|---|---|---|

# 青頭潛鴨

(普) qīng tóu qián yā

體長 length：41-47cm

Baer's Pochard | *Aythya baeri*

型鴨。脇部隱約有褐色和白色相間斑紋，尾下覆羽明顯白色。雄鳥頭墨綠色，虹膜白色，胸紅褐色。雌鳥全身大致褐色，虹膜褐色，嘴基有淡褐色斑塊。

Medium-sized duck. Both sexes have diffusely barred brown and white flanks, and prominent white undertail coverts. Male has dark green head, white iris and reddish brown breast. Female is mainly brown, with brownish iris and pale brown patches at the bill base.

1 adult female 雌成鳥：Tsim Bei Tsui 尖鼻咀：Aug-03, 03 年 8 月：Michelle and Peter Wong 江敏兒 · 黃理沛
2 adult female 雌成鳥：Tsim Bei Tsui 尖鼻咀：Aug-03, 03 年 8 月：Michelle and Peter Wong 江敏兒 · 黃理沛
3 adult female 雌成鳥：Tsim Bei Tsui 尖鼻咀：Aug-03, 03 年 8 月：Michelle and Peter Wong 江敏兒 · 黃理沛
4 adult female 雌成鳥：Lok Ma Chau 落馬洲：Martin Hale 夏敖天

| 春季過境遷徙鳥<br>Spring Passage Migrant | | | | 夏候鳥<br>Summer Visitor | | 秋季過境遷徙鳥<br>Autumn Passage Migrant | | | 冬候鳥<br>Winter Visitor | | |
|---|---|---|---|---|---|---|---|---|---|---|---|
| 1 | 2 | 3 | 4 | 5 | 6 | 7 | 8 | 9 | 10 | 11 | 12 |
| 留鳥<br>Resident | | | | 迷鳥<br>Vagrant | | | | 偶見鳥<br>Occasional Visitor | | | |

常見月份

# 白眼潛鴨

(普) bái yǎn qián yā

Ferruginous Duck | *Aythya nyroca*

體長 length：38-42cm

身形較青頭潛鴨小。雌雄羽色相似，全身褐色，尾下覆羽白色。雄鳥頭胸顏色較鮮明，虹膜白色。雌鳥顏色較暗淡。

Smaller than Baer's Pochard. Sexes similar. Plumage is mainly brownish with white undertail coverts. Male has a brighter head and breast, and prominent white iris. Female is duller.

1 adult male 雄成鳥；Mai Po 米埔；Owen Chiang 深藍
2 male 雄鳥；Lok Ma Chau 落馬洲；Martin Hale 夏敦天
3 adult male 雄成鳥；Mai Po 米埔；Feb-07, 07 年 2 月；Cherry Wong 黃卓研

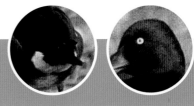

| 春季過境遷徙鳥 Spring Passage Migrant | | | 夏候鳥 Summer Visitor | | | 秋季過境遷徙鳥 Autumn Passage Migrant | | | 冬候鳥 Winter Visitor | | | |
|---|---|---|---|---|---|---|---|---|---|---|---|---|
| 1 | 2 | 3 | 4 | 5 | 6 | 7 | 8 | 9 | 10 | 11 | 12 | 常見月份 |
| 留鳥 Resident | | | | 迷鳥 Vagrant | | | | 偶見鳥 Occasional Visitor | | | | |

# 鳳頭潛鴨 fèng tóu qián yā

體長length：40-47cm

## Tufted Duck | *Aythya fuligula*

1

中 型鴨。嘴灰色，眼黃色，頭後有一小束飾羽，游泳輕盈迅速，尾貼近水面。雄鳥頭、胸、背、尾幾近黑色，白色的翼下和腹部對比鮮明。雌鳥全身大致深褐色，在秋季嘴基和尾下不時沾有白色。

Medium-sized duck. Both sexes have grey bill, yellow eyes and head tufts. Swims buoyantly with tail on the surface. Head, breast, back and tail almost black in male, contrasting with the white underwings and belly. Female is mainly brown, sometimes with white patch at the base of the bill and under the tail in autumn.

1 adult eclipse / 1st winter male 非繁殖羽成年 / 第一年冬天雄鳥：Nam Sang Wai 南生圍：Nov-07, 07 年 11 月：Aka Ho
2 adult male and female 雄成鳥和雌成鳥：Mai Po 米埔：Dec-03, 03 年 12 月：Jemi and John Holmes 孔思義、黃亞萍
3 adult eclipse male 非繁殖羽雄成鳥：Mai Po 米埔：Jan-09, 09 年 1 月：James Lam 林文華
4 adult male 雄成鳥：Mai Po 米埔：Jan-07, 07 年 1 月：Cherry Wong 黃卓研
5 adult male 雄成鳥：Mai Po 米埔：Jan-08, 08 年 1 月：Pang Chun Chiu 彭俊超
6 adult male and female 雄成鳥和雌成鳥：Mai Po 米埔：20-Jan-07, 07 年 1 月 20 日：Doris Chu 朱詠兒
7 adult male 雄成鳥：Mai Po 米埔：Jan-09, 09 年 1 月：James Lam 林文華
8 adult female 雌成鳥：Mai Po 米埔：Dec-06, 06 年 12 月：Cherry Wong 黃卓研

| | 春季過境遷徙鳥<br>Spring Passage Migrant | | | 夏候鳥<br>Summer Visitor | | | 秋李過境遷徙鳥<br>Autumn Passage Migrant | | | 冬候鳥<br>Winter Visitor | | |
|---|---|---|---|---|---|---|---|---|---|---|---|---|
| 常見月份 | 1 | 2 | 3 | 4 | 5 | 6 | 7 | 8 | 9 | 10 | 11 | 12 |
| | 留鳥<br>Resident | | | | 迷鳥<br>Vagrant | | | | 偶見鳥<br>Occasional Visitor | | | |

# 斑背潛鴨

(普)bān bèi qián yā

體長 length：40-51cm

Greater Scaup | *Aythya marila*

中型鴨。頭部較大渾圓，嘴大而闊，虹膜黃色。雄鳥頭部深黑色，帶綠色光澤，背部灰白，近看可見很多細紋。脇部白色，尾下黑色。雌鳥頭部明顯深褐色，嘴基可見淡色斑塊，上背灰褐帶深色細紋，脇部灰色沾褐。

Medium-sized duck. Both sexes have big and evenly round head, broad bill and yellow iris. Male has dark black head with green gloss and greyish white back. Many fine bars on the back visible at close range. White flanks and black undertail coverts. Female has dark brown head with prominent white oval patch at bill base. Upperparts are greyish brown with some narrow dark bars. Flanks are greyish and tinted brown.

[1] adult male and female 雄成鳥和雌成鳥；Mai Po 米埔；Jan-07, 07 年 1 月；Kelvin Yam 任德政
[2] female 雌鳥；San Tin 新田；Tam Yiu Leung 譚耀良
[3] male eclipse 非繁殖羽雄鳥；San Tin 新田；Tam Yiu Leung 譚耀良
[4] 1st winter male 第一年冬天雄鳥；Pui O 貝澳；Dec-08, 08 年 12 月；Ng Lin Yau 吳璉宥

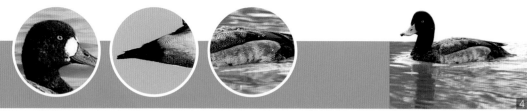

| 春季過境遷徙鳥 Spring Passage Migrant | | | | 夏候鳥 Summer Visitor | | 秋季過境遷徙鳥 Autumn Passage Migrant | | | 冬候鳥 Winter Visitor | | |
|---|---|---|---|---|---|---|---|---|---|---|---|
| 1 | 2 | 3 | 4 | 5 | 6 | 7 | 8 | 9 | 10 | 11 | 12 |

常見月份

| 留鳥 Resident | | | 迷鳥 Vagrant | | 偶見鳥 Occasional Visitor | | |
|---|---|---|---|---|---|---|---|

# 黑海番鴨

(普) hēi hǎi fān yā

體長 length：43-54cm

## Black Scoter | *Melanitta americana*

雄 鳥全身黑色，上嘴基部「腫脹」而呈鮮黃色。雌鳥和幼鳥大致褐色，臉和頸兩側淡色。

Male is overall black, with a bright yellow knob at the bill base. Female and juvenile are overall brown with pale cheeks and sides of neck.

① adult female 雌成鳥：Martin Hale 夏敖天
② adult female 雌成鳥：Mai Po 米埔：9-Dec-07, 07 年 12 月 9 日：Martin Hale 夏敖天
③ adult female 雌成鳥：Mai Po 米埔：9-Dec-07, 07 年 12 月 9 日：Martin Hale 夏敖天
④ adult female 雌成鳥：Martin Hale 夏敖天

| 春季過境遷徙鳥<br>Spring Passage Migrant | | | 夏候鳥<br>Summer Visitor | | | 秋季過境遷徙鳥<br>Autumn Passage Migrant | | | 冬候鳥<br>Winter Visitor | | |
|---|---|---|---|---|---|---|---|---|---|---|---|
| 1 | 2 | 3 | 4 | 5 | 6 | 7 | 8 | 9 | 10 | 11 | 12 |
| 留鳥<br>Resident | | | | 迷鳥<br>Vagrant | | | | 偶見鳥<br>Occasional Visitor | | | |

常見月份

# 白秋沙鴨 ⓟbái qiū shā yā

體長 length：35-44cm

## Smew | *Mergellus albellus*

中 型鴨。雄鳥全身黑白相間，眼周圍有大片黑斑，十分易認。雌鳥頭至後枕栗色，喉及頰部白色，其餘部分灰褐色。

Medium-sized duck. Male is a distinctive black and white with a large black patch around the eye. Female is greyish brown overall with dark brown head to nape and white throat and chins.

[1] adult female 雌成鳥；Mai Po 米埔；Dec-06, 06 年 12 月；Cherry Wong 黃卓研
[2] adult female 雌成鳥；San Tin 新田；Nov-18, 18 年 11 月；Kwok Tsz Ki 郭子祈
[3] adult female 雌成鳥；San Tin 新田；Jan-19, 19 年 1 月；Roman Lo 羅文凱
[4] adult female 雌成鳥；Mai Po 米埔；Dec-06, 06 年 12 月；Cherry Wong 黃卓研

| 春季過境遷徙鳥 Spring Passage Migrant | | | 夏候鳥 Summer Visitor | | | 秋季過境遷徙鳥 Autumn Passage Migrant | | | 冬候鳥 Winter Visitor | | |
|---|---|---|---|---|---|---|---|---|---|---|---|
| 1 | 2 | 3 | 4 | 5 | 6 | 7 | 8 | 9 | 10 | 11 | 12 |

常見月份

| 留鳥 Resident | 迷鳥 Vagrant | 偶見鳥 Occasional Visitor |
|---|---|---|
| • | | |

# 紅胸秋沙鴨

(普)hóng xiōng qiū shā yā

體長 length：52-58cm

## Red-breasted Merganser | *Mergus serrator*

嘴 長而帶鋸齒，枕部的羽毛長而鬆散，飛行時可見顯著的翼斑，但雌鳥前翼沒有白色。雄性頭部深綠色，上胸栗色，雌性頭帶褐色。

Long and saw-shaped bill with ragged crest, In flight, it shows prominent white wing patch, but female has no white colour on forewing. Male has dark greenish head and chestnut coloured upper breast. Female has brownish head.

1 adult female 雌成鳥：Tai Sang Wai 大生圍；15 Dec-12, 12 年 12 月 15 日；Beetle Cheng 鄭諾銘
2 adult female 雌成鳥：Po Toi 蒲台；7-Dec-06, 06 年 12 月 7 日；Geoff Welch
3 adult male 雄成鳥；16-Mar-07, 07 年 3 月 16 日；Geoff Welch

| 春季過境遷徙鳥<br>Spring Passage Migrant | | | | 夏候鳥<br>Summer Visitor | | 秋季過境遷徙鳥<br>Autumn Passage Migrant | | | 冬候鳥<br>Winter Visitor | | |
|---|---|---|---|---|---|---|---|---|---|---|---|
| 1 | 2 | 3 | 4 | 5 | 6 | 7 | 8 | 9 | 10 | 11 | 12 |

常見月份

| 留鳥<br>Resident | 迷鳥<br>Vagrant | 偶見鳥<br>Occasional Visitor |
|---|---|---|

# 普通夜鷹  普 pǔ tōng yè yīng

體長 length：24-27cm

## Grey Nightjar | *Caprimulgus jotaka*

其他名稱 Other names：Japanese Nightjar, Jungle Nightjar

雄 鳥毛色偏褐，尾近末端中間有斷開的白色橫帶，飛行時，從翼下見初級飛羽基部有小白點。雌鳥顏色偏灰。常伏在地上，叫聲為低音調而重複的「促─促─促─促」，約每秒鐘四個音節。

Male dark brown with white sub-terminal tail band that is broken in the middle. In flight, it has prominent white spots at the base of primaries. Female greyish colour. Usually rests on the ground. Call a chattering and repeated "chuck - chuck - chuck - chuck" at about four notes per second.

1 adult male 雄成鳥：Lai Chi Kok 荔枝角；Nov-08, 08 年 11 月；James Lam 林文華
2 adult male 雄成鳥：Lai Chi Kok 荔枝角；Nov-08, 08 年 11 月；James Lam 林文華
3 adult male 雄成鳥：Lai Chi Kok 荔枝角；Nov-08, 08 年 11 月；Sammy Sam and Winnie Wong 森美與雲妮
4 adult female 雌成鳥：Lamma Island 南丫島；5-Apr-07, 07 年 4 月 5 日；Geoff Smith
5 adult male 雄成鳥：Lai Chi Kok 荔枝角；Nov-08, 08 年 11 月；Sammy Sam and Winnie Wong 森美與雲妮

| | 春季過境遷徙鳥 Spring Passage Migrant | | 夏候鳥 Summer Visitor | | 秋季過境遷徙鳥 Autumn Passage Migrant | | 冬候鳥 Winter Visitor | | |
|---|---|---|---|---|---|---|---|---|---|
| 常見月份 | 1 | 2 | 3 | 4 | 5 | 6 | 7 | 8 | 9 | 10 | 11 | 12 |

| | 留鳥 Resident | | 迷鳥 Vagrant | | 偶見鳥 Occasional Visitor | |
|---|---|---|---|---|---|---|

# 林夜鷹 ⑱lín yè yīng

體長length：20-26cm

## Savanna Nightjar | *Caprimulgus affinis*

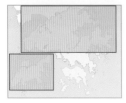

與普通夜鷹比較，林夜鷹尾部較短，褐色較濃和頭部有幼細斑點。雄鳥外側尾羽白色，飛行時初級飛羽有一片白斑。雌鳥尾羽顏色偏褐，翼斑較淡。叫聲為響亮的「chweep」，好像將塑膠在玻璃上摩擦的聲音。

Compared to Grey Nightjar, Savanna Nightjar has shorter tail, dark brownish overall with dense small marks on head. Male has white outer tail feathers and white wing patches which are especially visible in flight. Female has brown tail and duller wing patches. Call a "chweep", like the sound of plastic rubbing against glass.

1 male adult 雄性成年：Long Valley 塱原；Dec-07, 07 年 12 月； Wong Po Wai 黃寶偉
2 adult male 雄成鳥：Chau Tau 洲頭；May-06, 06 年 5 月；Jemi and John Holmes 孔思義、黃亞萍
3 adult male 雄成鳥：Ping Che 坪輋；Sep-08, 08 年 9 月；Sammy Sam and Winnie Wong 森美與雲妮
4 adult female 雌成鳥：Fanling 粉嶺；Sep-08, 08 年 9 月；Eling Lee 李佩玲
5 adult male 雄成鳥：Fanling 粉嶺；13-Sep-08, 08 年 9 月 13 日；Owen Chiang 深藍
6 juvenile 幼鳥：Tin Shui Wai 天水圍；Aug-08, 08 年 8 月；Sammy Sam and Winnie Wong 森美與雲妮
7 adult female 雌成鳥：Robin's Nest 紅花嶺；17-Oct-04, 04 年 10 月 17 日；Michelle and Peter Wong 江敏兒、黃理沛
8 adult female 雌成鳥：Robin's Nest 紅花嶺；Sep-06, 06 年 9 月；Jemi and John Holmes 孔思義、黃亞萍

| 春季過境遷徙鳥<br>Spring Passage Migrant | | | 夏候鳥<br>Summer Visitor | | | 秋季過境遷徙鳥<br>Autumn Passage Migrant | | | 冬候鳥<br>Winter Visitor | | |
|---|---|---|---|---|---|---|---|---|---|---|---|
| 1 | 2 | 3 | 4 | 5 | 6 | 7 | 8 | 9 | 10 | 11 | 12 |
| 留鳥<br>Resident | | | | 迷鳥<br>Vagrant | | | | 偶見鳥<br>Occasional Visitor | | | |

常見月份

雨燕科
Apodidae

# 短嘴金絲燕
(普) duǎn zuǐ jīn sī yàn

體長 length：13-14cm

## Himalayan Swiftlet | *Aerodramus brevirostris*

近 黑褐色的金絲燕，尾略呈叉形。腰色稍淡，顏色淺褐或偏灰色，與身體顏色對比不明顯。

A small dark brown swift with notched tail and paler rump. The rump colour varies from light brown to grey, sometimes not contrasting with the body.

1 4-Mar-05, 05 年 3 月 4 日：Jacky Yam 任永耀
2 24-Jan-08, 08 年 1 月 24 日：Martin Hale 夏敖天
3 4-Mar-05, 05 年 3 月 4 日：Jacky Yam 任永耀
4 3-Mar-05, 05 年 3 月 3 日：Jacky Yam 任永耀
5 4-Mar-05, 05 年 3 月 4 日：Jacky Yam 任永耀

| | 春季過境遷徙鳥<br>Spring Passage Migrant | | 夏候鳥<br>Summer Visitor | | 秋季過境遷徙鳥<br>Autumn Passage Migrant | | 冬候鳥<br>Winter Visitor | | |
|---|---|---|---|---|---|---|---|---|---|
| 常見月份 | 1 | 2 | 3 | 4 | 5 | 6 | 7 | 8 | 9 | 10 | 11 | 12 |
| | 留鳥<br>Resident | | | 迷鳥<br>Vagrant | | | 偶見鳥<br>Occasional Visitor | | |

# 白喉針尾雨燕 (普) bái hóu zhēn wěi yǔ yàn

體長 length：19-20cm

## White-throated Needletail | *Hirundapus caudacutus*

[1]

[2]

深 色大型雨燕，喉白色，三級飛羽有小塊白色，背上有銀白色馬鞍形斑塊。脅部至尾下覆羽有顯眼的U形白斑，翼比灰喉針尾雨燕長。

Large dark swift with diagnostic white throat and small white patch on tertials. Upperparts have silvery white wash saddle-like patch. An obvious U-shaped white pattern from flank to vent. Like Silver-backed, but with longer wings.

1 Po Toi 蒲台；16-Apr-09, 09 年 4 月 16 日；Geoff Welch
2 Beetle Cheng 鄭諾銘
3 Po Toi 蒲台；16-Apr-09, 09 年 4 月 16 日；Geoff Welch

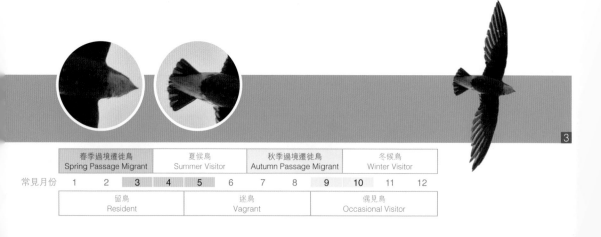

[3]

| 春季過境遷徙鳥<br>Spring Passage Migrant | | 夏候鳥<br>Summer Visitor | | 秋季過境遷徙鳥<br>Autumn Passage Migrant | | 冬候鳥<br>Winter Visitor | |
|---|---|---|---|---|---|---|---|

| 常見月份 | 1 | 2 | 3 | 4 | 5 | 6 | 7 | 8 | 9 | 10 | 11 | 12 |
|---|---|---|---|---|---|---|---|---|---|---|---|---|

| 留鳥<br>Resident | 迷鳥<br>Vagrant | 偶見鳥<br>Occasional Visitor |
|---|---|---|

# 灰喉針尾雨燕

(普) huī hóu zhēn wěi yǔ yàn

體長 length：20cm

## Silver-backed Needletail | *Hirundapus cochinchinensis*

其他名稱 Other names：白背針尾雨燕

① ② ③

體型大近黑色的針尾雨燕，脇部至尾下覆羽U形白斑顯眼，與白喉針尾雨燕的分別在喉部深灰色，身較修長，翼稍短。

Large needletail swift with white U-shaped patch from flank to vent. Compare to White-throated Needletail, it has a more dark greyish throat, slimmer body and shorter wings.

① Po Toi 蒲台：Apr-10, 10 年 4 月：Michelle and Peter Wong 江敏兒、黃理沛
② Po Toi 蒲台：Apr-10, 10 年 4 月：Michelle and Peter Wong 江敏兒、黃理沛
③ Po Toi 蒲台：Apr-10, 10 年 4 月：Michelle and Peter Wong 江敏兒、黃理沛
④ 15-Mar-07, 07 年 3 月 15 日：Geoff Welch
⑤ 15-Mar-07, 07 年 3 月 15 日：Geoff Welch

④

⑤

| 春季過境遷徙鳥<br>Spring Passage Migrant | | 夏候鳥<br>Summer Visitor | | 秋季過境遷徙鳥<br>Autumn Passage Migrant | | 冬候鳥<br>Winter Visitor | |
|---|---|---|---|---|---|---|---|
| 1 | 2 | 3 | 4 | 5 | 6 | 7 | 8 | 9 | 10 | 11 | 12 |

常見月份

| 留鳥<br>Resident | 迷鳥<br>Vagrant | 偶見鳥<br>Occasional Visitor |
|---|---|---|

雨燕科
Apodidae

# 白腰雨燕

(普) bái yāo yǔ yàn

體長 length：17-18cm

## Pacific Swift | *Apus pacificus*

其他名稱 Other names：大白腰雨燕

**大**型的深褐色雨燕，尾長而尾叉深，額和喉偏白，腰部白色。與小白腰雨燕的分別在於體型較大，顏色較淡，喉部顏色較深，腰部白色位置較窄。尾開叉。

Large dark brown swift with white rump and deeply forked tail. Pale throat and chin. Compare with House Swift, Pacific Swift is much larger in size, slimmer in appearance, paler colour, darker throat. White patch at rump is narrower, forked-tail.

1 Ma Tso Lung 馬草壟；9-Mar-08, 08 年 3 月 9 日：Sung Yik Hei 宋亦希
2 Mai Po 米埔：4-Apr-05, 05 年 4 月 4 日：Wong Hok Sze 王學思
3 Mai Po 米埔：4-Apr-05, 05 年 4 月 4 日：Wong Hok Sze 王學思
4 adult 成鳥；Ngau Tam Mei 牛潭尾；Apr-20, 20 年 4 月：Henry Lui 呂德恆
5 Ma Tso Lung 馬草壟；9-Mar-08, 08 年 3 月 9 日：Sung Yik Hei 宋亦希

| | 春季過境遷徙鳥 Spring Passage Migrant | | | 夏候鳥 Summer Visitor | | | 秋季過境遷徙鳥 Autumn Passage Migrant | | | 冬候鳥 Winter Visitor | | |
|---|---|---|---|---|---|---|---|---|---|---|---|---|
| 常見月份 | 1 | 2 | 3 | 4 | 5 | 6 | 7 | 8 | 9 | 10 | 11 | 12 |
| | 留鳥 Resident | | | | 迷鳥 Vagrant | | | | 偶見鳥 Occasional Visitor | | | |

# 小白腰雨燕 <small>(普)xiǎo bái yāo yǔ yàn</small>

## House Swift | *Apus nipalensis*

體長 length：15cm

1

小 型雨燕。全身黑色，但腰部白色，喉部偏白。愛成群飛行，飛行時像小型的船錨，尾張開時呈方形微凹，不斷「茲茲」鳴叫。在屋簷下築巢繁殖，香港中文大學圖書館有全港最大的繁殖群落。

Small-sized swift. Black overall except white rump and whitish throat. Often fly in twittering flocks. In flight, it looks like a small anchor, and tail square and slightly notched when spread. Call a continuous "si-si-". Nests built under eaves. The largest breeding colony in Hong Kong is at the Main Library of The Chinese University of Hong Kong.

[1] adult 成鳥：Lut Chau 甩洲；Jan-06, 06 年 1 月；Pippen Ho 何志剛
[2] adult 成鳥：Fanling Wai 粉嶺圍；Jun-04, 04 年 6 月；Henry Lui 呂德恆
[3] adult 成鳥：Long Valley 塱原；15-Nov-07, 07 年 11 月 15 日；Owen Chiang 深藍
[4] adult 成鳥：Tsim Bei Tsui 尖鼻咀；May-05, 05 年 5 月；Cherry Wong 黃卓研
[5] juvenile 幼鳥：Lantau Island 大嶼山；26-Jun-07, 07 年 6 月 26 日；Wong Hok Sze 王學思
[6] adult 成鳥：Tsim Bei Tsui 尖鼻咀；May-05, 05 年 5 月；Cherry Wong 黃卓研

| | 春季過境遷徙鳥<br>Spring Passage Migrant | | | 夏候鳥<br>Summer Visitor | | | 秋季過境遷徙鳥<br>Autumn Passage Migrant | | | 冬候鳥<br>Winter Visitor | | |
|---|---|---|---|---|---|---|---|---|---|---|---|---|
| 常見月份 | 1 | 2 | 3 | 4 | 5 | 6 | 7 | 8 | 9 | 10 | 11 | 12 |
| | 留鳥<br>Resident | | | | 迷鳥<br>Vagrant | | | | 偶見鳥<br>Occasional Visitor | | | |

# 褐翅鴉鵑
(普)hè chì yā juān

Greater Coucal | *Centropus sinensis*

體長 length：47-52cm

其他名稱 Other names：毛雞

**體**型大，成鳥全身黑色而帶光澤，翅膀鮮明栗色，嘴黑色，虹膜鮮紅色。幼鳥全身帶橫紋。常躲在樹叢下層，甚少飛行。發出低沉的「胡－胡－」聲。

Large adult has glossy black body with bright chestnut wings. Black bill and bright red iris. Juvenile is barred all over. Usually stay at under cover inside bushes. Call a repeated low pitched "woo woo" sound.

[1] breeding 繁殖羽；Cheung Chau 長洲；Nov-04, 04 年 11 月；Henry Lui 呂德恆
[2] breeding 繁殖羽；Cheung Chau 長洲；Jan-05, 05 年 1 月；Henry Lui 呂德恆
[3] adult 成鳥；Mai Po 米埔；Apr-20, 20年4月；Henry Lui 呂德恆
[4] juvenile 幼鳥；Mai Po 米埔；Sep-07, 07 年 9 月；Kinni Ho 何建業
[5] juvenile 幼鳥；Mai Po 米埔；Aug-08, 08 年 8 月；Lee Kai Hong 李啟康
[6] non-breeding adult 非繁殖羽成鳥；Cheung Chau 長洲；Jan-05, 05 年 1 月；Henry Lui 呂德恆
[7] juvenile 幼鳥；Mai Po 米埔；Sep-07, 07 年 9 月；James Lam 林文華

| | 春季過境遷徙鳥<br>Spring Passage Migrant | | | 夏候鳥<br>Summer Visitor | | | 秋季過境遷徙鳥<br>Autumn Passage Migrant | | | 冬候鳥<br>Winter Visitor | | |
|---|---|---|---|---|---|---|---|---|---|---|---|---|
| 常見月份 | 1 | 2 | 3 | 4 | 5 | 6 | 7 | 8 | 9 | 10 | 11 | 12 |
| | 留鳥<br>Resident | | | | 迷鳥<br>Vagrant | | | | 偶見鳥<br>Occasional Visitor | | | |

# 小鴉鵑

(普) xiǎo yā juān

體長 length：31-34cm

## Lesser Coucal | *Centropus bengalensis*

其他名稱 Other names：小毛雞

**外** 型和褐翅鴉鵑相似，但體型較小。嘴和腳黑色，虹膜深褐色。全身黑色，但翅膀和上背栗色。非繁殖期和幼鳥淡褐色，全身佈滿深色斑紋。叫聲為急促的「谷谷」聲。

Alike Greater Coucal but smaller in size. Bill and legs black. Dark brown iris. Black except chestnut wings and mantle. Non-breeding and juvenile are pale brown with dark stripes all over. Call a tight series of "hoot hoot hoot".

1 breeding 繁殖羽：Cheung Mok Jose Alberto 張振國
2 juvenile 幼鳥：Mai Po 米埔；Sep-07, 07 年 9 月；Kinni Ho 何建業
3 juvenile 幼鳥：Mai Po 米埔；Oct-07, 07 年 10 月；Lee Kai Hong 李啟康
4 juvenile 幼鳥：Ping Che 坪輋；18-May-09, 09 年 5 月 18 日；Martin Hale 夏敖天
5 juvenile 幼鳥：Kai Tak 啟德；Sep-17, 17 年 9 月；Beetle Cheng 鄭諾銘
6 adult 成鳥：Jul-10, 10 年 7 月；Christina Chan 陳燕明

| | 春季過境遷徙鳥<br>Spring Passage Migrant | | | 夏候鳥<br>Summer Visitor | | | 秋季過境遷徙鳥<br>Autumn Passage Migrant | | | 冬候鳥<br>Winter Visitor | | |
|---|---|---|---|---|---|---|---|---|---|---|---|---|
| 常見月份 | 1 | 2 | 3 | 4 | 5 | 6 | 7 | 8 | 9 | 10 | 11 | 12 |
| | 留鳥<br>Resident | | | | 迷鳥<br>Vagrant | | | | 偶見鳥<br>Occasional Visitor | | | |

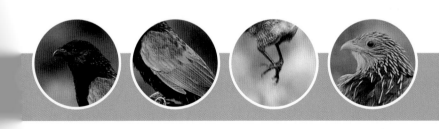

# 紅翅鳳頭鵑 (普)hóng chì fèng tóu juān

體長 length：38-46cm

## Chestnut-winged Cuckoo | *Clamator coromandus*

其他名稱 Other names：Red-winged Crested Cuckoo 紅翅鳳頭杜鵑

外 型獨特，有黑色的長冠羽和長尾。後頸有白帶、上體黑色且帶藍黑光澤。喉、上胸和翼栗色，下體偏白。日夜不停發出「必必」的叫聲。

Distinctive cuckoo with long black crest and long black tail. White band on hind neck. Dark upperparts with glossy blue colour. Chestnut throat, upper breast and wings. White belly. Persistent "peep-peep" call, day and night.

[1] adult 成鳥；Mai Po 米埔；Apr-07, 07 年 4 月；Felix Ng 伍昌齡
[2] adult 成鳥；Cheung Mok Jose Alberto 張振國
[3] adult 成鳥；Mai Po 米埔；Apr-07, 07 年 4 月；Felix Ng 伍昌齡
[4] adult 成鳥；Mai Po 米埔；Apr-07, 07 年 4 月；Cherry Wong 黃卓研
[5] adult 成鳥；Sai Kung 西貢；Jun-06, 06 年 6 月；Pippen Ho 何志剛
[6] adult 成鳥；Nam Chung 南涌；May-08, 08 年 5 月；Andy Cheung 張玉良

[1]

| | 春季過境遷徙鳥<br>Spring Passage Migrant | | | 夏候鳥<br>Summer Visitor | | | 秋季過境遷徙鳥<br>Autumn Passage Migrant | | | 冬候鳥<br>Winter Visitor | |
|---|---|---|---|---|---|---|---|---|---|---|---|
| 常見月份 | 1 | 2 | 3 | 4 | 5 | 6 | 7 | 8 | 9 | 10 | 11 | 12 |
| | 留鳥<br>Resident | | | | 迷鳥<br>Vagrant | | | | 偶見鳥<br>Occasional Visitor | | |

# 噪鵑 <sup>普</sup>záo juān

體長 length：39-46cm

## Asian Koel | *Eudynamys scolopaceus*

雄 鳥全身藍黑色，虹膜鮮紅色，嘴蘋果綠色。雌鳥和幼鳥相似，全身深褐色，帶淺色斑點。冬末至初夏鳴叫，叫聲為響亮而重複的粵音「戶－污－」。

Male is completely bluish black with bright red iris and apple green bill. Female and juvenile are similar, with dark brown body and pale spots. Calls heard from late winter to early summer which is loud and repeated "ko-el".

[1] adult male 雄成鳥；Mai Po 米埔；Mar-09, 09 年 3 月；Pippen Ho 何志剛
[2] adult female 雌成鳥；Mai Po 米埔；Sep-07, 07 年 9 月；Helen Chan 陳燕芳
[3] juvenile male 雄幼鳥；Mai Po 米埔；Aug-04, 04 年 8 月；Henry Lui 呂德恆
[4] adult female 雌成鳥；Sai Kung 西貢；Jun-07, 07 年 6 月；Sammy Sam and Winnie Wong 森美與雲妮
[5] juvenile male 雄幼鳥；Mai Po 米埔；Jun-07, 07 年 6 月；Cherry Wong 黃卓研
[6] adult male 雄成鳥；Mai Po 米埔；Sep-07, 07 年 9 月；Helen Chan 陳燕芳

| | 春季過境遷徙鳥 Spring Passage Migrant | | | 夏候鳥 Summer Visitor | | | 秋季過境遷徙鳥 Autumn Passage Migrant | | | 冬候鳥 Winter Visitor | | |
|---|---|---|---|---|---|---|---|---|---|---|---|---|
| 常見月份 | 1 | 2 | 3 | 4 | 5 | 6 | 7 | 8 | 9 | 10 | 11 | 12 |
| | 留鳥 Resident | | | | 迷鳥 Vagrant | | | | 偶見鳥 Occasional Visitor | | | |

# 八聲杜鵑 （普）bā shēng dù juān

## Plaintive Cuckoo | *Cacomantis merulinus*

體長 length：18-23.5cm

1

小型杜鵑。成鳥頭灰色，上體深灰褐色，下胸至腹部橙褐色，尾上深色，尾尖帶白點，尾下深色而有白色橫斑。幼鳥和赤色型雌鳥主要為紅褐色，有很多深色橫紋。獨特的八音節叫聲，前三聲長而慢，後五聲短而急、音調漸降。

Small-sized cuckoo. Grey head. Dark greyish brown upperparts, and brownish orange lower breast and belly. Dark uppertail with white spots at tips. Dark undertail with white bands. Juvenile and hepatic morph are mainly reddish brown with many dark bars. Distinctive call with eight notes, first three notes long, slow and rising, and last five notes short, fast and falling.

1 adult 成鳥：Long Valley 塱原：Jan-08, 08 年 1 月：Eling Lee 李佩玲
2 adult 成鳥：Long Valley 塱原：Feb-07, 07 年 2 月：Owen Chiang 深藍
3 adult 成鳥：Long Valley 塱原：Oct-07, 07 年 10 月：Andy Kwok 郭匯昌
4 juvenile 幼鳥：Long Valley 塱原：Oct-06, 06 年 10 月：Allen Chan 陳志雄
5 adult 成鳥：Long Valley 塱原：Feb-07, 07 年 2 月：James Lam 林文華
6 juvenile moulting into adult plumage 幼鳥轉換成羽：Long Valley 塱原：Feb-07, 07 年 2 月：James Lam 林文華
7 juvenile 幼鳥：Long Valley 塱原：Jan-08, 08 年 1 月：Eling Lee 李佩玲

| | 春季過境遷徙鳥<br>Spring Passage Migrant | | | 夏候鳥<br>Summer Visitor | | | 秋季過境遷徙鳥<br>Autumn Passage Migrant | | | 冬候鳥<br>Winter Visitor | | |
|---|---|---|---|---|---|---|---|---|---|---|---|---|
| 常見月份 | 1 | 2 | 3 | 4 | 5 | 6 | 7 | 8 | 9 | 10 | 11 | 12 |
| | 留鳥<br>Resident | | | | 迷鳥<br>Vagrant | | | | 偶見鳥<br>Occasional Visitor | | | |

# 烏鵑
⑪ wū juān

體長 length：24-25cm

## Square-tailed Drongo-Cuckoo | *Surniculus lugubris*

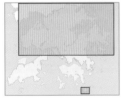

小型杜鵑。成鳥羽色全黑。頭、上背和上翼帶亮藍綠光澤。尾下覆羽和尾下有幼白間。幼鳥顏色偏褐，亮麗色澤較少，身上有白點。與卷尾相似，不過嘴部較幼和向下彎，翼較長，尾較短和呈淺叉狀，以及出現的生態環境不同。

Small-sized cuckoo. Adult mainly black in colour. Head, mantle and upperwings tinted glossy bluish green. White bars on undertails coverts and undertail. Juvenile has browner, less glossy plumage and white spots on the body. Resemble a Drongo, but the bill is much thinner and decurved, the wings are longer and the tail is shorter and less deeply forked. They occur in different habitats.

1 adult 成鳥：Po Toi 蒲台：Apr-07, 07 年 4 月：Matthew and TH Kwan 關朗曦、關子凱
2 adult 成鳥：Po Toi 蒲台：Apr-07, 07 年 4 月：Matthew and TH Kwan 關朗曦、關子凱
3 adult 成鳥：Po Toi 蒲台：Apr-07, 07 年 4 月：Martin Hale 夏敖天
4 adult 成鳥：Po Toi 蒲台：Oct-11, 11 年 10 月：Kinni Ho 何建業
5 adult 成鳥：Po Toi 蒲台：Oct-11, 11 年 10 月：Kinni Ho 何建業
6 juvenile 幼鳥：Tai Po Kau 大埔滘：Aug-04, 04 年 8 月：Michelle and Peter Wong 江敏兒、黃理沛

| 春季過境遷徙鳥 Spring Passage Migrant | | | 夏候鳥 Summer Visitor | | | 秋季過境遷徙鳥 Autumn Passage Migrant | | | 冬候鳥 Winter Visitor | | |
|---|---|---|---|---|---|---|---|---|---|---|---|
| 1 | 2 | 3 | 4 | 5 | 6 | 7 | 8 | 9 | 10 | 11 | 12 |

常見月份

| 留鳥 Resident | | | | 迷鳥 Vagrant | | | | 偶見鳥 Occasional Visitor | | | |
|---|---|---|---|---|---|---|---|---|---|---|---|

# 北鷹鵑
(普) běi yīng juān

體長 length：28-30cm

## Northern Hawk-Cuckoo | *Hierococcyx hyperythrus*

其他名稱 Other names：北方鷹鵑

嘴短黃色，嘴前端有一闊黑環，上嘴端向下彎。眼圈黃色，腳黃色。頭、背、翼及尾上藍灰色，後枕及三級飛羽有白斑，尾上有黑褐相間的橫帶。白色下體沾淺褐色。幼鳥下體有深色縱紋。叫聲為不斷重複的「止絕」聲。

Yellow and short bill with broad black ring near bill tip, upperbill decurved. Yellow eye-rings and legs. Head, mantle, wing and uppertail bluish grey. White patches on nape and tertials. Uppertail has alternative black and brown bars. White underparts tinted with pale brown. Juvenile has dark stripes on underparts. Repeated calls of "zhi...zhu...".

1 Tai Po Kau 大埔滘；Martin Hale 夏敖天
2 juvenile 幼鳥；Tai Po Kau 大埔滘；Aug-07, 07 年 8 月；Dick Lai 黎凱輝
3 Tai Po Kau 大埔滘；Tony Hung 洪教燊
4 juvenile 幼鳥；Tai Po Kau 大埔滘；Aug-07, 07 年 8 月；Dick Lai 黎凱輝

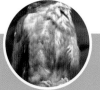

| 春季過境遷徙鳥 Spring Passage Migrant | | | | 夏候鳥 Summer Visitor | | | 秋季過境遷徙鳥 Autumn Passage Migrant | | 冬候鳥 Winter Visitor | | |
|---|---|---|---|---|---|---|---|---|---|---|---|
| 1 | 2 | 3 | 4 | 5 | 6 | 7 | 8 | 9 | 10 | 11 | 12 |

常見月份

| 留鳥 Resident | 迷鳥 Vagrant | 偶見鳥 Occasional Visitor |
|---|---|---|

杜鵑科
Cuculidae

# 大鷹鵑 （普）dà yīng juān

體長 length：38-40cm

Large Hawk-Cuckoo | *Hierococcyx sparverioides*

其他名稱 Other names：大鷹鵑

嘴 黃色，上嘴端向下彎，眼圈黃色，腳黃色。面頰淡灰色，頭、背、翼及尾上深灰褐色，尾部黑白和褐色橫紋相間。下體白色，喉至胸有深色縱紋，胸部紅褐色，十分明顯。幼鳥上體有紅褐色橫紋，下體偏褐而有深色縱紋。春、夏叫聲為不斷重複的「brain fever」聲。

Yellow bill, upperbill tip decurved. Yellow eye-rings and legs. Cheeks pale grey. Head, mantle, wings and uppertail greyish brown. Tail with alternative black, brown and white bars. White underparts with dark streaks. Breast has prominent reddish brown colour. Juvenile has reddish brown bars on upperparts, and dark streaks on pale brown underparts. Repeated calls of "brain fever" in spring and summer.

1 adult 成鳥；Shek Kong 石崗；Mar-09, 09 年 3 月；Allen Chan 陳志雄
2 adult 成鳥；Shek Kong 石崗；Mar-09, 09 年 3 月；Allen Chan 陳志雄
3 adult 成鳥；Shek Kong 石崗；Mar-07, 07 年 3 月；Martin Hale 夏敖天
4 juvenile 幼鳥；Kam Tin 錦田；May-07, 07 年 5 月；Andy Li 李偉仁
5 adult 成鳥；Kam Tin 錦田；Apr-07, 07 年 4 月；Pippen Ho 何志剛
6 adult 成鳥；Long Valley 塱原；May-08, 08 年 5 月；Chan Kai Wai 陳佳瑋
7 adult 成鳥；Long Valley 塱原；Apr-06, 06 年 4 月；Henry Lui 呂德恆

| | 春季過境遷徙鳥<br>Spring Passage Migrant | | 夏候鳥<br>Summer Visitor | | | 秋季過境遷徙鳥<br>Autumn Passage Migrant | | 冬候鳥<br>Winter Visitor | | | |
|---|---|---|---|---|---|---|---|---|---|---|---|
| 常見月份 | 1 | 2 | 3 | 4 | 5 | 6 | 7 | 8 | 9 | 10 | 11 | 12 |
| | 留鳥<br>Resident | | | 迷鳥<br>Vagrant | | | 偶見鳥<br>Occasional Visitor | | | | | |

杜鵑科
**Cuculidae**

# 霍氏鷹鵑

(普) huò shì yīng juān

體長 length：28-30cm

**Hodgson's Hawk-Cuckoo** | *Hierococcyx nisicolor*

頭部灰色，有黃色幼眼圈，嘴黑色，嘴端及嘴基沾黃。上體灰褐色雜有褐色紋，下體白色並滿佈粗黑縱紋，腳黃色。

Grey head with thin yellow eye rings. Black bill tinged with yellow on tip and base. Greyish-brown upperparts with some brown streaks. White underparts with lots of thick black stripes. Yellow legs.

1 adult female 雌成鳥：Tai Po Kau 大埔滘：Aug-16, 16 年 8 月：Ken Fung 馮漢城
2 adult 成鳥：Wu Kau Tang 烏蛟騰：1 May-13, 13 年 5 月 1 日：Beetle Cheng 鄭諾銘
3 adult female 雌成鳥：Tai Po Kau 大埔滘：Aug-16, 16 年 8 月：Ken Fung 馮漢城

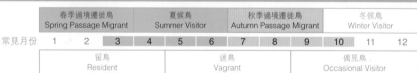

| | 春季過境遷徙鳥<br>Spring Passage Migrant | | 夏候鳥<br>Summer Visitor | | 秋季過境遷徙鳥<br>Autumn Passage Migrant | | | 冬候鳥<br>Winter Visitor | |
|---|---|---|---|---|---|---|---|---|---|
| 常見月份 | 1 | 2 | 3 | 4 | 5 | 6 | 7 | 8 | 9 | 10 | 11 | 12 |

| 留鳥<br>Resident | 迷鳥<br>Vagrant | 偶見鳥<br>Occasional Visitor |
|---|---|---|

# 小杜鵑

⓪ xiǎo dù juān

體長 length：25cm

## Lesser Cuckoo | *Cuculus poliocephalus*

①②

**體** 形細小的灰色杜鵑，形似中杜鵑及大杜鵑，上體灰色，白色腹部有較疏的黑色橫斑，臀部皮黃色。以五音節的叫聲最易與其他種類杜鵑區分。雌鳥似雄鳥，但有赤色形雌鳥。

Small-sized cuckoo similar in appearance to other Cuculus cuckoos. It has grey upperparts, widely spaced barring on white underparts and a buffy vent. Distinctive loud five-note call. Sexes alike, rufous female hepatic morph can be seen.

① juvenile 幼鳥；Po Toi 蒲台；17-Sep-06, 06 年 9 月 17 日；Michelle and Peter Wong 江敏兒 · 黃理沛
② juvenile 幼鳥；Po Toi 蒲台；17-Sep-06, 06 年 9 月 17 日；Michelle and Peter Wong 江敏兒 · 黃理沛
③ juvenile 幼鳥；Po Toi 蒲台；17-Sep-06, 06 年 9 月 17 日；Michelle and Peter Wong 江敏兒 · 黃理沛

③

| 春季過境遷徙鳥<br>Spring Passage Migrant | | | | 夏候鳥<br>Summer Visitor | | 秋季過境遷徙鳥<br>Autumn Passage Migrant | | | 冬候鳥<br>Winter Visitor | | |
|---|---|---|---|---|---|---|---|---|---|---|---|
| 1 | 2 | 3 | 4 | 5 | 6 | 7 | 8 | 9 | 10 | 11 | 12 |

常見月份

| 留鳥<br>Resident | 迷鳥<br>Vagrant | 偶見鳥<br>Occasional Visitor |
|---|---|---|

# 四聲杜鵑 (普)sì shēng dù juān

體長length：32-33cm

## Indian Cuckoo | *Cuculus micropterus*

中型杜鵑，頭灰色，嘴黑色，眼圈和腳黃色。上體灰褐色，下體白色帶深褐橫紋，尾羽末端有寬闊黑帶。雌鳥胸部略帶褐色。四月中旬時常發出響亮的叫聲，有如粵音「家婆打我」。

Medium-sized cuckoo. Grey head and greyish brown upperparts. Black bill, yellow eye-rings and legs. White underparts with dark brown bars. Broad black band on tips of tail feather. Female has brownish breast. Four note call from mid-April: 「Ko-ko-ta-to」(fourth note lower).

1 adult 成鳥；Mai Po 米埔；May-00, 00 年 5 月；Henry Lui 呂德恆
2 adult 成鳥；Mai Po 米埔；May-00, 00 年 5 月；Henry Lui 呂德恆
3 adult 成鳥；Long Valley 塱原；Jun-06, 06 年 6 月；Kinni Ho 何建業
4 Mai Po 米埔；5-May-07, 07 年 5 月 5 日；Owen Chiang 深藍

| | 春季過境遷徙鳥<br>Spring Passage Migrant | | | 夏候鳥<br>Summer Visitor | | | 秋季過境遷徙鳥<br>Autumn Passage Migrant | | | 冬候鳥<br>Winter Visitor | | |
|---|---|---|---|---|---|---|---|---|---|---|---|---|
| 常見月份 | 1 | 2 | 3 | 4 | 5 | 6 | 7 | 8 | 9 | 10 | 11 | 12 |
| | 留鳥<br>Resident | | | | 迷鳥<br>Vagrant | | | | 偶見鳥<br>Occasional Visitor | | | |

3 4

# 東方中杜鵑

(普) dōng fāng zhōng dù juān

體長 length：32-33cm

## Oriental Cuckoo | *Cuculus optatus*

[1]

[2][3]

雄 鳥頭胸及上體灰色，雌鳥上體褐色並全身滿佈黑色橫紋。下體白色並滿佈幼黑橫紋。幼鳥似雌鳥但上體較近黑色。

Male has grey head, breast and upperparts. Female has brown upperparts and black stripes all over its body. White lower parts with thin black stripes. Immature is similar to female with darker upperparts.

[1] adult female 雌成鳥；Po Toi 蒲台；Apr-13, 13 年 4 月；Allen Chan 陳志雄
[2] adult female 雌成鳥；Po Toi 蒲台；Apr-13, 13 年 4 月；Allen Chan 陳志雄
[3] adult female 雌成鳥；Po Toi 蒲台；Apr-13, 13 年 4 月；Allen Chan 陳志雄

| | 春季過境遷徙鳥<br>Spring Passage Migrant | | 夏候鳥<br>Summer Visitor | | 秋季過境遷徙鳥<br>Autumn Passage Migrant | | 冬候鳥<br>Winter Visitor | |
|---|---|---|---|---|---|---|---|---|
| 常見月份 | 1 | 2 | 3 | 4 | 5 | 6 | 7 | 8 | 9 | 10 | 11 | 12 |

| 留鳥<br>Resident | 迷鳥<br>Vagrant | 偶見鳥<br>Occasional Visitor |
|---|---|---|

# 大杜鵑

普 dà dù juān

體長 length：32-33cm

## Common Cuckoo | *Cuculus canorus*

中型杜鵑。虹膜及眼圈黃色，腳黃色。嘴黃色，嘴尖黑色。頭、上背至尾上覆羽及喉至胸部灰色，脇部、腹部至尾下覆羽白有黑色幼橫紋。翼及尾深灰色，尾外緣有白點。雄鳥為大家熟悉的「cuc-coo」叫聲，通常只在繁殖地才能聽到。

Medium-sized cuckoo. It has yellow iris, eye-rings and legs. Yellow bill with black tip. Head, mantle and uppertail coverts grey. Throat to breast is grey. Flanks, belly and undertail coverts are white with many fine dark bars. Dark grey wings, dark grey tail with spotting on outer edges. Male bird calls repeatly "cuc-coo" at its breeding site.

1 adult 成鳥；Po Toi 蒲台；4-Apr-07, 07 年 4 月 4 日；Geoff Welch

| 春季過境遷徙鳥<br>Spring Passage Migrant | | 夏候鳥<br>Summer Visitor | | 秋季過境遷徙鳥<br>Autumn Passage Migrant | | 冬候鳥<br>Winter Visitor | |
|---|---|---|---|---|---|---|---|
| 1 | 2 | 3 | 4 | 5 | 6 | 7 | 8 | 9 | 10 | 11 | 12 | 常見月份 |

| 留鳥<br>Resident | 迷鳥<br>Vagrant | 偶見鳥<br>Occasional Visitor |
|---|---|---|

# 原鴿 <sub>普</sub>yuán gē

體長 length：31-34cm

## Rock Dove | *Columba livia*

羽色多變，通常頭深灰色，頸部有綠色和紫色金屬光澤，翼尖和尾黑色，軀體藍灰色，並有兩條寬闊黑色翼帶。市區常見的多為野化及長期雜交的家鴿。叫聲為低沉的咕咕聲。

Highly variable plumage. Typical features include deep grey head, and metallic green and purple colour on the neck. Wing tip and tail black, body grey blue, with two broad black stripes on wing coverts. Most of the "wild" birds in urban areas originated from domestic pigeons with extensive in-breeding over time. Call is a low-pitched "hoo..hoo..".

1 adult 成鳥：Sai Kung 西貢：Oct-08, 08 年 10 月：Joyce Tang 鄧玉蓮
2 adult 成鳥：Kowloon Park 九龍公園：Aug-06, 06 年 8 月：James Lam 林文華
3 adult 成鳥：Kowloon Park 九龍公園：20-Apr-07, 07 年 4 月 20 日：Sonia and Kenneth Fung 馮啟文、蕭敏晶
4 adult 成鳥：Kowloon Park 九龍公園：Jul-07, 07 年 7 月：Chan Kai Wai 陳佳瑋
5 adult 成鳥：Sai Kung 西貢：Joyce Tang 鄧玉蓮
6 adult 成鳥：Kowloon Park 九龍公園：Mar-04, 04 年 3 月：Henry Lui 呂德恆
7 immature 未成年鳥：Kowloon Park 九龍公園：Aug-06, 06 年 8 月：James Lam 林文華

| | 春季過境遷徙鳥<br>Spring Passage Migrant | | 夏候鳥<br>Summer Visitor | | 秋季過境遷徙鳥<br>Autumn Passage Migrant | | 冬候鳥<br>Winter Visitor | | | |
|---|---|---|---|---|---|---|---|---|---|---|
| 常見月份 | 1 | 2 | 3 | 4 | 5 | 6 | 7 | 8 | 9 | 10 | 11 | 12 |
| | 留鳥<br>Resident | | | 迷鳥<br>Vagrant | | | 偶見鳥<br>Occasional Visitor | | | |

# 山斑鳩

(普) shān bān jiū
(粵) 鳩：音溝

體長 length：33-35cm

## Oriental Turtle Dove | *Streptopelia orientalis*

體型較珠頸斑鳩大，頸側有黑白相間的斑紋。背部至腰部藍灰色，嘴細而黑，腳紅色。翼上覆羽深灰而邊緣褐色，貌似鱗片。飛行時尾部深灰色，楔形，末端有白帶。

Bigger than Spotted Dove. Prominent patch on the side of the neck with black and white bands. Mantle to rump bluish grey, bill small and black. Red eye-rings. Upperwing coverts dark grey, appears scaly due to brownish feather fringes. In flight, it shows dark grey wedge-shaped tail, with a complete white band on the trailing edge.

1. adult 成鳥，Po Toi 蒲台：Nov-08, 08 年 11 月；James Lam 林文華
2. adult 成鳥，Mai Po 米埔，Nov-04, 04 年 11 月；Jemi and John Holmes 孔思義・黃亞萍
3. adult 成鳥，Long Valley 塱原；Nov-07, 07 年 11 月；Owen Chiang 深藍
4. adult 成鳥，Tsim Bei Tsui 尖鼻咀；Oct-07, 07 年 10 月；Kitty Koo 古愛婉
5. juvenile 幼鳥，Po Toi 蒲台；Oct-08, 08 年 10 月；James Lam 林文華
6. juvenile 幼鳥，Mai Po 米埔，Nov-06, 06 年 11 月；Pippen Ho 何志剛

| | 春季過境遷徙鳥<br>Spring Passage Migrant | | | 夏候鳥<br>Summer Visitor | | | 秋季過境遷徙鳥<br>Autumn Passage Migrant | | 冬候鳥<br>Winter Visitor | | |
|---|---|---|---|---|---|---|---|---|---|---|---|
| 常見月份 | 1 | 2 | 3 | 4 | 5 | 6 | 7 | 8 | 9 | 10 | 11 | 12 |

| 留鳥<br>Resident | 迷鳥<br>Vagrant | 偶見鳥<br>Occasional Visitor |
|---|---|---|

# 灰斑鳩

普 huī bān jiū
粵 鳩：音溝

體長 length：30-32cm

Eurasian Collared Dove | *Streptopelia decaocto*

全 身大致泥褐色，後頸有細黑頸帶，翼尖深色，尾部比相似的火斑鳩長。嘴黑色，腳紅色，羽色較珠頸斑鳩平淡。

Light ashy grey, with a narrow black bar on hind neck. Dark wing tips. Tail is longer than the Red Turtle Dove. Black bill and red legs. Plainer-looking than Spotted Dove.

[1] adult 成鳥：Mai Po 米埔：Mar-08, 08 年 3 月：Cheng Nok Ming 鄭諾銘
[2] adult 成鳥：Tsim Bei Tsui 尖鼻咀：Jan-06, 06 年 1 月：Henry Lui 呂德恆
[3] adult 成鳥：Mai Po 米埔：Sep-07, 07 年 9 月：Pippen Ho 何志剛
[4] adult 成鳥：Tsim Bei Tsui 尖鼻咀：Jan-05, 05 年 1 月：Cherry Wong 黃卓研
[5] adult 成鳥：Tsim Bei Tsui 尖鼻咀：30-Aug-03, 03 年 8 月 30 日：Michelle and Peter Wong 江敏兒、黃理沛
[6] adult 成鳥：Tsim Bei Tsui 尖鼻咀：30-Aug-03, 03 年 8 月 30 日：Michelle and Peter Wong 江敏兒、黃理沛
[7] adult 成鳥：Fung Lok Wai 豐樂圍：24-Dec-06, 06 年 12 月 24 日：Christina Chan 陳燕明

| 春季過境遷徙鳥 Spring Passage Migrant | | | 夏候鳥 Summer Visitor | | | 秋季過境遷徙鳥 Autumn Passage Migrant | | | 冬候鳥 Winter Visitor | | |
|---|---|---|---|---|---|---|---|---|---|---|---|
| 1 | 2 | 3 | 4 | 5 | 6 | 7 | 8 | 9 | 10 | 11 | 12 |
| 留鳥 Resident | | | | 迷鳥 Vagrant | | | | 偶見鳥 Occasional Visitor | | | |

常見月份

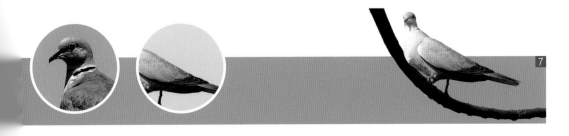

# 火斑鳩

(普) huǒ bān jiū
(粵) 鳩：音溝

體長 length：20.5-23cm

## Red Turtle Dove | *Streptopelia tranquebarica*

1

小型斑鳩。後頸有細黑頸帶，外側尾羽末端有寬闊白色邊緣，翼尖和尾上黑色，嘴黑色，腳紅色。雄鳥頭藍灰色，和紅褐色的身體對比鮮明。雌鳥大致黃褐色。

Small-sized dove. A narrow black bar on hind neck. White tips on outer tail feathers form a wide band. Wing tip and uppertail black in colour. Black bill and red legs. Male has bluish-grey head against deep pink body. Female is generally buffy brown.

1 adult male 雄成鳥；Lok Ma Chau 落馬洲；Sep-05, 05 年 9 月；Martin Hale 夏敖天
2 adult female 雌成鳥；Po Toi 蒲台；May-07, 07 年 5 月；Law Kam Man 羅錦文
3 adult female 雌成鳥；Po Toi 蒲台；May-07, 07 年 5 月；James Lam 林文華
4 adult female 雌成鳥；Po Toi 蒲台；Jun-07, 07 年 6 月；Michelle and Peter Wong 江敏兒、黃理沛
5 juvenile male 雄幼鳥；Long Valley 塱原；Oct-06, 06 年 10 月；Michelle and Peter Wong 江敏兒、黃理沛
6 juvenile 幼鳥；Fung Lok Wai 豐樂圍；Oct-03, 03 年 10 月；Jemi and John Holmes 孔思義、黃亞萍

| | 春季過境遷徙鳥<br>Spring Passage Migrant | | | 夏候鳥<br>Summer Visitor | | 秋季過境遷徙鳥<br>Autumn Passage Migrant | | 冬候鳥<br>Winter Visitor | | | |
|---|---|---|---|---|---|---|---|---|---|---|---|
| 常見月份 | 1 | 2 | 3 | 4 | 5 | 6 | 7 | 8 | 9 | 10 | 11 | 12 |
| | 留鳥<br>Resident | | | 迷鳥<br>Vagrant | | | 偶見鳥<br>Occasional Visitor | | | | | |

鳩鴿科
Columbidae

# 珠頸斑鳩

(普) zhū jǐng bān jiū
(粵) 鳩：音溝

體長 length：27.5-30cm

## Spotted Dove | *Spilopelia chinensis*

**常**見的斑鳩。後頸黑色且滿佈白點，頭灰色，全身褐色，嘴黑色，腳紅色。飛行時尾羽外側末端白色。雄鳥求偶時會鼓起喉頭，不斷向雌鳥鞠躬點頭。叫聲為低沉的「咕咕」聲。幼鳥大致淡灰褐色，後頸沒有黑斑和白點。

Common dove. Black hindneck with dense white spots. Grey head, overall plumage brown, black bill and red legs. White edge on tip of outer tail feather. In courtship, male inflates its cheek and bows to female. Call is a "hoo...hoo..." at low pitch. Juvenile bird is pale greyish brown overall, without marks on hind neck.

1 adult 成鳥；Mai Po 米埔；Sep-08, 08 年 9 月；Andy Kwok 郭匯昌
2 adult 成鳥；Kowloon Park 九龍公園；Aug-03, 03 年 8 月；Henry Lui 呂德恆
3 adult 成鳥；Kowloon Park 九龍公園；Apr-05, 05 年 4 月；Matthew and TH Kwan 關朗曦．關子凱
4 juvenile 幼鳥；North Point 北角；23-Jun-07, 07 年 6 月 23 日；Sonia and Kenneth Fung 馮啟文．蕭敏晶
5 adult 成鳥；Kowloon Park 九龍公園；Mar-06, 06 年 3 月；Owen Chiang 深藍
6 juvenile 幼鳥；Lok Ma Chau 落馬洲；Oct-06, 06 年 10 月；Martin Hale 夏敖天
7 adult 成鳥；Cheung Chau 長洲；Jan-08, 08 年 1 月；Kami Hui 許淑君

| | 春季過境遷徙鳥<br>Spring Passage Migrant | | | 夏候鳥<br>Summer Visitor | | | 秋季過境遷徙鳥<br>Autumn Passage Migrant | | | 冬候鳥<br>Winter Visitor | | |
|---|---|---|---|---|---|---|---|---|---|---|---|---|
| 常見月份 | 1 | 2 | 3 | 4 | 5 | 6 | 7 | 8 | 9 | 10 | 11 | 12 |
| | 留鳥<br>Resident | | | | 迷鳥<br>Vagrant | | | | 偶見鳥<br>Occasional Visitor | | | |

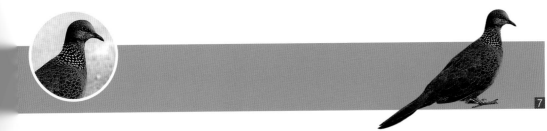

# 斑尾鵑鳩

(普) bān wěi juān jīu
(粵) 鳩：音溝

體長 length：37-41cm

## Barred Cuckoo-Dove | *Macropygia unchall*

身 軀修長的深褐色鳩，尾部頗長並滿佈深色橫紋。成鳥後頸有一片金屬綠色，雌鳥和幼鳥頭部至下體滿佈深色幼橫紋。

Dark brown dove with slender body. Relatively long tail with dark stripes. Adult has metallic green hindneck. Female and immature have thin dark stripes from head to underparts.

[1] adult female 雌成鳥；Tai Mei Tuk Catchwater 大美督引水道；Apr-20；Matthew Kwan 關朗曦
[2] adult male 雄成鳥；Siu Lek Yuen 小瀝源；May-19, 19 年 5 月；Ken Fung 馮漢城
[3] adult male 雄成鳥；Shek Kung；Dec-17, 17 年 12 月；Chan Jun Siu Brian 陳俊兆
[4] adult female 雌成鳥；KFBG 嘉道理農場暨植物園；Jan-14, 14 年 1 月；Kinni Ho 何建業
[5] adult female 雌成鳥；Natalie Chan 陳佩霞
[6] adult female 雌成鳥；KFBG 嘉道理農場暨植物園；Jan-14, 14 年 1 月；Kinni Ho 何建業

| | 春季過境遷徙鳥<br>Spring Passage Migrant | | | 夏候鳥<br>Summer Visitor | | 秋季過境遷徙鳥<br>Autumn Passage Migrant | | | 冬候鳥<br>Winter Visitor | | |
|---|---|---|---|---|---|---|---|---|---|---|---|
| 常見月份 | 1 | 2 | 3 | 4 | 5 | 6 | 7 | 8 | 9 | 10 | 11 | 12 |
| | 留鳥<br>Resident | | | | 迷鳥<br>Vagrant | | | | 偶見鳥<br>Occasional Visitor | | |

鳩鴿科
**Columbidae**

# 綠翅金鳩

(普) lù chì jīn jiū
(粵) 鳩：音溝

體長 length：23-27cm

## Common Emerald Dove | *Chalcophaps indica*

其他名稱 Other names：綠背金鳩

小型斑鳩，常躲在濃密的叢林中，受驚飛起時可見翼上深綠，背部近腰處有兩條白色粗橫紋。前額粉藍色，面頰至腹部灰褐色，嘴和腳紅色，翼尖和尾部深灰色。叫聲為「cu-oo」，在山間迴響。

Small-sized dove. Favours thick woodland. When flushed from forest footpath, it shows dark green upperwings and two white bars on the back near rump. Forehead and cap blueish-grey, cheek to belly pinkish brown, bill and legs red, wing tips and tail dark grey. Call a "cu-oo", heard among hills.

[1] adult male 雄成鳥：Kwai Chung 葵涌：Mar-07, 07 年 3 月：Pippen Ho 何志剛
[2] adult male and female 雄成鳥和雌成鳥：Kwai Chung 葵涌：Jul-07, 07 年 7 月：Owen Chiang 深藍
[3] adult male 雄成鳥：Ha Fa Shan 下花山：Feb-08, 08 年 2 月：Jasper Lee 李君哲
[4] juvenile 幼鳥：Tung Chung 東涌：May-08, 08 年 5 月：Kami Hui 許淑君
[5] Kwai Chung 葵涌：29-Jul-07, 07 年 7 月 29 日：Thomas Chan 陳土飛
[6] adult female 雌成鳥：Kwai Chung 葵涌：Apr-08, 08 年 4 月：Pippen Ho 何志剛
[7] Kwai Chung 葵涌：29-Jul-07, 07 年 7 月 29 日：Thomas Chan 陳土飛

| | 春季過境遷徙鳥<br>Spring Passage Migrant | | | 夏候鳥<br>Summer Visitor | | | 秋季過境遷徙鳥<br>Autumn Passage Migrant | | | 冬候鳥<br>Winter Visitor | | |
|---|---|---|---|---|---|---|---|---|---|---|---|---|
| 常見月份 | 1 | 2 | 3 | 4 | 5 | 6 | 7 | 8 | 9 | 10 | 11 | 12 |
| | 留鳥<br>Resident | | | | 迷鳥<br>Vagrant | | | | 偶見鳥<br>Occasional Visitor | | | |

# 橙胸綠鳩

(普)chéng xiōng lù jiū
(粵)鳩：音溝

體長 length：29cm

## Orange-breasted Green Pigeon | *Treron bicinctus*

身 軀主要綠色，飛羽深色。雄鳥上胸橙及粉紅色，後頸灰色。

Mainly green bird with darkish flying feathers. Male has orange and pink wash on upper breast, greyish neck.

1 Po Toi 蒲台：Feb-06, 06 年 2 月：Tam Yiu Leung 譚耀良
2 Po Toi 蒲台：2-Oct-06, 06 年 10 月 2 日：Chan Wing Kam 陳詠琴
3 18-Feb-06, 06 年 2 月 18 日：Michelle and Peter Wong 江敏兒，黃理沛
4 Po Toi 蒲台：2-Oct-06, 06 年 10 月 2 日：Chan Wing Kam 陳詠琴

| 春季過境遷徙鳥<br>Spring Passage Migrant | | | 夏候鳥<br>Summer Visitor | | | 秋季過境遷徙鳥<br>Autumn Passage Migrant | | | 冬候鳥<br>Winter Visitor | | |
|---|---|---|---|---|---|---|---|---|---|---|---|
| 1 | 2 | 3 | 4 | 5 | 6 | 7 | 8 | 9 | 10 | 11 | 12 |

常見月份

| 留鳥<br>Resident | 迷鳥<br>Vagrant | 偶見鳥<br>Occasional Visitor |
|---|---|---|

# 厚嘴綠鳩

(普)hòu zuǐ lǜ jiū
(粵)鳩：音溝

體長 length：24-31cm

## Thick-billed Green Pigeon | *Treron curvirostra*

嘴 厚呈淡綠色，嘴基紅色和明顯的灰藍色眼眶是最主要的特徵。
雄性背部及覆羽紫紅色，尾下覆羽栗色。雌鳥和幼鳥綠色為主。

Thick green bill with red at base and prominent greyish blue orbital skin are the main characteristics to separate from other green pigeons. Male has maroon mantle and cinnamon under tail coverts. Female and juveniles are mainly green.

1 adult female 雌成鳥：Lam Tsuen 林村；Jan-06, 06 年 1 月：Kinni Ho 何建業
2 adult female 雌成鳥：Ha Fa Shan 下花山；Dec-04, 04 年 12 月：Michelle and Peter Wong 江敏兒．黃理沛
3 adult female 雌成鳥：Lam Tsuen 林村；Dec-06, 06 年 12 月：Matthew and TH Kwan 關朗曦．關子凱
4 adult female 雌成鳥：Chai Wan 柴灣；Dec-04, 04 年 12 月：Martin Hale 夏敖天

| 春季過境遷徙鳥 Spring Passage Migrant | | | 夏候鳥 Summer Visitor | | | 秋季過境遷徙鳥 Autumn Passage Migrant | | | 冬候鳥 Winter Visitor | | |
|---|---|---|---|---|---|---|---|---|---|---|---|
| 1 | 2 | 3 | 4 | 5 | 6 | 7 | 8 | 9 | 10 | 11 | 12 |

常見月份

| 留鳥 Resident | 迷鳥 Vagrant | 偶見鳥 Occasional Visitor |
|---|---|---|

# 紅翅綠鳩

(普)hóng chì lǜ jiū
(粵)鳩：音溝

體長length：30-33cm

## White-bellied Green Pigeon | *Treron sieboldii*

**全**身綠色，微帶黃色，腹部灰綠色，嘴藍色。雄性覆羽帶紫紅色，前額有淡橙色。雌鳥全身綠色，腹部淡色帶淡黑斑，嘴灰藍色。

A Green Pigeon with greyish green belly and blue bill. Male has a maroon wash to the wing coverts, pale orange forehead. Female is all green except the belly, which is pale colour with dark marks, bill greyish blue.

1 adult female 雌成鳥：Shek Kong 石崗：Jan-07, 07 年 1 月：Owen Chiang 深藍
2 adult female 雌成鳥：Shek Kong 石崗：Dec-06, 06 年 12 月：Michelle and Peter Wong 江敏兒．黃理沛
3 adult female 雌成鳥：Shek Kong 石崗：Jan-07, 07 年 1 月：Allen Chan 陳志雄
4 adult female 雌成鳥：Shek Kong 石崗：Jan-07, 07 年 1 月：Martin Hale 夏敖天
5 adult female 雌成鳥：Shek Kong 石崗：Jan-07, 07 年 1 月：Owen Chiang 深藍

| 春季過境遷徙鳥<br>Spring Passage Migrant | | | | 夏候鳥<br>Summer Visitor | | | 秋季過境遷徙鳥<br>Autumn Passage Migrant | | 冬候鳥<br>Winter Visitor | | |
|---|---|---|---|---|---|---|---|---|---|---|---|
| 1 | 2 | 3 | 4 | 5 | 6 | 7 | 8 | 9 | 10 | 11 | 12 |
| 留鳥<br>Resident | | | | 迷鳥<br>Vagrant | | | | 偶見鳥<br>Occasional Visitor | | | |

常見月份

# 楔尾綠鳩

(普) xiē wěi lù jīu
(粤) 楔：音舌；鳩：音溝

體長 length：30-33cm

## Wedge-tailed Green Pigeon | *Treron sphenurus*

青綠色的中型鳩。翼及尾部深綠色，翼尖黑色，頭部至下體青黃色，雄鳥胸部沾橙色及肩部深褐色。飛行時可見尾部楔形，翼底灰色。

Bright green medium-sized Dove. Dark green wings and tail with black wing tips. Yellowish-green from head to underparts. Male's breast tinged with orange and has dark brown shoulders. Wedge-shaped tail shown in flight. Grey underwings.

[1] adult 成鳥：Sai Kung 西貢；Mar-16, 16 年 3 月；Martin Hale 夏敖天

| 春季過境遷徙鳥<br>Spring Passage Migrant | | | 夏候鳥<br>Summer Visitor | | | 秋季過境遷徙鳥<br>Autumn Passage Migrant | | | 冬候鳥<br>Winter Visitor | | |
|---|---|---|---|---|---|---|---|---|---|---|---|
| 1 | 2 | 3 | 4 | 5 | 6 | 7 | 8 | 9 | 10 | 11 | 12 |
| 留鳥<br>Resident | | | | 迷鳥<br>Vagrant | | | | 偶見鳥<br>Occasional Visitor | | | |

常見月份

# 灰腳秧雞

(普) huī jiǎo yāng jī
(粵) 秧：音央

體長 length：21-25cm

## Slaty-legged Crake | *Rallina eurizonoides*

其他名稱 Other names：Banded Crake

中型秧雞。嘴短而黑。頭和胸紅褐色，喉部有明顯白色斑塊。上背、上翼至尾部深褐色。下腹至尾下覆羽有很多黑白色相間的橫紋，腳深色。在樹林間濃密的矮叢活動。

Medium-sized rail. Bill short and black. Head and breast reddish brown, small white patch on throat. Mantle, upperwings to tail dark brown. Lower breast to undertail coverts has many black and white stripes, black legs. Prefers dense shrubs at forest floor.

1 adult 成鳥：Lai Chi Kok 荔枝角；Nov-07, 07 年 11 月：Michelle and Peter Wong 江敏兒、黃理沛
2 adult 成鳥：Lai Chi Kok 荔枝角；Nov-07, 07 年 11 月：James Lam 林文華
3 adult 成鳥：Lai Chi Kok 荔枝角；Michelle and Peter Wong 江敏兒、黃理沛
4 adult 成鳥：Lai Chi Kok 荔枝角；Nov-07, 07 年 11 月：Danny Ho 何國海

| 春季過境遷徙鳥 Spring Passage Migrant | | | 夏候鳥 Summer Visitor | | | 秋季過境遷徙鳥 Autumn Passage Migrant | | | 冬候鳥 Winter Visitor | | |
|---|---|---|---|---|---|---|---|---|---|---|---|
| 1 | 2 | 3 | 4 | 5 | 6 | 7 | 8 | 9 | 10 | 11 | 12 |

常見月份

| 留鳥 Resident | 迷鳥 Vagrant | 偶見鳥 Occasional Visitor |
|---|---|---|

# 灰胸秧雞

(普) hūi xiōng yāng jī
(粵) 秧：音央

體長 length：25-30cm

## Slaty-breasted Rail | *Gallirallus striatus*

其他名稱 Other names：藍胸秧雞, Banded Rail, Blue-breasted Banded Rail

中型秧雞，嘴直而紅褐色，嘴端沾灰色。前額至後頸栗色，上體暗褐色並有白色細橫紋，臉至上腹灰藍色，腹部白色，脇部有褐色橫紋。

Medium-sized rail. Bill straight and reddish, with greyish tip. Forehead to hindneck is chestnut in colour. Upperparts are darkish brown with narrow white bars. Face to upper breast is bluish grey, white belly. Brown bars on flanks.

[1] adult 成鳥；Tai O 大澳；Feb-08, 08 年 2 月；Ann To 陶偉意
[2] adult, race *albiventer* 成鳥, *albiventer* 亞種；Po Toi 蒲台；9-Apr-05, 05 年 4 月 9 日；Michelle and Peter Wong 江敏兒 · 黃理沛
[3] adult, race *gularis* 成鳥, *gularis* 亞種；Mai Po 米埔；22-Sep-06, 06 年 9 月 22 日；Owen Chiang 深藍
[4] juvenile 幼鳥；21-Oct-07, 07 年 10 月 21 日；Danny Ho 何國海

| 春季過境遷徙鳥<br>Spring Passage Migrant | | | 夏候鳥<br>Summer Visitor | | | 秋季過境遷徙鳥<br>Autumn Passage Migrant | | | 冬候鳥<br>Winter Visitor | | |
|---|---|---|---|---|---|---|---|---|---|---|---|
| 1 | 2 | 3 | 4 | 5 | 6 | 7 | 8 | 9 | 10 | 11 | 12 |
| 留鳥<br>Resident | | | | 迷鳥<br>Vagrant | | | | 偶見鳥<br>Occasional Visitor | | | |

常見月份

# 西方秧雞

(普) xī fāng yāng jī
(粵) 秧：音央

體長 length：25-28cm

## Western Water Rail | *Rallus aquaticus*

中型秧雞。上體褐色，臉至胸灰藍色，腹部黑白相間。與普通秧雞比較，臉全灰藍而沒有深褐貫眼紋，尾下覆羽沒有黑白橫紋。

Medium-sized rail. Brown upperparts. Greyish-blue from face to breast. Black and white stripes on belly. Distinguishes from Eastern Water Rail by fully greyish-blue face without brown eye-stripes, and lack of black and white stripes on undertail-coverts.

1 adult 成鳥：Mai Po 米埔；Dec-06, 06 年 12 月；Martin Hale 夏敖天
2 adult 成鳥：Mai Po 米埔；Dec-06, 06 年 12 月；Martin Hale 夏敖天

| 春季過境遷徙鳥<br>Spring Passage Migrant | | | 夏候鳥<br>Summer Visitor | | | 秋季過境遷徙鳥<br>Autumn Passage Migrant | | | 冬候鳥<br>Winter Visitor | | |
|---|---|---|---|---|---|---|---|---|---|---|---|
| 1 | 2 | 3 | 4 | 5 | 6 | 7 | 8 | 9 | 10 | 11 | 12 |
| 留鳥<br>Resident | | | | 迷鳥<br>Vagrant | | | | 偶見鳥<br>Occasional Visitor | | | |

常見月份

# 普通秧雞

(普) pǔ tōng yāng jī
(粵) 秧：音央

體長 length：23-29cm

## Eastern Water Rail | *Rallus indicus*

**中**型秧雞，嘴長，暗紅褐色微向下彎，上體褐色，有黑色短而粗的縱紋，胸部隱約可見棕色橫紋，臉及上腹石板灰色，下腹有黑、白、棕三色細橫紋相間。

Medium-sized rail. Long dull red bill is slightly decurved. Brownish mantle with short black streaks. The breast has unclear brown stripes. Face and upper belly are slaty grey. Lower belly has narrow black, white and brown stripes.

[1] juvenile 幼鳥；Pui O 貝澳；Jul-08, 08 年 7 月；Chan Kai Wai 陳佳瑋
[2] juvenile 幼鳥；Nam Sang Wai 南生圍；Dec-08, 08 年 12 月；Andy Kwok 郭匯昌
[3] juvenile 幼鳥；Pui O 貝澳；Mar-08, 08 年 3 月；Sonia and Kenneth Fung 馮啟文、蕭敏晶
[4] juvenile 幼鳥；Pui O 貝澳；Mar-08, 08 年 3 月；Cherry Wong 黃卓研
[5] juvenile 幼鳥；Pui O 貝澳；Mar-08, 08 年 3 月；Eling Lee 李佩玲
[6] juvenile 幼鳥；Pui O 貝澳；Mar-08, 08 年 3 月；Michelle and Peter Wong 江敏兒、黃理沛
[7] juvenile 幼鳥；Pui O 貝澳；Mar-08, 08 年 3 月；Chan Kin Chung Gary 陳建中

| | 春季過境遷徙鳥<br>Spring Passage Migrant | | 夏候鳥<br>Summer Visitor | | 秋季過境遷徙鳥<br>Autumn Passage Migrant | | 冬候鳥<br>Winter Visitor | |
|---|---|---|---|---|---|---|---|---|
| 常見月份 | 1 | 2 | 3 | 4 | 5 | 6 | 7 | 8 | 9 | 10 | 11 | 12 |

| 留鳥<br>Resident | 迷鳥<br>Vagrant | 偶見鳥<br>Occasional Visitor |
|---|---|---|

秧雞科
**Rallidae**

# 白胸苦惡鳥

(普) bái xiōng kǔ è niǎo

體長 length：28-33cm

## White-breasted Waterhen | *Amaurornis phoenicurus*

其他名稱 Other names：白胸秧雞

香港最常見的秧雞。臉至上腹白色，嘴部青黃色，上嘴基紅色。下腹至尾下覆羽橙褐色，背部深灰褐色。腳淡黃色。雛鳥全身長滿黑色絨毛，嘴和腳黑色，面頰偶有白點。經常發出像普通話「苦惡、苦惡」的叫聲，步行時頭前後擺動，尾巴上下搖動。

Commonest rail in Hong Kong. Face to upper belly white in color. Bill greenish yellow, with base of upper bill red. Lower belly to undertail coverts brownish orange. Upperparts dull brownish grey. Legs pale yellow. Chicks covered with dark down, bill and legs black, sometimes with white spots on chins. Loud "kuo-oa, kuo-oa" call. Head moves back-and-forth and tail flicks when walking.

1 adult 成鳥；Mai Po 米埔；Sept-06, 06 年 9 月；Owen Chiang 深藍
2 adult with chicks 成鳥和雛鳥；Kam Tin 錦田；May-07, 07 年 5 月；Aka Ho
3 adult with chicks 成鳥和雛鳥；Kam Tin 錦田；May-07, 07 年 5 月；Herman Ip 葉紀江
4 adult 成鳥；Mai Po 米埔；Apr-06, 06 年 4 月；Tam Yip Shing 譚業成
5 juvenile 幼鳥；Mai Po 米埔；Jul-07, 07 年 7 月；James Lam 林文華
6 chick 雛鳥；Maipo 米埔；Aug-08, 08 年 8 月；Thomas Chan 陳土飛
7 adult 成鳥；Mai Po 米埔；Apr-06, 06 年 4 月；Henry Lui 呂德恆

| | 春季過境遷徙鳥<br>Spring Passage Migrant | | | 夏候鳥<br>Summer Visitor | | | 秋季過境遷徙鳥<br>Autumn Passage Migrant | | | 冬候鳥<br>Winter Visitor | | |
|---|---|---|---|---|---|---|---|---|---|---|---|---|
| 常見月份 | 1 | 2 | 3 | 4 | 5 | 6 | 7 | 8 | 9 | 10 | 11 | 12 |
| | 留鳥<br>Resident | | | | 迷鳥<br>Vagrant | | | | 偶見鳥<br>Occasional Visitor | | | |

秧雞科
**Rallidae**

# 紅腳苦惡鳥 (普)hóng jiǎo kǔ è niǎo

體長length：26-28cm

Brown Crake | *Amaurornis akool*

**中** 型秧雞。上體褐色，下體前部灰藍色。嘴灰黃色，腳胭脂紅色，腳趾頗長。幼鳥偏褐色。

Medium-sized rail. Upperparts brown. Front part of underparts greyish blue. Bill greyish yellow. Crimson legs with long toes. Juveniles are brownish.

① adult 成鳥：Nam Chung 南涌；15-Feb-09, 09 年 2 月 15 日：Kinni Ho 何建業
② adult 成鳥：Nam Chung 南涌；Feb-09, 09 年 2 月：Danny Ho 何國海
③ adult 成鳥：4-Feb-09, 09 年 2 月 4 日：Christina Chan 陳燕明

| 春季過境遷徙鳥<br>Spring Passage Migrant | | 夏候鳥<br>Summer Visitor | | 秋季過境遷徙鳥<br>Autumn Passage Migrant | | 冬候鳥<br>Winter Visitor | |
|---|---|---|---|---|---|---|---|

| 常見月份 | 1 | 2 | 3 | 4 | 5 | 6 | 7 | 8 | 9 | 10 | 11 | 12 |
|---|---|---|---|---|---|---|---|---|---|---|---|---|

| 留鳥<br>Resident | 迷鳥<br>Vagrant | 偶見鳥<br>Occasional Visitor |
|---|---|---|

# 小田雞

（普）xiǎo tián jī

體長 length：17-19cm

Baillon's Crake | *Porzana pusilla*

小 型秧雞。上體栗色，有黑色縱紋及白斑。上胸淺灰色，腹部有白色橫紋。見於水淹耕地及沼澤地。

Small-sized Rail. Reddish brown upperparts, bluish grey underparts with dark and white barring on rear body. Lives in flooded cultivation and marshland.

1 adult 成鳥；Shing Mun Valley Park 城門谷公園；May-09, 09 年 5 月；Wong Shui Chi 黃瑞芝
2 adult 成鳥；Shing Mun Valley Park 城門谷公園；May-09, 09 年 5 月；Helen Chan 陳燕芳
3 juvenile 幼鳥；Long Valley 塱原；Oct-07, 07 年 10 月；Andy Kwok 郭匯昌
4 juvenile 幼鳥；Long Valley 塱原；Oct-07, 07 年 10 月；Andy Cheung 張玉良
5 juvenile 幼鳥；Long Valley 塱原；27-Oct-07, 07 年 10 月 27 日；Michelle and Peter Wong 江敏兒、黃理沛
6 juvenile 幼鳥；Long Valley 塱原；27-Oct-07, 07 年 10 月 27 日；Michelle and Peter Wong 江敏兒、黃理沛

| 春季過境遷徙鳥<br>Spring Passage Migrant | | | 夏候鳥<br>Summer Visitor | | | 秋季過境遷徙鳥<br>Autumn Passage Migrant | | | 冬候鳥<br>Winter Visitor | | |
|---|---|---|---|---|---|---|---|---|---|---|---|
| 1 | 2 | 3 | 4 | 5 | 6 | 7 | 8 | 9 | 10 | 11 | 12 |

常見月份

| 留鳥<br>Resident | 迷鳥<br>Vagrant | 偶見鳥<br>Occasional Visitor |
|---|---|---|

# 斑脇田雞

(普) bān xié tián jī
(粵) 脇：音脅

體長 length：20-22cm

## Band-bellied Crake | *Porzana paykullii*

1

居 於淡水濕地草甸或稻田的中型秧雞。頭冠和上體深褐色，偏黃的短嘴，腳紅，頭側及胸部紅栗色，兩脇和尾下覆羽有較細的黑白條紋，翼覆羽有白橫斑。

Medium-sized rail, lives in freshwater swamp or paddy field. Upperparts and head deep brown, yellowish short bill, red legs, sides of head and underparts reddish chestnut. Flanks and tail coverts have some black and white strips, while wing coverts are with white spots.

[1] Lung Fu Shan 龍虎山：May-09, 09 年 5 月：Pippen Ho 何志剛
[2] Lung Fu Shan 龍虎山：May-09, 09 年 5 月：Yue Pak Wai 余柏維
[3] Lung Fu Shan 龍虎山：May-09, 09 年 5 月：Michelle and Peter Wong 江敏兒、黃理沛
[4] Lung Fu Shan 龍虎山：May-09, 09 年 5 月：Pippen Ho 何志剛
[5] Lung Fu Shan 龍虎山：8-May-09, 09 年 5 月 8 日：Wong Shui Chi 黃瑞芝
[6] Lung Fu Shan 龍虎山：May-09, 09 年 5 月：Michelle and Peter Wong 江敏兒、黃理沛
[7] Lung Fu Shan 龍虎山：May-09, 09 年 5 月：Pippen Ho 何志剛
[8] Lung Fu Shan 龍虎山：May-09, 09 年 5 月：Sung Yik Hei 宋亦希

| | 春季過境遷徙鳥<br>Spring Passage Migrant | | | 夏候鳥<br>Summer Visitor | | | 秋季過境遷徙鳥<br>Autumn Passage Migrant | | | 冬候鳥<br>Winter Visitor | | |
|---|---|---|---|---|---|---|---|---|---|---|---|---|
| 常見月份 | 1 | 2 | 3 | 4 | 5 | 6 | 7 | 8 | 9 | 10 | 11 | 12 |
| | 留鳥<br>Resident | | | | 迷鳥<br>Vagrant | | | | 偶見鳥<br>Occasional Visitor | | | |

# 紅胸田雞

(普) hóng xiōng tián jī

體長 length：21-23cm

## Ruddy-breasted Crake | *Porzana fusca*

其他名稱 Other names：Ruddy Crake

中型秧雞。嘴短而黑。上體褐色，臉至胸部紅棕色，喉部淡色，下腹黑白色橫紋相間，腳紅色。

Medium-sized rail. Short bill. Brown upperparts, face to breast brownish red, throat pale in colour. Lower belly has fine black and white barring, red legs.

1 Long Valley 塱原：Oct-07, 07 年 10 月：Michelle and Peter Wong 江敏兒、黃理沛
2 Long Valley 塱原：Apr-07, 07 年 4 月：Cherry Wong 黃卓研

| 春季過境遷徙鳥<br>Spring Passage Migrant | | | 夏候鳥<br>Summer Visitor | | | 秋季過境遷徙鳥<br>Autumn Passage Migrant | | | 冬候鳥<br>Winter Visitor | | |
|---|---|---|---|---|---|---|---|---|---|---|---|
| 1 | 2 | 3 | 4 | 5 | 6 | 7 | 8 | 9 | 10 | 11 | 12 |

常見月份

| | 留鳥<br>Resident | | | 迷鳥<br>Vagrant | | | 偶見鳥<br>Occasional Visitor | |
|---|---|---|---|---|---|---|---|---|

# 白眉田雞
⟨普⟩bái méi tián jī

體長 length：15-20cm

## White-browed Crake | *Porzana cinerea*

小 型田雞。黑色貫眼紋和白色眉紋對比鮮明。上體黃褐色，下體灰白色，脇至尾下覆羽淡橙褐色，腳淡黃色。

Small crake. Black eye-stripes contrast with white supercilia. Yellowish-brown upperparts with greyish-white underparts. Pale brownish-orange from flanks to undertail-coverts. Pale yellow legs.

[1] adult 成鳥：Long Valley 塱原：Sep-14, 14 年 9 月：Allen Chan 陳志雄
[2] adult 成鳥：Long Valley 塱原：Sep-14, 14 年 9 月：Allen Chan 陳志雄
[3] adult 成鳥：Long Valley 塱原：Sep-14, 14 年 9 月：Allen Chan 陳志雄
[4] adult 成鳥：Long Valley 塱原：Sep-14, 14 年 9 月：Allen Chan 陳志雄

| 春季過境遷徙鳥 Spring Passage Migrant | | | 夏候鳥 Summer Visitor | | | 秋季過境遷徙鳥 Autumn Passage Migrant | | | 冬候鳥 Winter Visitor | | |
|---|---|---|---|---|---|---|---|---|---|---|---|
| 1 | 2 | 3 | 4 | 5 | 6 | 7 | 8 | 9 | 10 | 11 | 12 |

常見月份

| 留鳥 Resident | 迷鳥 Vagrant | 偶見鳥 Occasional Visitor |
|---|---|---|

# 董雞 <sup>普</sup>dǒng jī

體長length：42-43cm

## Watercock | *Gallicrex cinerea*

1

大型秧雞。站立時身體挺直，嘴粗大。繁殖期雄鳥身黑色，額板鮮紅色突起，腳紅色。雌鳥及非繁殖鳥，體羽黃褐色有暗褐色斑，腳黃綠色，嘴黃色。

Large-sized rail. Upright posture and thick bill. Breeding male is black, with bright red frontal plate and red feet. Female and non-breeding birds are brownish yellow with dull brown spots, yellowish green legs and yellow bill.

① Long Valley 塱原；Martin Hale 夏敖天
② adult male 雄成鳥；Long Valley 塱原；16-May-06, 06 年 5 月 16 日；Owen Chiang 深藍

2

| | 春季過境遷徙鳥<br>Spring Passage Migrant | | | 夏候鳥<br>Summer Visitor | | 秋季過境遷徙鳥<br>Autumn Passage Migrant | | 冬候鳥<br>Winter Visitor | | |
|---|---|---|---|---|---|---|---|---|---|---|
| 常見月份 | 1 | 2 | 3 | 4 | 5 | 6 | 7 | 8 | 9 | 10 | 11 | 12 |
| | 留鳥<br>Resident | | | 迷鳥<br>Vagrant | | 偶見鳥<br>Occasional Visitor | | |

# 紫水雞

(普) zǐ shuǐ jī

體長 length：38-50cm

## Grey-headed Swamphen | *Porphyrio poliocephalus*

顏色艷麗的大型水雞。粗厚的嘴全紅色，額部有紅色骨板，腳紅色，身體羽毛紫藍色，臉淡色。

Colourful large rail. Think red bill with red skin on forehead. Red legs. Purplish-blue body with pale face.

[1] adult 成鳥：Lok Ma Chau 落馬洲；Dec-14, 14 年 12 月；Jemi and John Holmes 孔思義、黃亞萍
[2] adult 成鳥：Mai Po 米埔；Jun-18, 18 年 6 月；Kwok Tsz Ki 郭子祈
[3] adult 成鳥：Lok Ma Chau 落馬洲；Dec-14, 14 年 12 月；Jemi and John Holmes 孔思義、黃亞萍

| 春季過境遷徙鳥 Spring Passage Migrant | | | 夏候鳥 Summer Visitor | | | 秋季過境遷徙鳥 Autumn Passage Migrant | | | 冬候鳥 Winter Visitor | | |
|---|---|---|---|---|---|---|---|---|---|---|---|
| 1 | 2 | 3 | 4 | 5 | 6 | 7 | 8 | 9 | 10 | 11 | 12 |
| 留鳥 Resident | | | | 迷鳥 Vagrant | | | | 偶見鳥 Occasional Visitor | | | |

常見月份

# 黑水雞

普 hēi shuǐ jī

體長 length：30-38cm

## Common Moorhen | *Gallinula chloropus*

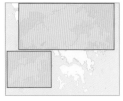

大 型秧雞，雌雄同色。體羽主要黑色，上體沾褐色。嘴小而紅，額板紅色，嘴端黃色，腳青黃色。翼上有一道明顯白紋，脇部隱約可見白色細橫紋，尾下覆羽兩側白色。

Large-sized rail. Sexes alike. Plumage is mainly black, upperparts stained with brown. Bill small and red, red frontal plate, yellow bill tip, greenish yellow legs. Distinctive white stripe above each folded wing. Conspicuous white narrow stripes under flanks. White colour on both sides of undertail coverts.

1 adult 成鳥：HK Wetland Park 香港濕地公園：Nov-07, 07 年 11 月：Aka Ho
2 adult with juvenile 成鳥和幼鳥：Mai Po 米埔：Oct-04, 04 年 10 月：Henry Lui 呂德恆
3 adult 成鳥：Mai Po 米埔：Apr-03, 03 年 4 月：Henry Lui 呂德恆
4 adult 成鳥：Mai Po 米埔：Dec-06, 06 年 12 月：Danny Ho 何國海
5 juvenile 幼鳥：Long Valley 塱原：Dec-08, 08 年 12 月：Yue Pak Wai 余柏維
6 juvenile 幼鳥：Mong Tseng 輞井：3-Apr-04, 04 年 4 月 3 日：Doris Chu 朱詠兒
7 juvenile 幼鳥：HK Wetland Park 香港濕地公園：Nov-07, 07 年 11 月：Aka Ho
8 juvenile 幼鳥：Mai Po 米埔：Oct-04, 04 年 10 月：Henry Lui 呂德恆

| | 春季過境遷徙鳥 Spring Passage Migrant | | 夏候鳥 Summer Visitor | | 秋季過境遷徙鳥 Autumn Passage Migrant | | 冬候鳥 Winter Visitor | | | |
|---|---|---|---|---|---|---|---|---|---|---|
| 常見月份 | 1 | 2 | 3 | 4 | 5 | 6 | 7 | 8 | 9 | 10 | 11 | 12 |
| | 留鳥 Resident | | | 迷鳥 Vagrant | | | 偶見鳥 Occasional Visitor | | | |

# 骨頂雞

普 gǔ dǐng jī

**Eurasian Coot** | *Fulica atra*

體長 length：36-39cm

大型秧雞。體羽黑色，虹膜深紅色，嘴及額板鮮明白色，腳暗綠色。常成群出現，經常游泳，間中潛入水中。近年冬季數量明顯下跌。

Large-sized rail. Black plumage, deep red iris. Contrasting white bill and forehead. Legs dark green. Gregarious, often swims and sometimes dives in water. In recent years, wintering population decreased significantly.

[1] adult 成鳥；Mai Po 米埔；Nov-06, 06 年 11 月；Cherry Wong 黃卓研
[2] adult 成鳥；Mai Po 米埔；Apr-08, 08 年 4 月；Chan Kai Wai 陳佳瑋
[3] adult 成鳥；Mai Po 米埔；15-Feb-06, 06 年 2 月 15 日；Owen Chiang 深藍

| 春季過境遷徙鳥 Spring Passage Migrant | | | 夏候鳥 Summer Visitor | | | 秋季過境遷徙鳥 Autumn Passage Migrant | | | 冬候鳥 Winter Visitor | | |
|---|---|---|---|---|---|---|---|---|---|---|---|
| 1 | 2 | 3 | 4 | 5 | 6 | 7 | 8 | 9 | 10 | 11 | 12 |
| 留鳥 Resident | | | | 迷鳥 Vagrant | | | | 偶見鳥 Occasional Visitor | | | |

常見月份

# 白鶴 <small>普 bái hè</small>

體長 length：140cm

## Siberian Crane | *Leucogeranus leucogeranus*

大型水鳥，成鳥靜立時全身白色，前額及臉部裸露皮膚橙紅色，嘴及腳淡紅，初級飛羽黑色。幼鳥全身淡褐色，虹膜黃色，腹白色。香港首個記錄是2002年12月於米埔錄得一隻幼鳥。

Large-sized waterbird. Adult is overall white when at rest. Forehead and naked face are with reddish orange colour. Bill and legs are pale red. Primaries black in colour. Juvenile is overall light brown, with yellow iris, white belly. The first Hong Kong record was a juvenile at Mai Po in December 2002.

1 adult 成鳥：Mai Po 米埔；Jan-17, 17 年 1 月；Henry Lui 呂德恆
2 juvenile 幼鳥：Mai Po 米埔；Dec-02, 02 年 12 月；Jemi and John Holmes 孔思義・黃亞萍
3 adult 成鳥：Mai Po 米埔；Jan- 17, 17 年 1 月；Kinni Ho 何建業
4 immature 未成年鳥：Mai Po 米埔；11-Dec-02, 02 年 12 月 11 日；Lo Kar Man 盧嘉孟
5 Mai Po 米埔；Martin Hale 夏敖天

| 春季過境遷徙鳥<br>Spring Passage Migrant | | | 夏候鳥<br>Summer Visitor | | | 秋季過境遷徙鳥<br>Autumn Passage Migrant | | | 冬候鳥<br>Winter Visitor | | |
|---|---|---|---|---|---|---|---|---|---|---|---|
| 1 | 2 | 3 | 4 | 5 | 6 | 7 | 8 | 9 | 10 | 11 | 12 |

常見月份

| 留鳥<br>Resident | 迷鳥<br>Vagrant | 偶見鳥<br>Occasional Visitor |
|---|---|---|

# 小鸊鷉

普 xiǎo pì tī
粵 鸊鷉：音僻梯

體長 length：25-29cm

## Little Grebe | *Tachybaptus ruficollis*

其他名稱 Other names：水葫蘆 Common Grebe, Red-throated Grebe

1

小 型鸊鷉。外貌像鴨，嘴尖細。雌雄同色，頭頂明顯深色，虹膜和嘴角淺色。非繁殖期大致淡褐色，繁殖期面頰和頸則轉棗紅色。幼鳥頭頸長滿黑色條紋，成長時條紋逐漸變淡。喜愛單隻或小群游泳，叫聲像急促的馬嘶叫聲，不時潛入水中捕捉食物。

Small grebe that resembles a duckling, but has a small and pointed bill. Sexes alike, with prominent dark cap, light-coloured gape patches and iris. Looks buffy brown in non-breeding season. The cheeks and neck turn claret in breeding season. Juveniles have striped head and neck. The stripes fade during growth. Usually single or in small groups. Voice is a high-pitched whinny. Often dives under water to catch food.

1 breeding 繁殖羽；Mai Po 米埔；Jul-06, 06 年 7 月；Law Kam Man 羅錦文
2 non-breeding adult 非繁殖羽成鳥；Mai Po 米埔；Oct-06, 06 年 10 月；Andy Kwok 郭匯昌
3 non-breeding adult 非繁殖羽成鳥；Mai Po 米埔；25-Jan-09, 09 年 1 月 25 日；Owen Chiang 深藍
4 non-breeding adult 非繁殖羽成鳥；Mai Po 米埔；Dec-08, 08 年 12 月；Joyce Tang 鄧玉蓮
5 adult and juvenile 成鳥和幼鳥；Mai Po 米埔；Apr-07, 07 年 4 月；James Lam 林文華
6 juvenile 幼鳥；Mai Po 米埔；Sep-00, 00 年 9 月；Henry Lui 呂德恆
7 breeding 繁殖羽；Lok Ma Chau 落馬洲；Martin Hale 夏敖天

| 春季過境遷徙鳥<br>Spring Passage Migrant | | | 夏候鳥<br>Summer Visitor | | | 秋季過境遷徙鳥<br>Autumn Passage Migrant | | | 冬候鳥<br>Winter Visitor | | |
|---|---|---|---|---|---|---|---|---|---|---|---|
| 1 | 2 | 3 | 4 | 5 | 6 | 7 | 8 | 9 | 10 | 11 | 12 |
| 留鳥<br>Resident | | | | 迷鳥<br>Vagrant | | | | 偶見鳥<br>Occasional Visitor | | | |

常見月份

# 鳳頭鸊鷉

（普）fèng tóu pì tī
（粵）鸊鷉：音僻梯

體長 length：46-61cm

## Great Crested Grebe | *Podiceps cristatus*

其他名稱 Other names：冠鸊鷉

1

大型鸊鷉。外貌像鴨。頸細長，常垂直水面。嘴帶粉紅色，背部深褐色，頸和面頰鮮明白色，頭頂黑色。繁殖期頭頂長有飾羽，頸和脇部羽毛帶紅褐色。喜歡數隻一起在海岸較深水處游泳，不時潛入水中覓食。

Large grebe. Neck is thin and long, usually held vertically. Bill is pink, and the back is dark brown. Black cap contrasts sharply with white cheeks and neck. Wears striking head ornaments. In breeding, plumage shows reddish brown neck and flanks. Usually swims in groups of several individuals at deep waters near coastal area. Dives under water to catch food.

[1] breeding 繁殖羽：Nam Sang Wai 南生圍：Jan-09, 09 年 1 月：James Lam 林文華
[2] breeding 繁殖羽：Tsim Bei Tsui 尖鼻咀：Jan-07, 07 年 1 月：Lee Kai Hong 李啟康
[3] breeding 繁殖羽：Mai Po 米埔：Apr-03, 03 年 4 月：Henry Lui 呂德恆
[4] breeding 繁殖羽：Nam Sang Wai 南生圍：Feb-09, 09 年 2 月：Sammy Sam and Winnie Wong 森美與雲妮
[5] breeding 繁殖羽：Mai Po 米埔：Mar-08, 08 年 3 月：Pippen Ho 何志剛
[6] breeding 繁殖羽：Tsim Bei Tsui 尖鼻咀：20-Jan-07, 07 年 1 月 20 日：Michelle and Peter Wong 江敏兒、黃理沛
[7] breeding 繁殖羽：Wang Chau 橫洲：Jan-07, 07 年 1 月：Danny Ho 何國海

| 春季過境遷徙鳥<br>Spring Passage Migrant | | | 夏候鳥<br>Summer Visitor | | | 秋季過境遷徙鳥<br>Autumn Passage Migrant | | 冬候鳥<br>Winter Visitor | | | |
|---|---|---|---|---|---|---|---|---|---|---|---|
| 1 | 2 | 3 | 4 | 5 | 6 | 7 | 8 | 9 | 10 | 11 | 12 |
| 留鳥<br>Resident | | | | 迷鳥<br>Vagrant | | | | 偶見鳥<br>Occasional Visitor | | | |

常見月份

# 黑頸鸊鷉

(普) hēi jǐng pì tī
(粵) 鸊鷉：音僻梯

體長 length：28-34cm

## Black-necked Grebe | *Podiceps nigricollis*

其他名稱 Other names：Eared Grebe

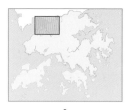

主要為黑白兩色。頭部黑色，面頰、喉及頸側白色，虹膜紅色，嘴微向上翹。上體黑色，有淡色斑紋，下體白色。繁殖期有明顯褐黃色耳羽，頸和胸轉深色。

Black and white grebe with black head. Cheeks, throat and both sides of the neck white. Red iris, bill slightly upcurved. Upperparts black with pale stripes. Underparts white. Breeding birds have distinctive brownish yellow ear coverts, dark neck and breast.

1 non-breeding 非繁殖羽；Mai Po 米埔；12-Nov-03, 03 年 11 月 12 日：Martin Hale 夏敖天
2 non-breeding 非繁殖羽；Mai Po 米埔；12-Nov-03, 03 年 11 月 12 日：Martin Hale 夏敖天
3 non-breeding 非繁殖羽；Mai Po 米埔；12-Nov-03, 03 年 11 月 12 日：Martin Hale 夏敖天
4 non-breeding 非繁殖羽；Mai Po 米埔；Michelle and Peter Wong 江敏兒、黃理沛

| 春季過境遷徙鳥 Spring Passage Migrant | | | 夏候鳥 Summer Visitor | | | 秋季過境遷徙鳥 Autumn Passage Migrant | | | 冬候鳥 Winter Visitor | | |
|---|---|---|---|---|---|---|---|---|---|---|---|
| 1 | 2 | 3 | 4 | 5 | 6 | 7 | 8 | 9 | 10 | 11 | 12 |
| 留鳥 Resident | | | | 迷鳥 Vagrant | | | | 偶見鳥 Occasional Visitor | | | |

常見月份

三趾鶉科
**Turnicidae**

# 黃腳三趾鶉

(普) huáng jiǎo sān zhǐ chún
(圖) 鶉：音春

體長 length：15-18cm

## Yellow-legged Buttonquail | *Turnix tanki*

其他名稱 Other names：Common Buttonquail

**渾**圓的褐色鶉，上嘴深褐而下嘴黃色，黃色腳，覆羽有黑點，胸部沾棕色橫紋。

Brown round-shaped quail. Dark brown upper bill with yellow lower bill. Yellow legs. Black spots on wing-coverts. Breast tinged with brown stripes.

1 adult 成鳥；Long Valley 塱原；Oct-18, 18 年 10 月；Leo Sit 薛國華
2 adult 成鳥；Long Valley 塱原；Oct-18, 18 年 10 月；Leo Sit 薛國華
3 adult 成鳥；Long Valley 塱原；Oct-18, 18 年 10 月；Leo Sit 薛國華
4 adult 成鳥；Long Valley 塱原；Oct-18, 18 年 10 月；Leo Sit 薛國華

| 春季過境遷徙鳥<br>Spring Passage Migrant | | | 夏候鳥<br>Summer Visitor | | | 秋季過境遷徙鳥<br>Autumn Passage Migrant | | 冬候鳥<br>Winter Visitor | | |
|---|---|---|---|---|---|---|---|---|---|---|---|
| 1 | 2 | 3 | 4 | 5 | 6 | 7 | 8 | 9 | 10 | 11 | 12 |

常見月份

| 留鳥<br>Resident | 迷鳥<br>Vagrant | 偶見鳥<br>Occasional Visitor |
|---|---|---|

# 棕三趾鶉

(普) zōng sān zhǐ chún
(粵) 鶉：音春

體長length：13.5-17.5cm

Barred Buttonquail | *Turnix suscitator*

其他名稱 Other names：Common Buttonquail

體型細小而渾圓，尾短，虹膜明顯奶白色，腳灰色。上體栗紅色具黑色及褐色縱紋，脇部橙褐色，腹及尾下覆羽淡色。雄鳥上體黑斑明顯，胸及腹部白色並有黑色短橫紋。雌鳥喉部黑色，喉下面部分的斑紋較雄鳥多。

Small-sized, short tail, round body, prominent milky white iris and grey legs. Rufous chestnut upperparts, with black and buff streaks, rufous orange flanks, and paler belly and vent. Male has obvious white throat, and black streaks on whitish breast.

Female has prominent black throat with black bands below, more boldly patterned than those of male.

[1] male 雄鳥：Cloudy Hill 九龍坑山；Jemi and John Holmes 孔思義．黃亞萍
[2] Chu Cho Yan & Chu Chui Ping 朱祖仁．朱翠萍
[3] male 雄鳥：Mai Po 米埔；Marcus Ho 何萬邦

| 春季過境遷徙鳥 Spring Passage Migrant | | | 夏候鳥 Summer Visitor | | | 秋季過境遷徙鳥 Autumn Passage Migrant | | 冬候鳥 Winter Visitor | | | |
|---|---|---|---|---|---|---|---|---|---|---|---|
| 1 | 2 | 3 | 4 | 5 | 6 | 7 | 8 | 9 | 10 | 11 | 12 |
| 留鳥 Resident | | | | 迷鳥 Vagrant | | | | 偶見鳥 Occasional Visitor | | | |

常見月份

石鴴科
**Burhinidae**

# 大石鴴
(普) dà shí héng
(粵) 鴴：音恆

體長length：41.5-54cm

Great Stone-Curlew | *Esacus recurvirostris*

`1` `2`

♨

頭 大嘴粗，下嘴端微向上翹，頭上有黑白色斑。上體及胸呈灰褐色及無斑紋，下體白色。飛行時初級飛羽及次級飛羽黑色並具白色粗斑紋。香港遠離其分布範圍。

Big head with heavy bill, with lower bill slightly upturned. Diagnostic black and white pattern on head. Unstreaked greyish brown upperparts and breast, with rest of the underparts in white. Primaries and secondaries black with board white patches in flight. Hong Kong is far from its distribution areas.

1 Mai Po 米埔：24-Jun-09, 09 年 6 月 24 日；Michelle and Peter Wong 江敏兒、黃理沛
2 Mai Po 米埔：24-Jun-09, 09 年 6 月 24 日；Michelle and Peter Wong 江敏兒、黃理沛
3 Mai Po 米埔：24-Jun-09, 09 年 6 月 24 日；Kinni Ho 何建業
4 Mai Po 米埔：24-Jun-09, 09 年 6 月 24 日；Kinni Ho 何建業
5 Mai Po 米埔：24-Jun-09, 09 年 6 月 24 日；Michelle and Peter Wong 江敏兒、黃理沛
6 Mai Po 米埔：24-Jun-09, 09 年 6 月 24 日；Michelle and Peter Wong 江敏兒、黃理沛
7 Mai Po 米埔：24-Jun-09, 09 年 6 月 24 日；Kinni Ho 何建業

| | 春季過境遷徙鳥<br>Spring Passage Migrant | | | 夏候鳥<br>Summer Visitor | | | 秋季過境遷徙鳥<br>Autumn Passage Migrant | | | 冬候鳥<br>Winter Visitor | | |
|---|---|---|---|---|---|---|---|---|---|---|---|---|
| 常見月份 | 1 | 2 | 3 | 4 | 5 | 6 | 7 | 8 | 9 | 10 | 11 | 12 |
| | 留鳥<br>Resident | | | | 迷鳥<br>Vagrant | | | | 偶見鳥<br>Occasional Visitor | | | |

# 蠣鷸
(普)lì yù
(粵)蠣鷸：音例核

體長length：40-47.5cm

## Eurasian Oystercatcher | *Haematopus ostralegus*

1

頭部、上體和翼上覆羽黑色，胸部和下體白色。橙紅色的嘴長而直，虹膜紅色，腳橙紅色，獨特易認。

Distinctive in the wild. Black head, upperparts and upperwing coverts. White breast and underparts. Bill is long and straight, reddish orange in colour. Red iris, and reddish orange legs.

1 Mai Po 米埔；9-Apr-09, 09 年 4 月 9 日；James Lam 林文華
2 adult and 1st winter 成鳥和第一年冬天；Mai Po 米埔：Apr-09, 09 年 4 月；James Lam 林文華
3 Mai Po 米埔；9-Apr-09, 09 年 4 月 9 日；Winnie Wong and Sammy Sam 森美與雲妮
4 Mai Po 米埔；9-Apr-09, 09 年 4 月 9 日；Allen Chan 陳志雄
5 Mai Po 米埔；9-Apr-09, 09 年 4 月 9 日；Winnie Wong and Sammy Sam 森美與雲妮
6 Mai Po 米埔；9-Apr-09, 09 年 4 月 9 日；James Lam 林文華
7 1st winter 第一年冬天；Mai Po 米埔；Jan-10, 10 年 1 月；Michelle and Peter Wong 江敏兒，黃理沛

| 春季過境遷徙鳥 Spring Passage Migrant | | | 夏候鳥 Summer Visitor | | 秋季過境遷徙鳥 Autumn Passage Migrant | | 冬候鳥 Winter Visitor | | |
|---|---|---|---|---|---|---|---|---|---|
| 1 | 2 | 3 | 4 | 5 | 6 | 7 | 8 | 9 | 10 | 11 | 12 |

常見月份

| 留鳥 Resident | 迷鳥 Vagrant | 偶見鳥 Occasional Visitor |
|---|---|---|

# 黑翅長腳鷸

(普) hēi chì cháng jiǎo yù
(粵) 鷸：音核

體長 length：35-40cm

**Black-winged Stilt** | *Himantopus himantopus*

其他名稱 Other names：黑翅長腳鴴

雌 雄同色。腳甚長呈粉紅色，嘴黑色，既長且直。背部及翅膀黑色，頭、頸、尾及下體白色，頭頂至後頸有時沾灰。幼鳥頭頂和後頸有時呈淡灰褐色，背及翼上羽毛有白邊，腳部顏色較暗。

Sexes alike. Very long pink legs. Narrow and straight black bill. Mantle and wings black. Head, neck, tail and underparts white. Crown and hindneck sometimes marked with grey. Juvenile has light greyish brown on crown and nape. White edges on mantle and wing coverts. Duller legs.

[1] adult male 雄成鳥：Nam Sang Wai 南生圍：Jan-06, 06 年 1 月：Jemi and John Holmes 孔思義、黃亞萍
[2] juvenile 幼鳥：Po Toi 蒲台：Oct-07, 07 年 10 月：Joyce Tang 鄧玉蓮
[3] adult 成鳥：Mai Po 米埔：Apr-04, 04 年 4 月：Henry Lui 呂德恆
[4] chick 雛鳥：Mai Po 米埔：Jun-07, 07 年 6 月：Cherry Wong 黃卓研
[5] juvenile moulting into 1st winter plumage 幼鳥：Long Valley 塑原：Nov-07, 07 年 11 月：Isaac Chan 陳家強
[6] adult female 雌成鳥：Long Valley 塑原：Dec-06, 06 年 12 月：Matthew and TH Kwan 關朗曦、關子凱
[7] juvenile 幼鳥：Mai Po 米埔：Jul-04, 04 年 7 月：Jemi and John Holmes 孔思義、黃亞萍

| 春季過境遷徙鳥<br>Spring Passage Migrant | | | 夏候鳥<br>Summer Visitor | | | 秋季過境遷徙鳥<br>Autumn Passage Migrant | | | 冬候鳥<br>Winter Visitor | | |
|---|---|---|---|---|---|---|---|---|---|---|---|
| 1 | 2 | 3 | 4 | 5 | 6 | 7 | 8 | 9 | 10 | 11 | 12 |
| 留鳥<br>Resident | | | | 迷鳥<br>Vagrant | | | | 偶見鳥<br>Occasional Visitor | | | |

常見月份

# 反嘴鷸

(普) fǎn zuǐ yù
(粵) 鷸：音核

體長 length：42-45cm

## Pied Avocet | *Recurvirostra avosetta*

其他名稱 Other names：反嘴鴴, Avocet

1

黑白分明。嘴黑色細長，末端向上彎，腳灰色。頭頂至後頸、翼角和翼尖，以及翼上覆羽皆為黑色，其他部分大致白色。

Unmistakable black-and-white bird, with long, narrow up-curved bill. Grey legs. Black from crown to hindneck, on wing tips and carpal joints, and on upperwing coverts. Other parts are white.

1 Mai Po 米埔；Samson So 蘇毅雄
2 adult 成鳥；Nam Sang Wai 南生圍；8-Dec-07, 07 年 12 月 8 日；Doris Chu 朱詠兒
3 1st winter 第一年冬天；Nam Sang Wai 南生圍；Nov-03, 03 年 11 月；Jemi and John Holmes 孔思義、黃亞萍
4 adult in courtship 成鳥，正在交配；6-May-07, 07 年 5 月 6 日；Michelle and Peter Wong 江敏兒、黃理沛
5 Mai Po 米埔；Marcus Ho 何萬邦
6 Long Valley 塱原；Dec-06, 06 年 12 月；Matthew and TH Kwan 關朗曦、關子凱
7 Mai Po 米埔；Mar-07, 07 年 3 月；Bill Man 文權溢
8 1st winter 第一年冬天；Mai Po 米埔；Nov-04, 04 年 11 月；Pippen Ho 何志剛

| | 春季過境遷徙鳥<br>Spring Passage Migrant | | 夏候鳥<br>Summer Visitor | | 秋季過境遷徙鳥<br>Autumn Passage Migrant | | 冬候鳥<br>Winter Visitor | | |
|---|---|---|---|---|---|---|---|---|---|---|---|---|
| 常見月份 | 1 | 2 | 3 | 4 | 5 | 6 | 7 | 8 | 9 | 10 | 11 | 12 |
| | 留鳥<br>Resident | | | 迷鳥<br>Vagrant | | | | 偶見鳥<br>Occasional Visitor | | | | |

# 鳳頭麥雞

(普) fèng tóu mài jī

體長 length：28-31cm

## Northern Lapwing | *Vanellus vanellus*

其他名稱 Other names：Lapwing

長 長的冠羽非常易認。嘴黑色，腳暗紅色。上體暗綠有光澤，喉及腹部白色，有一條黑色粗胸帶，容易辨認。飛行時翼底白色，初級和次級飛羽黑色。

Distinctive long crest. Black bill. Dull pinkish legs. Upperparts dark green with a metallic gloss. Throat and belly white, with a thick black breast band. Underwings white with black primaries and secondaries, readily seen in flight.

1 non-breeding adult 非繁殖羽成鳥；Mai Po 米埔；Owen Chiang 深藍
2 juvenile 幼鳥；Mai Po 米埔；15-Nov-08, 08 年 11 月 15 日；Jacky Chan 陳家華
3 non-breeding adult 非繁殖羽成鳥；Mai Po 米埔；15-Jan-05, 05 年 1 月 15 日；Michelle and Peter Wong 江敏兒．黃理沛
4 juvenile 幼鳥；Mai Po 米埔；Dec-03, 03 年 12 月；Jemi and John Holmes 孔思義．黃亞萍
5 juvenile 幼鳥；Mai Po 米埔；Nov-08, 08 年 11 月；Ng Lin Yau 吳璉宥
6 juvenile 幼鳥；Mai Po 米埔；4-Dec-05, 05 年 12 月 4 日；Doris Chu 朱詠兒

| 春季過境遷徙鳥 Spring Passage Migrant | | | 夏候鳥 Summer Visitor | | | 秋季過境遷徙鳥 Autumn Passage Migrant | | | 冬候鳥 Winter Visitor | | |
|---|---|---|---|---|---|---|---|---|---|---|---|
| 1 | 2 | 3 | 4 | 5 | 6 | 7 | 8 | 9 | 10 | 11 | 12 |
| 留鳥 Resident | | | | 迷鳥 Vagrant | | | | 偶見鳥 Occasional Visitor | | | |

常見月份

# 灰頭麥雞

(普) huī tóu mài jī

體長 length：34-37cm

## Grey-headed Lapwing | *Vanellus cinereus*

頭灰色，腹部白色，嘴黃色，嘴端黑色，虹膜紅色，頸及上胸灰褐色，有幼細黑色胸帶。飛行時翼面有明顯黑、白和褐色配搭。繁殖期有明顯的粗黑色胸帶，頸及上胸灰色，眼圈黃色鮮明。

Grey head, white belly, yellow bill with black tip. Red iris, greyish brown neck and upper breast with a narrow black border. Shows distinctive black, white and brown in flight. In breeding plumage, it has thicker black breast band, with neck and upperbreast also turn grey. Prominent yellow eye-rings.

[1] non-breeding adult 非繁殖羽成鳥；Kam Tin 錦田；Dec-04, 04 年 12 月；Cherry Wong 黃卓研
[2] breeding 繁殖羽；Kam Tin 錦田；Mar-08, 08 年 3 月；Herman Ip 葉紀江
[3] breeding 繁殖羽；Kam Tin 錦田；0 Dec 07, 07 年 12 月 9 日；Doris Chu 朱詠兒
[4] breeding 繁殖羽；Kam Tin 錦田；Owen Chiang 深藍
[5] juvenile 幼鳥；Mai Po 米埔；Nov-06, 06 年 11 月；Allen Chan 陳志雄
[6] breeding 繁殖羽；Kam Tin 錦田；Dec-08, 08 年 12 月；Andy Kwok 郭匯昌
[7] breeding 繁殖羽；Mai Po 米埔；Mar-08, 08 年 3 月；Michelle and Peter Wong 江敏兒、黃理沛
[8] non-breeding adult 非繁殖羽成鳥；Mai Po 米埔；Aug-04, 04 年 8 月；Henry Lui 呂德恆

| | 春季過境遷徙鳥 Spring Passage Migrant | | | | 夏候鳥 Summer Visitor | | | 秋季過境遷徙鳥 Autumn Passage Migrant | | 冬候鳥 Winter Visitor | | |
|---|---|---|---|---|---|---|---|---|---|---|---|---|
| 常見月份 | 1 | 2 | 3 | 4 | 5 | 6 | 7 | 8 | 9 | 10 | 11 | 12 |
| | 留鳥 Resident | | | | 迷鳥 Vagrant | | | 偶見鳥 Occasional Visitor | | | | |

# 歐金鴴

普 ōu jīn héng
粵 鴴：音恆

體長 length：26-29cm

## European Golden Plover | *Pluvialis apricaria*

全 身羽毛黃褐色，翅膀羽毛邊緣圍有白點。和相似的太平洋金斑鴴比較，體型稍大，頭和身軀看上去較矮胖，腳比例上較短，翼底較白。

Yellowish-brown body. White fringes on wing-coverts. Resembles Pacific Golden Plover but slightly larger in size. Looks shorter and fatter overall. Legs are proportionally shorter. Underwing comparatively whiter.

1 adult 成鳥；Mai Po 米埔；Oct-18, 18 年 10 月；Kinni Ho 何建業
2 adult 成鳥；Mai Po 米埔；Oct-15, 15 年 10 月；Jemi and John Holmes 孔思義、黃亞萍
3 adult 成鳥；Mai Po 米埔；Oct-15, 15 年 10 月；Jemi and John Holmes 孔思義、黃亞萍
4 adult 成鳥；Mai Po 米埔；Oct-15, 15 年 10 月；Jemi and John Holmes 孔思義、黃亞萍

| | 春季過境遷徙鳥<br>Spring Passage Migrant | | | 夏候鳥<br>Summer Visitor | | | 秋季過境遷徙鳥<br>Autumn Passage Migrant | | 冬候鳥<br>Winter Visitor | | |
|---|---|---|---|---|---|---|---|---|---|---|---|
| 常見月份 | 1 | 2 | 3 | 4 | 5 | 6 | 7 | 8 | 9 | 10 | 11 | 12 |
| | 留鳥<br>Resident | | | 迷鳥<br>Vagrant | | | 偶見鳥<br>Occasional Visitor | | | | | |

# 太平洋金斑鴴

(普) tài píng yáng jīn bān héng
(粵) 鴴：音恆

體長 length：23-26cm

## Pacific Golden Plover | *Pluvialis fulva*

其他名稱 Other names：金鴴

**繁**殖期上體有金黃色、黑色和白色斑點，與黑色下體之間有一條寬闊的白帶。非繁殖期及幼鳥沒有白帶，下體黑色，羽色變淡。嘴比灰斑鴴纖細及較短。

Breeding birds have gold-spangled upperparts with black and white spots, marked out from the black underparts by a broad white line. On non-breeding bird and juvenile, the white line and black underparts are absent, upperparts paler. Bill is shorter and thinner than Grey Plover.

1 breeding 繁殖羽；Mai Po 米埔；Apr-08, 08 年 4 月；James Lam 林文華
2 non-breeding adult moulting into breeding plumage 非繁殖羽成鳥轉換繁殖羽；Mai Po 米埔；Apr-08, 08 年 4 月；Joyce Tang 鄧玉蓮
3 breeding 繁殖羽；Mai Po 米埔；Apr-08, 08 年 4 月；Wallace Tse 謝鑑超
4 breeding 繁殖羽；Mai Po 米埔；Apr-06, 06 年 4 月；Martin Hale 夏敖天
5 juvenile 幼鳥；Kam Tin 錦田；Oct-06, 06 年 10 月；Eling Lee 李佩玲
6 non-breeding adult 非繁殖羽成鳥；Mai Po 米埔；2-Apr-05, 05 年 4 月 2 日；Doris Chu 朱詠兒
7 breeding 繁殖羽；Mai Po 米埔；Apr-06, 06 年 4 月；Tam Yip Shing 譚業成
8 non-breeding adult 非繁殖羽成鳥；Mai Po 米埔；Apr-08, 08 年 4 月；James Lam 林文華

| | 春季過境遷徙鳥<br>Spring Passage Migrant | | | 夏候鳥<br>Summer Visitor | | 秋季過境遷徙鳥<br>Autumn Passage Migrant | | 冬候鳥<br>Winter Visitor | | | |
|---|---|---|---|---|---|---|---|---|---|---|---|
| 常見月份 | 1 | 2 | 3 | 4 | 5 | 6 | 7 | 8 | 9 | 10 | 11 | 12 |
| | 留鳥<br>Resident | | | | 迷鳥<br>Vagrant | | | 偶見鳥<br>Occasional Visitor | | | | |

# 灰斑鴴

(普) huī bān héng
(粵) 鴴：音恆

體長 length：27-31cm

Grey Plover | *Pluvialis squatarola*

其他名稱 Other names：灰鴴

[1]

繁殖期上體有黑色、灰色和白色斑點，下體黑色，其間有寬闊白帶分隔。非繁殖期及幼鳥沒有白帶，下體亦無黑色，毛色灰褐。飛行時脇羽黑色，尾上覆羽白色。體型比太平洋金斑鴴大，嘴較長。

Breeding bird has black, grey and white spots on underparts, marked out from the black underparts by a broad white line. On non-breeding bird and juvenile, the white line and black underparts are absent, plumage is greyish. Shows black axillaries and white uppertail coverts in flight. Bigger and longer-billed than Pacific Golden Plover.

[1] non-breeding adult 非繁殖羽成鳥；Mai Po 米埔；May-08, 08 年 5 月；Cherry Wong 黃卓研
[2] non-breeding 非繁殖羽；Mai Po 米埔；24-Feb-07, 07 年 2 月 24 日；Michelle and Peter Wong 江敏兒、黃理沛
[3] non-breeding 非繁殖羽；Mai Po 米埔；Nov-06, 06 年 11 月；Frankie Chu 朱錦滿
[4] non-breeding adult 非繁殖羽成鳥；Mai Po 米埔；Geoff Carey 賈知行
[5] juvenile 幼鳥；Mai Po 米埔；8-May-05, 05 年 5 月 8 日；Michelle and Peter Wong 江敏兒、黃理沛

| | 春季過境遷徙鳥 Spring Passage Migrant | | | 夏候鳥 Summer Visitor | | | 秋季過境遷徙鳥 Autumn Passage Migrant | | | 冬候鳥 Winter Visitor | | |
|---|---|---|---|---|---|---|---|---|---|---|---|---|
| 常見月份 | 1 | 2 | 3 | 4 | 5 | 6 | 7 | 8 | 9 | 10 | 11 | 12 |
| | | 留鳥 Resident | | | | 迷鳥 Vagrant | | | | 偶見鳥 Occasional Visitor | | |

# 劍鴴

(普) jiàn héng
(粵) 鴴：音恆

體長 length：18-20cm

## Common Ringed Plover | *Charadrius hiaticula*

中型鴴類。頭褐色，額白色，喉部白色並有完整的白色頸圈。面頰和耳羽深褐色；上胸至背部、上體褐色，下體白色。繁殖羽額上有粗黑橫帶。腳橙紅色，嘴黑色。繁殖羽嘴橙紅色而嘴端黑，飛行時有顯眼白色翼帶。

Medium-sized plover. Brownish head, white forehead, white throat and complete neck ring. Cheek and ear coverts dark brown. Upper breast to mantle and upperparts brown, and underparts white in colour. During breeding, it has thick black band on top of white forehead, reddish orange legs, black bill, reddish orange bill with black tip. Prominent white wing band in flght.

[1] non-breeding adult 非繁殖羽成鳥；Mai Po 米埔；Oct-01, 01 年 10 月；Jemi and John Holmes 孔思義、黃亞萍

| 春季過境遷徙鳥<br>Spring Passage Migrant | | | 夏候鳥<br>Summer Visitor | | | 秋季過境遷徙鳥<br>Autumn Passage Migrant | | 冬候鳥<br>Winter Visitor | | | |
|---|---|---|---|---|---|---|---|---|---|---|---|
| 1 | 2 | 3 | 4 | 5 | 6 | 7 | 8 | 9 | 10 | 11 | 12 |
| 留鳥<br>Resident | | | | 迷鳥<br>Vagrant | | | | 偶見鳥<br>Occasional Visitor | | | |

常見月份

# 長嘴鴴

(普)cháng zuǐ héng
(粵)鴴：音恆

體長 length：18-21cm

## Long-billed Plover | *Charadrius placidus*

**體**型比金眶鴴大。嘴較長，腳淡黃色，耳羽褐色，前額白斑比較明顯，後額有寬闊黑帶，眼後有淺色粗眼眉，白頸圈完整並有黑色及灰褐色邊。頭頂和上體灰褐色，腹部白色。飛行時有細白色翼帶。幼鳥上體羽毛末端淺色。

Bigger than Little Ringed Plover and has longer bill. Legs pale yellow, ear coverts brown. Obvious white patch on forehead, thick black band on forecrown. Thick light-coloured supercilium behind the eyes, complete white collar with black and greyish brown border. Greyish brown crown and upperparts, white belly. Thin white lines on wings in flight. Juvenile has pale feather tip on upperparts.

[1] juvenile 幼鳥：Kam Tin 錦田：Jemi and John Holmes 孔思義，黃亞萍

| 春季過境遷徙鳥 Spring Passage Migrant | | | 夏候鳥 Summer Visitor | | | 秋季過境遷徙鳥 Autumn Passage Migrant | | | 冬候鳥 Winter Visitor | | |
|---|---|---|---|---|---|---|---|---|---|---|---|
| 1 | 2 | 3 | 4 | 5 | 6 | 7 | 8 | 9 | 10 | 11 | 12 |
| 留鳥 Resident | | | | 迷鳥 Vagrant | | | | 偶見鳥 Occasional Visitor | | | |

常見月份

# 金眶鴴

(普) jīn kuàng héng
(粵) 眶鴴：音框恆

Little Ringed Plover | *Charadrius dubius*

體長 length：14-17cm

1

嘴部大致黑色，眼圈鮮明黃色，喉部白色並有完整的白色頸圈。繁殖羽有黑色胸帶，眼後白眉向上延伸至頭頂，嘴基有時沾紅。非繁殖期頭及胸帶的黑色變成褐色。

Black bill, obvious yellow eye-rings, white throat and complete white neck ring. In breeding plumage, it has completely black breast band and white supercilium behind eye extending to forecrown. During non-breeding season, black colour on breast band and head turns brown.

1 breeding 繁殖羽；Shan Pui River 山貝河；Dec-05, 05 年 12 月；Jemi and John Holmes 孔思義、黃亞萍
2 juvenile 幼鳥；Shan Pui River 山貝河；Dec-05, 05 年 12 月；Jemi and John Holmes 孔思義、黃亞萍
3 breeding 繁殖羽；Long Valley 塱原；Mar-07, 07 年 3 月；Allen Chan 陳志雄
4 non-breeding adult 非繁殖羽成鳥；Long Valley 塱原；Mar-08, 08 年 3 月；Ken Fung 馮漢城
5 juvenile moulting into breeding plumage 幼鳥轉換繁殖羽；Mai Po 米埔；19-Aug-07, 07 年 8 月 19 日；Michelle and Peter Wong 江敏兒、黃理沛
6 juvenile 幼鳥；Mai Po 米埔；Feb-07, 07 年 2 月；Cherry Wong 黃卓研
7 adult 非繁殖羽成鳥；Pui O 貝澳；Mar-08, 08 年 3 月；Cherry Wong 黃卓研

| 春季過境遷徙鳥 Spring Passage Migrant | | | | 夏候鳥 Summer Visitor | | 秋季過境遷徙鳥 Autumn Passage Migrant | | | 冬候鳥 Winter Visitor | | |
|---|---|---|---|---|---|---|---|---|---|---|---|
| 1 | 2 | 3 | 4 | 5 | 6 | 7 | 8 | 9 | 10 | 11 | 12 |
| 留鳥 Resident | | | | 迷鳥 Vagrant | | | | 偶見鳥 Occasional Visitor | | | |

常見月份

# 環頸鴴

普 huán jǐng héng
粵 鴴：音恆

體長 length：15-17.5cm

Kentish Plover | *Charadrius alexandrinus*

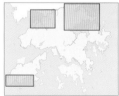

脚 深色，嘴黑色。胸帶在胸前斷開，與金眶鴴有別。飛行時有明顯的白色翼帶，尾部兩側白色。繁殖時，後額有黑色斑，頭頂至後枕栗色，胸帶黑色，在胸前及後頸斷開。非繁殖時，頭和胸帶黑色部分變為褐色，後枕栗色部分變為淡褐色。

Dark legs and black bill. Broken breast band distinguishes it from Little Ringed Plover. White wing bar and white outer tail feathers are conspicuous in flight. Breeding plumage shows black patch behind forehead, chestnut crown and nape, black breast band broken at the breast and hindneck. In non-breeding season, the black colour on head and breast turns brown, and chestnut colour on crown and nape becomes pale brown.

[1] breeding male 繁殖羽雄鳥：Mai Po 米埔：Mar-04, 04 年 3 月：Jemi and John Holmes 孔思義、黃亞萍
[2] adult female 雌成鳥：Mai Po 米埔：4-Mar-05, 05 年 3 月 4 日：Wong Hok Sze 王學思
[3] breeding male 繁殖羽雄鳥：Mai Po 米埔：20-Mar-04, 04 年 3 月 20 日：Lo Kar Man 盧嘉孟
[4] adult female 雌成鳥：Mai Po 米埔：26-Mar-09, 09 年 3 月 26 日：Martin Hale 夏敖天
[5] breeding male 繁殖羽雄鳥：Mai Po 米埔：26-Mar-09, 09 年 3 月 26 日：Martin Hale 夏敖天
[6] breeding male 繁殖羽雄鳥：Mai Po 米埔：20-Mar-04, 04 年 3 月 20 日：Lo Kar Man 盧嘉孟
[7] 1st winter 第一年冬天：Mai Po 米埔：Oct-06, 06 年 10 月：Joyce Tang 鄧玉蓮
[8] juvenile moulting into breeding male 雄幼鳥轉換繁殖羽：Mai Po 米埔：Daniel CK Chan 陳志光

| | 春季過境遷徙鳥<br>Spring Passage Migrant | | | 夏候鳥<br>Summer Visitor | | | 秋季過境遷徙鳥<br>Autumn Passage Migrant | | | 冬候鳥<br>Winter Visitor | | |
|---|---|---|---|---|---|---|---|---|---|---|---|---|
| 常見月份 | 1 | 2 | 3 | 4 | 5 | 6 | 7 | 8 | 9 | 10 | 11 | 12 |
| | | 留鳥<br>Resident | | | | 迷鳥<br>Vagrant | | | | 偶見鳥<br>Occasional Visitor | | |

鴴科
Charadriidae

# 蒙古沙鴴

[普] měng gǔ shā héng
[粵] 鴴：音恆

體長 length：18-21cm

Lesser Sand Plover | *Charadrius mongolus*

其他名稱 Other names：Mongolian Plover

[1]

 🏖️ 🌊

體型較大的鴴。嘴黑色，短而纖細；腳偏綠，可和鐵嘴沙鴴區別。繁殖期頭頂前端至後枕沾栗色，粗栗色胸帶伸延至上腹，胸帶上黑色細紋將白色喉部分隔。非繁殖期頭胸的栗色變成與上體相若的灰褐色，有淺色眼眉。

Large-sized plover. Compared with Greater Sand Plover, it has shorter, finer black bill and greenish legs. In breeding plumage, the area between forecrown and nape is tinted with chestnut, and thick chestnut breast band extends to upper belly. A thin black line separates breast band and white throat. In non-breeding season, chestnut colour on head and breast turns greyish brown, resembling the colour of the upperparts, and pale supercilium also appears.

[1] breeding male 繁殖羽雄鳥；Mai Po 米埔；May-04, 04 年 5 月；Jemi and John Holmes 孔思義、黃亞萍
[2] juvenile 幼鳥；Sai Kung 西貢；Oct-08, 08 年 10 月；Freeman Yue 余柏維
[3] juvenile 幼鳥；Ma On Shan 馬鞍山；Oct-08, 08 年 10 月；Ken Fung 馮漢城
[4] juvenile 幼鳥；Ma On Shan 馬鞍山；Oct-08, 08 年 10 月；Ken Fung 馮漢城
[5] juvenile 幼鳥；Ma On Shan 馬鞍山；Oct-08, 08 年 10 月；Ken Fung 馮漢城
[6] juvenile 幼鳥；Sai Kung 西貢；Oct-08, 08 年 10 月；Joyce Tang 鄧玉蓮
[7] breeding 繁殖羽；Mai Po 米埔；Samson So 蘇毅雄

| 春季過境遷徙鳥<br>Spring Passage Migrant | | | 夏候鳥<br>Summer Visitor | | | 秋季過境遷徙鳥<br>Autumn Passage Migrant | | | 冬候鳥<br>Winter Visitor | | |
|---|---|---|---|---|---|---|---|---|---|---|---|
常見月份 | 1 | 2 | 3 | 4 | 5 | 6 | 7 | 8 | 9 | 10 | 11 | 12 |

| 留鳥<br>Resident | 迷鳥<br>Vagrant | 偶見鳥<br>Occasional Visitor |
|---|---|---|

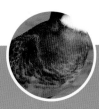

# 鐵嘴沙鴴

(普) tiě zuǐ shā héng
(粵) 鴴：音恆

體長 length：20-25cm

## Greater Sand Plover | *Charadrius leschenaultii*

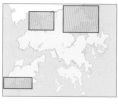

**體**型較大的鴴。嘴黑色，較蒙古沙鴴長和粗；腳偏黃色且較長。繁殖羽頭頂前端至後枕沾栗色，胸帶栗色且較蒙古沙鴴幼細。非繁殖羽頭胸上的栗色變成與上體相若的灰褐色，有淺色眼眉。

Large-sized plover. Compared with Lesser Sand Plover, it has longer, thicker black bill, and longer yellowish legs. In breeding plumage, the area between forecrown and nape is tinted chestnut, and the chestnut breast band is narrower than Lesser Sand Plover. In non-breeding plumage, chestnut colour on head and breast turn greyish brown, resembling the colour of the upperparts, and pale supercilium also appears.

1 breeding male 繁殖羽雄鳥；Pui O 貝澳；Apr-04, 04 年 4 月：Henry Lui 呂德恆
2 breeding male, race *atrifrons* 繁殖羽雄鳥, *atrifrons* 亞種；Mai Po 米埔；Apr-06, 06 年 4 月：Tam Yip Shing 譚業成
3 juvenile 幼鳥；8-Apr-07, 07 年 4 月 8 日：Michelle and Peter Wong 江敏兒、黃理沛
4 adult moulting into breeding plumage 成鳥轉換繁殖羽；Mai Po 米埔；May-06, 06 年 5 月：Martin Hale 夏敖天
5 non-breeding adult 非繁殖羽成鳥；Mai Po 米埔；May-06, 06 年 5 月：Martin Hale 夏敖天
6 breeding female 繁殖羽雌鳥；Mai Po 米埔；Apr-07, 07 年 4 月：James Lam 林文華
7 juvenile 幼鳥；Mai Po 米埔；May-09, 09 年 5 月：Kitty Koo 古愛婉

| 春季過境遷徙鳥<br>Spring Passage Migrant | | | 夏候鳥<br>Summer Visitor | | 秋季過境遷徙鳥<br>Autumn Passage Migrant | | | 冬候鳥<br>Winter Visitor | | | |
|---|---|---|---|---|---|---|---|---|---|---|---|
| 常見月份 | | | | | | | | | | | |
| 1 | 2 | 3 | 4 | 5 | 6 | 7 | 8 | 9 | 10 | 11 | 12 |
| 留鳥<br>Resident | | | | 迷鳥<br>Vagrant | | | | 偶見鳥<br>Occasional Visitor | | | |

# 東方鴴

(普) dōng fāng héng
(粵) 鴴：音恆

體長 length：22-25.5cm

## Oriental Plover | *Charadrius veredus*

其他名稱 Other names：紅胸鴴

1

體型較大的鴴。上體褐色，嘴黑色，佇立時翅膀很長。繁殖時胸部紅褐色，下有黑色細紋將白色腹部分隔。雄鳥頭部白色，頭頂沾有褐色；雌鳥頭部褐色斑較多。非繁殖期頭褐色，胸部淡黃褐色。

Large-sized plover. Brown upperparts, black bill, long wing at rest. In breeding plumage, it shows a reddish brown breast with thin black line at the bottom, above the white belly. Male has white head and brownish crown; female has more brown colour patches on head. In non-breeding plumage, it has brown head and pale yellowish brown breast.

1 breeding male 繁殖羽雄鳥；Tai Sang Wai 大生圍；14-Mar-09, 09 年 3 月 14 日；Kwok Ka Ki 郭加祈
2 breeding female 繁殖羽雌鳥；Tai Sang Wai 大生圍；14-Mar-09, 09 年 3 月 14 日；Kwok Ka Ki 郭加祈
3 breeding male 繁殖羽雄鳥；Tai Sang Wai 大生圍；14-Mar-09, 09 年 3 月 14 日；Kwok Ka Ki 郭加祈
4 juvenile 幼鳥；Mai Po 米埔；Apr-08, 08 年 4 月；Winnie Wong and Sammy Sam 森美與雲妮
5 breeding male 繁殖羽雄鳥；Tsim Bei Tsui 尖鼻咀；Martin Hale 夏敖天
6 juvenile 幼鳥；Mai Po 米埔；Apr-08, 08 年 4 月；Winnie Wong and Sammy Sam 森美與雲妮
7 breeding female 繁殖羽雌鳥；Tai Sang Wai 大生圍；14-Mar-09, 09 年 3 月 14 日；Kwok Ka Ki 郭加祈
8 breeding female 繁殖羽雌鳥；Tai Sang Wai 大生圍；14-Mar-09, 09 年 3 月 14 日；Kwok Ka Ki 郭加祈

| | 春季過境遷徙鳥<br>Spring Passage Migrant | | 夏候鳥<br>Summer Visitor | | 秋季過境遷徙鳥<br>Autumn Passage Migrant | | 冬候鳥<br>Winter Visitor | | |
|---|---|---|---|---|---|---|---|---|---|
| 常見月份 | 1 | 2 | 3 | 4 | 5 | 6 | 7 | 8 | 9 | 10 | 11 | 12 |
| | 留鳥<br>Resident | | | 迷鳥<br>Vagrant | | | 偶見鳥<br>Occasional Visitor | | |

# 彩鷸

(普)căi yù
(粵)鷸：音核

體長 length：23-28cm

## Greater Painted-snipe | *Rostratula benghalensis*

[1]

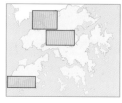

中型涉禽。有別於一般雀鳥，雌鳥的色彩比雄鳥鮮艷。眼圈至眼後明顯白色，腹部白色向上伸延至上背。嘴粗長而嘴端微向下彎，腳綠色。雄鳥頭、胸、上體和翼上覆羽褐色並有黃色斑紋。雌鳥頭頸深栗紅色，胸部沾黑，背及翼上覆羽深橄欖褐色。

Medium-sized wader. Unlike most other birds, female is more colourful than male. Obvious white eye-rings extend behind the eyes. White belly extends upwards to mantle. Long thick bill with slightly decurved tip. Green legs. Head, breast, upperparts and upperwing coverts of male are well-camouflaged brown with yellow patches. Female has dark chestnut red head and neck, breast strippled with black. Mantle and wing coverts dark olive brown.

[1] adult female with male 雌成鳥和雄鳥；Mai Po 米埔；Jun-05, 05 年 6 月；Cherry Wong 黃卓研
[2] adult female 雌成鳥；Shek Kong 石崗；Mar-07, 07 年 3 月；James Lam 林文華
[3] adult male 雄成鳥；Mai Po 米埔；May-08, 08 年 5 月；Danny Ho 何國海
[4] adult male 雄成鳥；Shek Kong 石崗；Mar-07, 07 年 3 月；Martin Hale 夏敖天
[5] adult male with chick 雄成鳥；Mai Po 米埔；May-08, 08 年 5 月；Cherry Wong 黃卓研
[6] juvenile 幼鳥；Mai Po 米埔；28-May-09, 09 年 5 月 28 日；James Lam 林文華
[7] adult female 雌成鳥；Mai Po 米埔；Jun-05, 05 年 6 月；Cherry Wong 黃卓研
[8] adult male 雄成鳥；Mai Po 米埔；14-May-06, 06 年 5 月 14 日；Doris Chu 朱詠兒

| | 春季過境遷徙鳥<br>Spring Passage Migrant | | | 夏候鳥<br>Summer Visitor | | | 秋季過境遷徙鳥<br>Autumn Passage Migrant | | | 冬候鳥<br>Winter Visitor | | |
|---|---|---|---|---|---|---|---|---|---|---|---|---|
| 常見月份 | 1 | 2 | 3 | 4 | 5 | 6 | 7 | 8 | 9 | 10 | 11 | 12 |
| | 留鳥<br>Resident | | | | 迷鳥<br>Vagrant | | | | 偶見鳥<br>Occasional Visitor | | | |

# 水雉
(普) shuǐ zhì
(粵) 雉：音字

體長 length：39-58cm

## Pheasant-tailed Jacana | *Hydrophasianus chirurgus*

**外** 型獨特的水鳥。腳及趾長，頭至前頸白色，頭頂深色，黑色貫眼紋沿頸側一直伸延至胸帶，後頸金黃色，上體及覆羽褐色，翼底白色，翼尖黑色，腹部白色。繁殖期貫眼紋消失，腹部變成黑色，有特長的黑色尾羽。常在浮葉植物上行走。

Distinctive waterbird. Long legs and toes. Head to foreneck white in colour. Dark cap, black eyestripe extends along the side of the neck to breast band. Golden yellow hindneck, brown upperparts and wing coverts. White underwing with black tip. White belly. When breeding, eyestripe disappears, belly turns black, with elongated black tail. Often walks on floating-leave plant.

1 non-breeding adult 非繁殖羽成鳥：Long Valley 塱原；Nov-08，08 年 11 月；Michelle and Peter Wong 江敏兒、黃理沛
2 juvenile 幼鳥：Long Valley 塱原；Nov-07，07 年 11 月；Joyce Tang 鄧玉蓮
3 juvenile moulting into adult plumage 幼鳥轉換成羽：Lok Ma Chau 落馬洲；Oct-06，06 年 10 月；Martin Hale 夏敖天
4 adult moulting into breeding plumage 成鳥轉換繁殖羽：Mai Po 米埔；Apr-08，08 年 4 月；Ng Lin Yau 吳璉宥
5 juvenile 幼鳥：Lok Ma Chau 落馬洲；Sep-06，06 年 9 月；Allen Chan 陳志雄
6 immature 未成年鳥：Kam Tin 錦田；11-Nov-00，00 年 11 月 11 日；Lo Kar Man 盧嘉孟
7 juvenile 幼鳥：Mai Po 米埔；Owen Chiang 深藍
8 juvenile 幼鳥：Mai Po 米埔；Oct-06，06 年 10 月；Hoon Kwok Wai 洪國偉

| | 春季過境遷徙鳥<br>Spring Passage Migrant | | | 夏候鳥<br>Summer Visitor | | 秋季過境遷徙鳥<br>Autumn Passage Migrant | | 冬候鳥<br>Winter Visitor | | |
|---|---|---|---|---|---|---|---|---|---|---|
| 常見月份 | 1 | 2 | 3 | 4 | 5 | 6 | 7 | 8 | 9 | 10 | 11 | 12 |
| | 留鳥<br>Resident | | | | 迷鳥<br>Vagrant | | | 偶見鳥<br>Occasional Visitor | | |

# 中杓鷸

(普) zhōng sháo yù
(粵) 杓鷸：音桌核

體長 length：40-46cm

## Eurasian Whimbrel | *Numenius phaeopus*

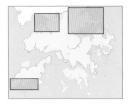

外 型與白腰杓鷸相似，但體型較小，向下彎的嘴亦較短。上體深
灰褐色有白色斑點，頭頂有深褐色粗側冠紋。飛行時腰部白色。

Resembles Eurasian Curlew but smaller in size, with shorter decurved bill. Upperparts dark greyish brown with white spots. Prominent dark brown lateral crown stripes. Shows white rump in flight.

1 adult 成鳥；Mai Po 米埔；Apr-08, 08 年 4 月；Joyce Tang 鄧玉蓮
2 Mai Po 米埔；Sep-07, 07 年 9 月；Chan Kin Chung Gary 陳建中
3 juvenile 幼鳥；Mai Po 米埔；Sep-07, 07 年 9 月；Kitty Koo 古愛婉
4 adult 成鳥；Mai Po 米埔；Oct-06, 06 年 10 月；Andy Kwok 郭匯昌
5 adult 成鳥；Mai Po 米埔；Owen Chiang 深藍
6 adult 成鳥；Mai Po 米埔；May-08, 08 年 5 月；Cherry Wong 黃卓研
7 Mai Po 米埔；Sep-07, 07 年 9 月；Kitty Koo 古愛婉

| 春季過境遷徙鳥<br>Spring Passage Migrant | | | 夏候鳥<br>Summer Visitor | | | 秋季過境遷徙鳥<br>Autumn Passage Migrant | | | 冬候鳥<br>Winter Visitor | | |
|---|---|---|---|---|---|---|---|---|---|---|---|
| 1 | 2 | 3 | 4 | 5 | 6 | 7 | 8 | 9 | 10 | 11 | 12 |
| 留鳥<br>Resident | | | | 迷鳥<br>Vagrant | | | | 偶見鳥<br>Occasional Visitor | | | |

常見月份

鷸科
Scolopacidae

# 小杓鷸

(普) xiǎo sháo yù
(粵) 杓鷸：音桌核

體長 length：28-34cm

## Little Curlew | *Numenius minutus*

其他名稱 Other names：Little Whimbrel

看 似細小的中杓鷸，嘴短，嘴端微向下彎。上體深褐色有白色斑點，有明淺黃褐色眼眉，頭頂有一條淡褐色細冠紋。飛行時腰部褐色。

Resembles a small Whimbrel. Short bill with slightly decurved tip. Upperparts darkish brown colour with white spots, with buff supercilium and narrow crown stripe. Obvious brown rump in flight.

1 adult 成鳥：Kam Tin 錦田：23-Apr-05, 05 年 4 月 23 日：Michelle and Peter Wong 江敏兒、黃理沛
2 adult 成鳥：Kam Tin 錦田：23-Apr-05, 05 年 4 月 23 日：Michelle and Peter Wong 江敏兒、黃理沛
3 adult 成鳥：Kam Tin 錦田：23-Apr-05, 05 年 4 月 23 日：Michelle and Peter Wong 江敏兒、黃理沛

| 春季過境遷徙鳥 Spring Passage Migrant | | | 夏候鳥 Summer Visitor | | | 秋季過境遷徙鳥 Autumn Passage Migrant | | | 冬候鳥 Winter Visitor | | |
|---|---|---|---|---|---|---|---|---|---|---|---|
| 常見月份 1 | 2 | 3 | 4 | 5 | 6 | 7 | 8 | 9 | 10 | 11 | 12 |

| 留鳥 Resident | 迷鳥 Vagrant | 偶見鳥 Occasional Visitor |
|---|---|---|

# 紅腰杓鷸

(普)hóng yāo sháo yù
(粵)杓鷸：音桌核

體長length：53-66cm

## Far Eastern Curlew | *Numenius madagascariensis*

其他名稱 Other names：Eastern Curlew, Australian Curlew

近 似白腰杓鷸，嘴部比白腰杓鷸長，腰部深黃褐色有斑紋，佇立時呈更濃的褐色；腹部及尾下覆羽黃褐色。飛行時上體和腰部深黃褐色明顯可見，翼下覆羽有濃密橫紋。常單隻混在白腰杓鷸群中。

Average longer-billed than Eurasian Curlew, with dark brownish-barred buff rump, belly and undertail coverts. In flight, it shows brown rump, brown upperparts and dense bars on underwings. Usually single birds are found in flocks of Eurasian Curlews.

1 adult 成鳥：Mai Po 米埔：May-07, 07 年 5 月；Cherry Wong 黃卓研
2 breeding 繁殖羽：Mai Po 米埔：27-Mar-09, 09 年 3 月 27 日；Martin Hale 夏敖天
3 adult 成鳥（right, 右面）：Mai Po 米埔：May-07, 07 年 5 月；Cherry Wong 黃卓研
4 Mai Po 米埔：10-Apr-09, 09 年 4 月 10 日；Hoon Kwok Wai 洪國偉

| 春季過境遷徙鳥<br>Spring Passage Migrant | | 夏候鳥<br>Summer Visitor | | 秋季過境遷徙鳥<br>Autumn Passage Migrant | | 冬候鳥<br>Winter Visitor | |
|---|---|---|---|---|---|---|---|
| 1 | 2 | 3 | 4 | 5 | 6 | 7 | 8 | 9 | 10 | 11 | 12 | 常見月份 |
| 留鳥<br>Resident | | | 迷鳥<br>Vagrant | | | 偶見鳥<br>Occasional Visitor | |

# 白腰杓鷸

(普) bái yāo sháo yù
(粵) 杓鷸：音桌核

體長 length：50-60cm

**Eurasian Curlew** | *Numenius arquata*

**體**型大，嘴長而向下彎。上體淡褐色，有黑褐色縱紋，腹及尾下覆羽白色。飛行時可見明顯的白色翼下覆羽及腰，尾上有深褐色橫紋。

Large-sized wader, with very long decurved bill. Upperparts light brown with heavy blackish brown streaks. White belly and undertail coverts. In flight, it shows prominent white underwings, white rump and white tail with dark brown bars.

[1] adult 成鳥；Mai Po 米埔；Feb-07, 07 年 2 月；Cherry Wong 黃卓研
[2] Mai Po 米埔；Oct-04, 04 年 10 月；Jemi and John Holmes 孔思義，黃亞萍
[3] race *orientalis*, *orientalis* 亞種；Mai Po 米埔；Oct-07, 07 年 10 月；Isaac Chan 陳家強
[4] juvenile 幼鳥；Mai Po 米埔；May-06, 06 年 5 月；Andy Kwok 郭匯昌
[5] race *orientalis*, *orientalis* 亞種；Mai Po 米埔；Oct-06, 06 年 10 月；Martin Hale 夏敖天

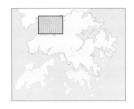

| 春季過境遷徙鳥<br>Spring Passage Migrant | | | 夏候鳥<br>Summer Visitor | | | 秋季過境遷徙鳥<br>Autumn Passage Migrant | | | 冬候鳥<br>Winter Visitor | | |
|---|---|---|---|---|---|---|---|---|---|---|---|
| 1 | 2 | 3 | 4 | 5 | 6 | 7 | 8 | 9 | 10 | 11 | 12 |

常見月份

| 留鳥<br>Resident | 迷鳥<br>Vagrant | 偶見鳥<br>Occasional Visitor |
|---|---|---|

# 斑尾塍鷸

(普) bān wěi chéng yù
(粵) 塍鷸：音成核

體長 length：37-41cm

## Bar-tailed Godwit | *Limosa lapponica*

中型涉禽，嘴端微向上彎。繁殖期除尾羽及尾下覆羽外，全身深紅褐色，上體羽毛邊緣白色。尾羽白色並有數條橫紋。非繁殖期羽色灰褐。

Medium-sized wader. Bill tip slightly upturned. In breeding plumage, body turns deep reddish brown, with white tail and undertail coverts; white fringes on upperparts feathers. Uppertail white with blackish bars. Greyish-brown in non-breeding plumage.

[1] breeding and non-breeding 繁殖羽和非繁殖羽；Mai Po 米埔；Cheung Ho Fai 張浩輝
[2] non-breeding 非繁殖羽；Mai Po 米埔；Lo Kar Man 盧嘉孟
[3] breeding male 繁殖羽雄鳥；Mai Po 米埔；Mar-08, 08 年 3 月；Pippen Ho 何志剛
[4] non-breeding adult 非繁殖羽成鳥；Nam Sang Wai 南生圍；Oct-06, 06 年 10 月；Kelvin Yam 任德政

| 春季過境遷徙鳥<br>Spring Passage Migrant | | 夏候鳥<br>Summer Visitor | | 秋季過境遷徙鳥<br>Autumn Passage Migrant | | 冬候鳥<br>Winter Visitor | |
|---|---|---|---|---|---|---|---|
| 1 | 2 | 3 | 4 | 5 | 6 | 7 | 8 | 9 | 10 | 11 | 12 |

常見月份

| 留鳥<br>Resident | 迷鳥<br>Vagrant | 偶見鳥<br>Occasional Visitor |
|---|---|---|

# 黑尾塍鷸

(普) hēi wěi chéng yù
(粵) 塍鷸：音成核

體長 length：36-44cm

## Black-tailed Godwit | *Limosa limosa*

中型涉禽，較斑尾塍鷸小。腳長，嘴直而長，尖端黑色。尾羽白色，末端有寬闊黑色橫帶，飛行時特別清楚。繁殖羽頸及胸橙褐色，背部灰褐色且有黑、褐色斑紋。非繁殖羽全身偏灰。

Medium-sized wader. Smaller than Bar-tailed Godwit. Long legs. Bill long and straight with black tip. Tail white with broad black terminal band which shows clearly in flight. Neck and breast turn orange brown in breeding plumage, with black and brown patches on back. Generally grey in non-breeding plumage.

1 Mai Po 米埔；Samson So 蘇毅雄
2 breeding 繁殖羽；Mai Po 米埔；1-May-05, 05 年 5 月 1 日；Michelle and Peter Wong 江敏兒、黃理沛
3 non-breeding adult 非繁殖羽成鳥；Mai Po 米埔；Mar-08, 08 年 3 月；Pippen Ho 何志剛
4 breeding 繁殖羽；Mai Po 米埔；10-Apr-05, 05 年 4 月 10 日：Doris Chu 朱詠兒
5 breeding 繁殖羽；Mai Po 米埔；Apr-04, 04 年 4 月：Jemi and John Holmes 孔思義、黃亞萍

| | 春季過境遷徙鳥<br>Spring Passage Migrant | | | 夏候鳥<br>Summer Visitor | | | 秋季過境遷徙鳥<br>Autumn Passage Migrant | | | 冬候鳥<br>Winter Visitor | | |
|---|---|---|---|---|---|---|---|---|---|---|---|---|
| 常見月份 | 1 | 2 | 3 | 4 | 5 | 6 | 7 | 8 | 9 | 10 | 11 | 12 |
| | | 留鳥<br>Resident | | | | 迷鳥<br>Vagrant | | | | 偶見鳥<br>Occasional Visitor | | |

# 大濱鷸

(普) dà bīn yù
(粵) 鷸:音核

Great Knot | *Calidris tenuirostris*

體長 length：26-28cm

外 型比紅腹濱鷸大和豐滿，黑色嘴部亦較長，近末端處微向下彎。胸部斑點濃密，飛行時腰部白色，沒有明顯翼帶。繁殖期上背有明顯的V型栗色斑紋，胸部斑點較大及濃密，非繁殖期上體偏灰。

Larger and heavier than the Red Knot, the black bill is longer and slightly decurved at the tip. Heavy black spots on the breast. In flight, it shows a white rump and no conspicuous wing bar. Breeding bird has chestnut V-shaped patch on the mantle, and spots on breast appear larger and more dense. Non-breeding bird is greyer on the upperparts.

[1] breeding 繁殖羽；Mai Po 米埔；28-Mar-05, 05 年 3 月 28 日；Michelle and Peter Wong 江敏兒、黃理沛
[2] non-breeding adult moniting into breeding plumage 非繁殖成鳥轉換繁殖羽；Mai Po 米埔；Captain Wong 黃倫昌
[3] breeding 繁殖羽；Mai Po 米埔；20-Mar-04, 04 年 3 月 20 日；Lo Kar Man 盧嘉孟
[4] adult 成鳥；Ma On Shan 馬鞍山；Sep-17, 17 年 9 月；Ken Fung 馮漢城
[5] adult 成鳥；Ma On Shan 馬鞍山；Sep-17, 17 年 9 月；Ken Fung 馮漢城
[6] non-breeding adult 非繁殖羽成鳥；Mai Po 米埔；Apr-04, 04 年 4 月；Jemi and John Holmes 孔思義、黃亞萍

| 春季過境遷徙鳥 Spring Passage Migrant | | | | 夏候鳥 Summer Visitor | | 秋季過境遷徙鳥 Autumn Passage Migrant | | 冬候鳥 Winter Visitor | | | |
|---|---|---|---|---|---|---|---|---|---|---|---|
| 1 | 2 | 3 | 4 | 5 | 6 | 7 | 8 | 9 | 10 | 11 | 12 |

常見月份

| 留鳥 Resident | 迷鳥 Vagrant | 偶見鳥 Occasional Visitor |
|---|---|---|

# 翻石鷸

(普) fān shí yù
(粵) 鷸：音核

體長 length：21-26cm

## Ruddy Turnstone | *Arenaria interpres*

頭和胸部黑白斑駁，嘴短，腳短呈橙色。繁殖期上體變成黑、白和紅棕三色，非繁殖羽和幼鳥上體均偏褐色。

Distinctive mix of black and white on head and breast. The short bill and short orange legs are easily distinguishable. In breeding plumage, it shows black, white and rufous colour on upperparts, which is brownish on non-breeding and juvenile birds.

1 breeding male 繁殖羽雄鳥；Mai Po 米埔：Apr-07, 07 年 4 月；Andy Kwok 郭匯昌
2 breeding female 繁殖羽雌鳥；Mai Po 米埔：Apr-08, 08 年 4 月；James Lam 林文華
3 breeding female 繁殖羽雌鳥；Mai Po 米埔：Apr-07, 07 年 4 月；Lee Kai Hong 李啟康
4 breeding female 繁殖羽雌鳥；Mai Po 米埔：May-04, 04 年 5 月；Jemi and John Holmes 孔思義、黃亞萍
5 breeding male 繁殖羽雄鳥；Mai Po 米埔：Apr-97, 97 年 4 月；Martin Hale 夏敖天
6 breeding male 繁殖羽雄鳥；Mai Po 米埔：Apr-07, 07 年 4 月；Lee Kai Hong 李啟康

| 春季過境遷徙鳥<br>Spring Passage Migrant | | | 夏候鳥<br>Summer Visitor | | 秋季過境遷徙鳥<br>Autumn Passage Migrant | | 冬候鳥<br>Winter Visitor | | |
|---|---|---|---|---|---|---|---|---|---|
| 常見月份 | 1 | 2 | 3 | 4 | 5 | 6 | 7 | 8 | 9 | 10 | 11 | 12 |

| 留鳥<br>Resident | 迷鳥<br>Vagrant | 偶見鳥<br>Occasional Visitor |
|---|---|---|

# 紅腹濱鷸

(普) hóng fù bīn yù
(粵) 鷸：音核

體長 length：23-25cm

Red Knot | *Calidris canutus*

1

2

腳

短矮胖的水鳥，黑色的嘴短而粗，輕微向下彎。飛行時腰部污白，翼帶不明顯。繁殖期胸腹沾鮮栗色，非繁殖期則為均勻的灰褐色；幼鳥胸部微帶褐色，背部羽毛有魚鱗狀淺色斑紋。

Short-legged and plump wader, with thick, black, slightly decurved bill. In flight, it shows dirty white rump and inconspicuous wing bars. Breeding bird has chestnut colour breast and belly, while non-breeding bird is greyish overall brown. Juvenile bird is brownish on the breast, and upperparts have scaly feathers with pale fringes.

1 breeding 繁殖羽；Mai Po 米埔；Apr-09, 09 年 4 月；James Lam 林文華
2 breeding 繁殖羽；Mai Po 米埔；Apr-09, 09 年 4 月；Serene Wong 王維萍

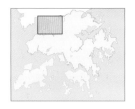

| 春季過境遷徙鳥<br>Spring Passage Migrant | | 夏候鳥<br>Summer Visitor | | 秋季過境遷徙鳥<br>Autumn Passage Migrant | | 冬候鳥<br>Winter Visitor | |
|---|---|---|---|---|---|---|---|

常見月份

| 1 | 2 | 3 | 4 | 5 | 6 | 7 | 8 | 9 | 10 | 11 | 12 |
|---|---|---|---|---|---|---|---|---|---|---|---|

| 留鳥<br>Resident | 迷鳥<br>Vagrant | 偶見鳥<br>Occasional Visitor |
|---|---|---|

# 流蘇鷸

(普) liú sū yù
(粵) 鷸：音核

體長 length：26-32cm

Ruff | *Calidris pugnax*

其他名稱 Other names：Reeve

頸 粗長，嘴短黑色而微向下彎，背部羽毛邊緣白色，腿部顏色多變。繁殖期雄鳥的頸、上胸和背部有鮮明紅褐和黑色。非繁殖期羽色灰褐為主，背部有大片的鱗狀羽毛，嘴基羽毛淡白色。幼鳥羽色也相似，但胸部微沾淡褐色。

Long thick neck, short and slightly decurved bill. Feathers of upperparts have white fringes. Leg colour varies. Breeding male has rufous and black colour on neck, upperbreast and upperparts. Non-breeding plumage mainly greyish brown, characterised by large scale-like feathers on mantle, and whitish area at the base of bill. Juvenile is similar to non-breeding adult, but with pale brown tinted on the breast.

1 non-breeding adult male moulting into breeding plumage 非繁殖羽雄鳥轉換繁殖羽；Mai Po 米埔；Aug-07, 07 年 8 月；Kinni Ho 何建業
2 1st winter moulting into breeding female 第一年冬天雌鳥轉換繁殖羽；Mai Po 米埔；Apr-04, 04 年 4 月；Jemi and John Holmes 孔思義、黃亞萍
3 1st winter moulting into breeding female 第一年冬天雌鳥轉換繁殖羽；Mai Po 米埔；Mar-05, 05 年 3 月；Cherry Wong 黃卓研
4 1st winter moulting into breeding female 第一年冬天雌鳥轉換繁殖羽；Mai Po 米埔；Mar-05, 05 年 3 月；Henry Lui 呂德恆
5 1st winter moulting into breeding female 第一年冬天雌鳥轉換繁殖羽；Mai Po 米埔；Apr-04, 04 年 4 月；Jemi and John Holmes 孔思義、黃亞萍

| 春季過境遷徙鳥<br>Spring Passage Migrant | | 夏候鳥<br>Summer Visitor | | 秋季過境遷徙鳥<br>Autumn Passage Migrant | | 冬候鳥<br>Winter Visitor | |
|---|---|---|---|---|---|---|---|
| 1 | 2 | 3 | 4 | 5 | 6 | 7 | 8 | 9 | 10 | 11 | 12 |

常見月份

| 留鳥<br>Resident | 迷鳥<br>Vagrant | 偶見鳥<br>Occasional Visitor |
|---|---|---|

鷸科
Scolopacidae

# 闊嘴鷸

(普) kuò zuǐ yù
(粵) 鷸：音核

體長 length：16-18cm

## Broad-billed Sandpiper | *Calidris falcinellus*

**嘴** 直、粗厚而末端向下彎。有白色眼眉和側冠紋，並在前端相連，有如兩道眼眉。繁殖期和幼鳥上體深褐色，羽毛邊緣白色，非繁殖羽帶灰色，胸部有淡色縱紋。

Bill straight, thick and bill tip decurved. White supercilium and lateral crown stripe meet at the front like "double eye-brows". Breeding bird and juvenile have dark brownish mantle feathers with white fringes. Non-breeding plumage is comparatively greyish, with pale stripes on breast.

1 breeding 繁殖羽；Mai Po 米埔；May-08, 08 年 5 月；Cherry Wong 黃卓研
2 breeding 繁殖羽；Mai Po 米埔；Apr-08, 08 年 4 月；Jemi and John Holmes 孔思義、黃亞萍
3 breeding 繁殖羽；Mai Po 米埔；Apr-08, 08 年 4 月；Jemi and John Holmes 孔思義、黃亞萍
4 breeding 繁殖羽；Mai Po 米埔；21-Apr-07, 07 年 4 月 21 日；Michelle and Peter Wong 江敏兒、黃理沛
5 breeding 繁殖羽；Mai Po 米埔；21-Apr-07, 07 年 4 月 21 日；Michelle and Peter Wong 江敏兒、黃理沛

| 春季過境遷徙鳥 Spring Passage Migrant | 夏候鳥 Summer Visitor | 秋季過境遷徙鳥 Autumn Passage Migrant | 冬候鳥 Winter Visitor |
|---|---|---|---|

| 常見月份 | 1 | 2 | 3 | 4 | 5 | 6 | 7 | 8 | 9 | 10 | 11 | 12 |
|---|---|---|---|---|---|---|---|---|---|---|---|---|

| 留鳥 Resident | 迷鳥 Vagrant | 偶見鳥 Occasional Visitor |
|---|---|---|

# 尖尾濱鷸

⊜ jiān wěi bīn yù
⊜ 鷸：音核

體長 length：17-22cm

## Sharp-tailed Sandpiper | *Calidris acuminata*

其他名稱 Other names：尖尾鷸

頭 頂深栗色，與淡色眼眉成對比；上體羽毛深色，與淺色羽緣對比鮮明。嘴短而微向下彎，基部淺色。繁殖羽胸部銹褐色，有濃密V型斑紋伸延至脇部；上體為鮮鐵銹色，頭頂較鮮明。非繁殖羽顏色較暗淡，下體斑點較不明顯。幼鳥顏色鮮明，胸部沾橙色。

Dark rufous crown contrasts with pale supercilium. Dark upperparts with pale fringes. Short bill and bill tip is slightly decurved, bill base is pale. In breeding plumage, breast rusty with heavy V-shaped marks extending to flanks. Upperparts orange, with brighter crown. Non-breeding bird is duller with less prominent markings on the underparts. Juvenile is brightly coloured and the breast tinted with orange colour.

1 breeding 繁殖羽；Po Toi 蒲台；24-Apr-05, 05 年 4 月 24 日；Michelle and Peter Wong 江敏兒、黃理沛
2 breeding 繁殖羽；Mai Po 米埔；Apr-06, 06 年 4 月；Martin Hale 夏敖天
3 breeding 繁殖羽；Po Toi 蒲台；24-Apr-05, 05 年 4 月 24 日；Michelle and Peter Wong 江敏兒、黃理沛
4 breeding 繁殖羽；Po Toi 蒲台；24-Apr-05, 05 年 4 月 24 日；Michelle and Peter Wong 江敏兒、黃理沛

| 春季過境遷徙鳥<br>Spring Passage Migrant | | 夏候鳥<br>Summer Visitor | | 秋季過境遷徙鳥<br>Autumn Passage Migrant | | 冬候鳥<br>Winter Visitor | |
|---|---|---|---|---|---|---|---|

| 1 | 2 | 3 | 4 | 5 | 6 | 7 | 8 | 9 | 10 | 11 | 12 | 常見月份 |
|---|---|---|---|---|---|---|---|---|---|---|---|---|

| 留鳥<br>Resident | 迷鳥<br>Vagrant | 偶見鳥<br>Occasional Visitor |
|---|---|---|

鷸科
Scolopacidae

# 彎嘴濱鷸

(普) wān zuǐ bīn yù
(粵) 鷸：音核

體長 length：18-23cm

Curlew Sandpiper | *Calidris ferruginea*

中型水鳥，嘴黑色，向下彎。與黑腹濱鷸相比，腳較長、嘴較彎。繁殖期頭、頸、胸和腹為磚紅色，背部夾雜紅褐、黑和白色斑點。非繁殖期羽色單調，頭、頸和上胸有灰褐色細縱紋。

Medium-sized wader. Black bill slightly decurved. Compared with Dunlin, it has longer legs and a decurved bill. In breeding plumage, head, neck, breast and belly are brick red; upperparts have small rufous, black and white patches. Non-breeding plumage is duller, with thin brownish grey stripes on the head, neck and upper breast.

1 breeding 繁殖羽：Mai Po 米埔；Apr-08, 08 年 4 月：James Lam 林文華
2 non-breeding adult 非繁殖羽成鳥：Mai Po 米埔；Apr-06, 06 年 4 月：Martin Hale 夏敖天
3 breeding 繁殖羽：Mai Po 米埔；Apr-06, 06 年 4 月：Martin Hale 夏敖天
4 juvenile 幼鳥：Mai Po 米埔；Sep-05, 05 年 9 月：Jemi and John Holmes 孔思義、黃亞萍
5 breeding 繁殖羽：Mai Po 米埔；Apr-06, 06 年 4 月：Tam Yip Shing 譚業成
6 Mai Po 米埔；Apr-04, 04 年 4 月：Jemi and John Holmes 孔思義、黃亞萍
7 breeding 繁殖羽：Mai Po 米埔；Apr-04, 04 年 4 月：Jemi and John Holmes 孔思義、黃亞萍
8 breeding 繁殖羽：Mai Po 米埔；May-04, 04 年 5 月：Jemi and John Holmes 孔思義、黃亞萍

| | 春季過境遷徙鳥 Spring Passage Migrant | | 夏候鳥 Summer Visitor | | 秋季過境遷徙鳥 Autumn Passage Migrant | | 冬候鳥 Winter Visitor | |
|---|---|---|---|---|---|---|---|---|---|---|---|---|
| 常見月份 | 1 | 2 | 3 | 4 | 5 | 6 | 7 | 8 | 9 | 10 | 11 | 12 |
| | 留鳥 Resident | | | 迷鳥 Vagrant | | | 偶見鳥 Occasional Visitor | | |

# 青腳濱鷸

(普) qīng jiǎo bīn yù
(粵) 鷸：音核

體長 length：13-15cm

## Temminck's Stint | *Calidris temminckii*

1

**體**型細小，羽色和磯鷸相似但較單調，腳黃綠色，頭、頸、胸和背有均勻的灰褐色，腹部白色。繁殖時背部灰褐色，羽毛邊緣黃褐色。

Small, plain-looking wader resembling a Common Sandpiper. Yellowish green legs. Greyish brown head, neck, breast and upperparts, white belly. In breeding plumage, upperparts greyish brown with brownish yellow fringes.

[1] breeding 繁殖羽：Mai Po 米埔；Apr-08, 08 年 4 月：Allen Chan 陳志雄
[2] breeding 繁殖羽：Mai Po 米埔；Apr-08, 08 年 4 月：Allen Chan 陳志雄
[3] non-breeding adult 非繁殖羽成鳥；Fung Lok Wai 豐樂圍：Dec-03, 03 年 11 月：Jemi and John Holmes 孔思義、黃亞萍
[4] non-breeding adult 非繁殖羽成鳥；Fung Lok Wai 豐樂圍：Dec-04, 04 年 12 月：Jemi and John Holmes 孔思義、黃亞萍
[5] non-breeding adult 非繁殖羽成鳥；Mai Po 米埔；Dec-00, 00 年 12 月：Allen Chan 陳志雄
[6] non-breeding adult 非繁殖羽成鳥；Mai Po 米埔；Dec-08, 08 年 12 月：Allen Chan 陳志雄

| 春季過境遷徙鳥 Spring Passage Migrant | | | | 夏候鳥 Summer Visitor | | 秋季過境遷徙鳥 Autumn Passage Migrant | | 冬候鳥 Winter Visitor | | | |
|---|---|---|---|---|---|---|---|---|---|---|---|
| 常見月份 1 | 2 | 3 | 4 | 5 | 6 | 7 | 8 | 9 | 10 | 11 | 12 |
| 留鳥 Resident | | | | 迷鳥 Vagrant | | | | 偶見鳥 Occasional Visitor | | | |

# 長趾濱鷸

(普) cháng zhǐ bīn yù
(粵) 鷸：音核

體長 length：13-16cm

Long-toed Stint | *Calidris subminuta*

1

🌊 🌱

**體** 型細小，羽色似尖尾濱鷸，具黃綠色的長腳和腳趾長。繁殖期及幼鳥頭頂棕紅色，和白色眉紋成強烈對比；三級飛羽深色，有較寬的橙棕色邊緣。非繁殖期主要為灰色和白色。站姿挺直。

Small-sized wader. Resembles a small Sharp-tailed Sandpiper. Long and yellowish green legs, with long toes. Breeding and juvenile birds have reddish brown cap, with contrasting white supercilium, dark tertials with thick orange rufous fringes. Non-breeding bird is mainly grey and white. Upright stance.

1 breeding 繁殖羽：Mai Po 米埔：Apr-07, 07 年 4 月：Pippen Ho 何志剛
2 breeding 繁殖羽：Fung Lok Wai 豐樂圍：Apr-04, 04 年 4 月：Jemi and John Holmes 孔思義、黃亞萍
3 breeding 繁殖羽：Mai Po 米埔：Apr-07, 07 年 4 月：Martin Hale 夏敦天
4 breeding 繁殖羽：Fung Lok Wai 豐樂圍：Apr-04, 04 年 4 月：Jemi and John Holmes 孔思義、黃亞萍
5 breeding 繁殖羽：Fung Lok Wai 豐樂圍：Apr-04, 04 年 4 月：Jemi and John Holmes 孔思義、黃亞萍
6 breeding 繁殖羽：Fung Lok Wai 豐樂圍：Apr-04, 04 年 4 月：Jemi and John Holmes 孔思義、黃亞萍

| 春季過境遷徙鳥 Spring Passage Migrant | | | 夏候鳥 Summer Visitor | | | 秋季過境遷徙鳥 Autumn Passage Migrant | | | 冬候鳥 Winter Visitor | | |
|---|---|---|---|---|---|---|---|---|---|---|---|
| 常見月份 1 | 2 | 3 | 4 | 5 | 6 | 7 | 8 | 9 | 10 | 11 | 12 |

| 留鳥 Resident | 迷鳥 Vagrant | 偶見鳥 Occasional Visitor |
|---|---|---|

# 勺嘴鷸

（普）sháo zuǐ yù
（粵）鷸：音核

體長 length：14-16cm

## Spoon-billed Sandpiper | *Calidris pygmaea*

嘴端扁平如勺子，前額帶白色，是識別的特徵。常混在紅胸濱鷸群中，覓食經常低頭。繁殖期羽色與紅胸濱鷸相似，非殖繁期和幼鳥主要為上體灰褐和下體白色。

Characterised by its spoon-shaped bill and whitish forehead. Often mixes with flocks of Red-necked Stints and keeps lowering its head when feeding. Breeding plumage resembles Red-necked Stint. Non-breeder and juvenile are mainly greyish brown at the upperparts and white at the underparts.

[1] non-breeding adult 非繁殖羽成鳥；Mai Po 米埔；19-Apr-09, 09 年 4 月 19 日；Michelle and Peter Wong 江敏兒、黃理沛
[2] non-breeding adult 非繁殖羽成鳥；Mai Po 米埔；19-Apr-09, 09 年 4 月 19 日；Michelle and Peter Wong 江敏兒、黃理沛
[3] non-breeding adult moulting into breeding plumage 非繁殖羽成鳥轉換繁殖羽；Mai Po 米埔；Apr-08, 08 年 4 月；Jemi and John Holmes 孔思義、黃亞萍
[4] non-breeding adult moulting into breeding plumage 非繁殖羽成鳥轉換繁殖羽；Mai Po 米埔；Apr-08, 08 年 4 月；Jemi and John Holmes 孔思義、黃亞萍
[5] non-breeding adult moulting into breeding plumage 非繁殖羽成鳥轉換繁殖羽；Mai Po 米埔；Apr-08, 08 年 4 月；Michelle and Peter Wong 江敏兒、黃理沛
[6] non-breeding adult 非繁殖羽成鳥；Mai Po 米埔；Apr-05, 05 年 4 月；Cherry Wong 黃卓研

| 春季過境遷徙鳥<br>Spring Passage Migrant | | | 夏候鳥<br>Summer Visitor | | | 秋季過境遷徙鳥<br>Autumn Passage Migrant | | | 冬候鳥<br>Winter Visitor | | |
|---|---|---|---|---|---|---|---|---|---|---|---|
| 1 | 2 | 3 | 4 | 5 | 6 | 7 | 8 | 9 | 10 | 11 | 12 |

常見月份

| 留鳥<br>Resident | 迷鳥<br>Vagrant | 偶見鳥<br>Occasional Visitor |
|---|---|---|

# 紅頸濱鷸

（普）hóng jǐng bīn yù
（粵）鷸：音核

體長 length：13-16cm

## Red-necked Stint | *Calidris ruficollis*

其他名稱 Other names：Rufous-necked Sandpiper, 紅胸濱鷸

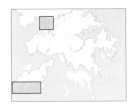

**體**型細小，嘴和腳明顯黑色，幼鳥腳部有時沾黃。繁殖期面頰、頸、胸部和上體有十分顯眼的紅棕色。非繁殖期毛色暗淡。

Small-sized wader, with prominent black bill and legs, but juvenile may have yellowish legs. Breeding bird has reddish cheeks, necks, breast and upperparts. Non-breeding bird is overall paler.

1 adult moulting into breeding plumage 成鳥轉換繁殖羽：Mai Po 米埔：Apr-06, 06 年 4 月：Martin Hale 夏敖天
2 breeding 繁殖羽：Mai Po 米埔：Apr-07, 07 年 4 月：Cherry Wong 黃卓研

| 春季過境遷徙鳥<br>Spring Passage Migrant | | 夏候鳥<br>Summer Visitor | | 秋季過境遷徙鳥<br>Autumn Passage Migrant | | 冬候鳥<br>Winter Visitor | |
|---|---|---|---|---|---|---|---|

| 1 | 2 | 3 | 4 | 5 | 6 | 7 | 8 | 9 | 10 | 11 | 12 | 常見月份 |
|---|---|---|---|---|---|---|---|---|---|---|---|---|

| 留鳥<br>Resident | 迷鳥<br>Vagrant | 偶見鳥<br>Occasional Visitor |
|---|---|---|

# 三趾濱鷸

(普) sān zhǐ bīn yù
(粵) 鷸：音核

體長 length：20-21cm

Sanderling | *Calidris alba*

其他名稱 Other names：三趾鷸

香 港所見主要為獨特易認的灰白色非繁殖羽，嘴和腳黑色，頭頂至背部為淡灰色，身體其他部分白色。跟其他涉禽不同，三趾濱鷸沒有後趾。

Mostly seen in Hong Kong are birds with distinctive whitish-grey non-breeding plumage. Bill and legs are black in colour, with pale grey upperparts and white underparts. Unlike other waders. Sanderling lacks a hind claw.

1 1st winter 第一年冬天；Po Toi 蒲台；Apr-08, 08 年 4 月；Ken Fung 馮漢城
2 1st winter 第一年冬天；Po Toi 蒲台；Apr-08, 08 年 4 月；Ken Fung 馮漢城
3 1ct winter 第一年冬天；Po Toi 蒲台；Apr-08, 08 年 4 月；Andy Kwok 郭匯昌
4 1st winter 第一年冬天；Po Toi 蒲台；Apr-08, 08 年 4 月；Ken Fung 馮漢城
5 1st winter 第一年冬天；Po Toi 蒲台；Apr-08, 08 年 4 月；Cherry Wong 黃卓研
6 1st winter 第一年冬天；Po Toi 蒲台；Apr-08, 08 年 4 月；Ken Fung 馮漢城
7 1st winter 第一年冬天；Po Toi 蒲台；Jun-08, 08 年 6 月；Chan Kai Wai 陳佳瑋
8 1st winter 第一年冬天；Po Toi 蒲台；Apr-08, 08 年 4 月；Ken Fung 馮漢城

| 春季過境遷徙鳥 Spring Passage Migrant | | | 夏候鳥 Summer Visitor | | | 秋季過境遷徙鳥 Autumn Passage Migrant | | | 冬候鳥 Winter Visitor | | |
|---|---|---|---|---|---|---|---|---|---|---|---|
| 1 | 2 | 3 | 4 | 5 | 6 | 7 | 8 | 9 | 10 | 11 | 12 |
| 留鳥 Resident | | | | 迷鳥 Vagrant | | | | 偶見鳥 Occasional Visitor | | | |

常見月份

# 黑腹濱鷸

(普) hēi fù bīn yù
(粵) 鷸：音核

體長 length：16-22cm

Dunlin | *Calidris alpina*

♨

中型水鳥，黑色嘴直長而末端稍向下彎。與彎嘴濱鷸相比，腳較短、嘴較直。繁殖期背部沾紅棕色，腹部明顯黑色。非繁殖期上體為單調的灰褐色，頸部有細縱紋，下體白色。

Medium-sized wader. Long and straight bill with bill tip slightly decurved. Compared with Curlew Sandpiper, it has shorter legs and straight bill. In breeding plumage, upperparts rufous brown, with conspicuous black belly. Non-breeding plumage is dull greyish brown on the upperparts, with thin brownish grey stripes on the neck. Underparts white.

1 non-breeding adult 非繁殖羽成鳥：Luk Keng 鹿頸；Oct-05, 05 年 10 月；Allen To 杜偉倫
2 non-breeding adult 非繁殖羽成鳥：Luk Keng 鹿頸；Oct-05, 05 年 10 月；Allen To 杜偉倫
3 non-breeding adult moulting into breeding plumage 非繁殖羽成鳥轉換繁殖羽；Mai Po 米埔；Dec-08, 08 年 12 月；Allen Chan 陳志雄
4 non-breeding adult moulting into breeding plumage 非繁殖羽成鳥轉換繁殖羽；Mai Po 米埔；Dec-08, 08 年 12 月；Allen Chan 陳志雄

| 春季過境遷徙鳥 Spring Passage Migrant | | | 夏候鳥 Summer Visitor | | 秋季過境遷徙鳥 Autumn Passage Migrant | | | 冬候鳥 Winter Visitor | | | |
|---|---|---|---|---|---|---|---|---|---|---|---|
| 1 | 2 | 3 | 4 | 5 | 6 | 7 | 8 | 9 | 10 | 11 | 12 |

常見月份

| 留鳥 Resident | 迷鳥 Vagrant | 偶見鳥 Occasional Visitor |
|---|---|---|

# 小濱鷸

(普) xiǎo bīn yù
(粵) 鷸：音核

體長 length：12-14cm

## Little Stint | *Calidris minuta*

體 型細小，近似紅胸濱鷸，但軀體後部稍長，黑色的腳較長，嘴較長且嘴端較尖細。繁殖期羽色兼有紅胸濱鷸和長趾濱鷸的特徵：頭、頸、胸側和背部羽毛邊緣皆為橙棕色。喉部白色。三級飛羽深色，有較寬的橙棕色邊緣。

Small-sized wader. Resembles Red-necked Stint but with slightly longer rear part, longer black legs and longer bill with finer tip. In breeding plumage, it has the characteristics of both Red-necked and Long-toed Stints: orange rufous colour on head, neck, both sides of the breast, and fringes on mantle feathers. White throat. Dark tertials, with thick orange rufous fringes.

1 breeding 繁殖羽：Mai Po 米埔：18-Apr-09, 09 年 4 月 18 日：Cherry Wong 黃卓研
2 breeding 繁殖羽：Mai Po 米埔：Shirley Lam 林鳳兒

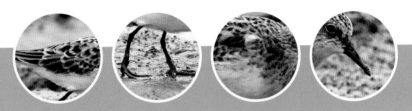

| 春季過境遷徙鳥 Spring Passage Migrant | | 夏候鳥 Summer Visitor | | 秋季過境遷徙鳥 Autumn Passage Migrant | | 冬候鳥 Winter Visitor | |
|---|---|---|---|---|---|---|---|

| 1 | 2 | 3 | 4 | 5 | 6 | 7 | 8 | 9 | 10 | 11 | 12 | 常見月份 |
|---|---|---|---|---|---|---|---|---|---|---|---|---|

| 留鳥 Resident | 迷鳥 Vagrant | 偶見鳥 Occasional Visitor |
|---|---|---|

# 飾胸鷸

(普) shì xiōng yù
(粵) 鷸：音核

體長 length：18-20cm

## Buff-breasted Sandpiper | *Calidris subruficollis*

臉部至胸部淡粉紅色，頭形較方，淡色的臉顯得黑色眼睛特別突出，黑色嘴幼，腳橙黃色。上體羽毛黑色而帶黃褐色邊，腹部近白色，翼底白色。

Pale pink from face to breast. Square-shaped head. Black eyes look prominent on its pale face. Thin black bill. Yellowish-orange legs. Black upperparts with yellowish-brown fringes on feathers. Whitish belly. White underwing.

[1] immature 未成年鳥；San Tin 新田；Dec-16, 16 年 12 月；Wong Wong Yung 王煌容
[2] immature 未成年鳥；San Tin 新田；Dec-16, 16 年 12 月；Kwok Tsz Ki 郭子祈
[3] immature 未成年鳥；San Tin 新田；Dec-16, 16 年 12 月；Jemi and John Holmes 孔思義、黃亞萍
[4] immature 未成年鳥；San Tin 新田；22 Dec-16, 16 年 12 月 22 日；Beetle Cheng 鄭諾銘

| | 春季過境遷徙鳥<br>Spring Passage Migrant | | | 夏候鳥<br>Summer Visitor | | | 秋季過境遷徙鳥<br>Autumn Passage Migrant | | | 冬候鳥<br>Winter Visitor | | |
|---|---|---|---|---|---|---|---|---|---|---|---|---|
| 常見月份 | 1 | 2 | 3 | 4 | 5 | 6 | 7 | 8 | 9 | 10 | 11 | 12 |

| 留鳥<br>Resident | 迷鳥<br>Vagrant | 偶見鳥<br>Occasional Visitor |
|---|---|---|

# 斑胸濱鷸

(普)bān xiōng bīn yù
(粵)鷸：音核

體長 length：19-23cm

Pectoral Sandpiper | *Calidris melanotos*

**體**型跟尖尾濱鷸相似，但沒有栗色頭頂。眼眉短及不明顯。嘴尖微向下及有淺色基部。上體羽毛深色及有淺褐色羽毛邊緣。胸部有濃密縱紋，縱紋於中部尖出，與白色腹部作出明顯分界。脇部沒有或只有極少量縱紋。非繁殖羽顏色較暗灰。

Resembles Sharp-tailed Sandpiper but no rufous tinge on crown. It has a short and inconspicuous supercilium. Bill is slightly decurved at tip with pale base. Upperparts feathers are dark with pale brown fringes. Dense streaks on breast come to a point in the centre and with a distinctive sharp demarcation to the white belly. Very few streaks on flanks. Non-breeding plumage is duller.

1 breeding 繁殖羽；Mai Po 米埔：25-Apr-05, 05 年 4 月 25 日；Michelle and Peter Wong 江敏兒、黃理沛
2 breeding 繁殖羽；Mai Po 米埔：25-Apr-05, 05 年 4 月 25 日；Michelle and Peter Wong 江敏兒、黃理沛
3 Mai Po 米埔：23-Apr-04, 04 年 4 月 23 日；Tam Yiu Leung 譚耀良
4 Mai Po 米埔：23-Apr-04, 04 年 4 月 23 日；Tam Yiu Leung 譚耀良
5 breeding 繁殖羽；Mai Po 米埔：25-Apr-05, 05 年 4 月 25 日；Michelle and Peter Wong 江敏兒、黃理沛

| 春季過境遷徙鳥<br>Spring Passage Migrant | | | 夏候鳥<br>Summer Visitor | | | 秋季過境遷徙鳥<br>Autumn Passage Migrant | | | 冬候鳥<br>Winter Visitor | | |
|---|---|---|---|---|---|---|---|---|---|---|---|
| 1 | 2 | 3 | 4 | 5 | 6 | 7 | 8 | 9 | 10 | 11 | 12 |

常見月份

| 留鳥<br>Resident | 迷鳥<br>Vagrant | 偶見鳥<br>Occasional Visitor |
|---|---|---|

# 半蹼鷸

(普) bàn pǔ yù
(粵) 蹼鷸：音樸核

體長 length：33-36cm

## Asian Dowitcher | *Limnodromus semipalmatus*

其他名稱 Other names：Asiatic Dowitcher

外 形像小型的塍鷸，但嘴全黑、粗長而渾圓，末端微微隆起；脅部有很多細橫斑。繁殖期頭至胸部橙褐色，有淡色眼眉。非繁殖期全身大致灰褐色，背部深色羽毛有明顯粗白邊。幼鳥胸部沾有黃褐色。覓食時嘴部一下一下頻密地直插泥中。

Resembles small-sized godwits, but the long black bill is thick, round and with a bulbous tip. Flanks heavily barred. In breeding plumage the head and breast are brownish orange, with pale supercilium. Non-breeding bird is buff and overall brown, and upperpart feathers are dark with prominent white fringes. Juvenile has buff breast. Stabs bill rapidly into mud when feeding.

1 breeding 繁殖羽：Mai Po 米埔；10-Apr-05, 05 年 4 月 10 日：Doris Chu 朱詠兒
2 juvenile 幼鳥：Mai Po 米埔；6-May-07, 07 年 5 月 6 日：Michelle and Peter Wong 江敏兒、黃理沛
3 breeding 繁殖羽：Mai Po 米埔；23-Apr-06, 06 年 4 月 23 日：Doris Chu 朱詠兒
4 breeding 繁殖羽：Mai Po 米埔；Apr-08, 08 年 4 月：Cherry Wong 黃卓研
5 breeding 繁殖羽：Mai Po 米埔；Apr-08, 08 年 4 月：James Lam 林义華

| 春季過境遷徙鳥 Spring Passage Migrant | | 夏候鳥 Summer Visitor | | 秋季過境遷徙鳥 Autumn Passage Migrant | | 冬候鳥 Winter Visitor | |
|---|---|---|---|---|---|---|---|

| 常見月份 | 1 | 2 | 3 | 4 | 5 | 6 | 7 | 8 | 9 | 10 | 11 | 12 |
|---|---|---|---|---|---|---|---|---|---|---|---|---|

| 留鳥 Resident | 迷鳥 Vagrant | 偶見鳥 Occasional Visitor |
|---|---|---|

# 長嘴鷸
(普) cháng zuǐ yù
(粵) 鷸：音核

體長 length：24-30cm

## Long-billed Dowitcher | *Limnodromus scolopaceus*

外 形近似半蹼鷸，但體型較小。嘴長而深褐色，嘴基顏色較淡；腳淡綠色。繁殖羽頭、頸、胸和腹深棕紅色；背部羽毛深色，有白色及棕紅色羽尖。非繁殖期整體深灰褐色，腹部顏色較淡，有深色斑紋。幼鳥羽色接近非繁殖期但褐色較濃。

Resembles Asian Dowitcher but much smaller in size. Bill long and dark brown, bill base is paler. Legs are pale green. In breeding plumage, head, neck, breast and belly dark rufous; upperparts darker in colour, with white and rufous tips. In non-breeding plumage, it is overall dark greyish brown, and belly is paler with dark bars. Juveniles resembles non-breeding bird but browner.

1 breeding 繁殖羽：Mai Po 米埔：11-Apr-09, 09 年 4 月 11 日：Yue Pak Wai 余柏維
2 non-breeding 非繁殖羽：Mai Po 米埔：Oct-08, 08 年 10 月：Jemi and John Holmes 孔思義、黃亞萍
3 non-breeding 非繁殖羽：Mai Po 米埔：25-Mar-03, 03 年 3 月 25 日：KK Hui 許光杰
4 breeding 繁殖羽：Mai Po 米埔：Apr-09, 09 年 4 月：Cherry Wong 黃卓研
5 breeding 繁殖羽：Mai Po 米埔：Apr-09, 09 年 4 月：Cherry Wong 黃卓研

| 春季過境遷徙鳥<br>Spring Passage Migrant | | 夏候鳥<br>Summer Visitor | | 秋季過境遷徙鳥<br>Autumn Passage Migrant | | 冬候鳥<br>Winter Visitor | |
|---|---|---|---|---|---|---|---|

| 1 | 2 | 3 | 4 | 5 | 6 | 7 | 8 | 9 | 10 | 11 | 12 | 常見月份 |
|---|---|---|---|---|---|---|---|---|---|---|---|---|

| 留鳥<br>Resident | 迷鳥<br>Vagrant | 偶見鳥<br>Occasional Visitor |
|---|---|---|

# 丘鷸

(普) qiū yù
(粵) 鷸：音核

體長 length：33-35cm

## Eurasian Woodcock | *Scolopax rusticola*

其他名稱 Other names：Woodcock

**體** 型大，外形矮壯似沙錐，頭部呈三角形。頭頂至後頸有深色粗橫斑，雙翼圓闊。在山溪、矮樹叢和開闊地方出現。主要在夜間覓食，日間則躲在林區邊緣休息。

Large-sized, stout appearance similar to a snipe, with triangular head. Darkish thick bars from crown to hindneck, and the wings are round and broad. Occurs at shrubland and open areas near streams on hillsides. Forages at night, and hides in forest edge during daytime.

1　Pui O 貝澳；Mar-08, 08 年 3 月；Eling Lee 李佩玲
2　Pui O 貝澳；Mar-08, 08 年 3 月；Kitty Koo 古愛嫦
3　Tai Po Kau 大埔滘；7-Nov-07, 07 年 11 月 7 日；Michelle and Peter Wong 江敏兒、黃理沛
4　Pui O 貝澳；Mar-08, 08 年 3 月；James Lam 林文華
5　Pui O 貝澳；Mar-08, 08 年 3 月；Eling Lee 李佩玲
6　Lai Chi Kok 荔枝角；Nov-07, 07 年 11 月；Ng Lin Yau 吳璉宥
7　Shek Wu Wai 石湖圍；Oct-04, 04 年 10 月；Jemi and John Holmes 孔思義、黃亞萍

| 春季過境遷徙鳥 Spring Passage Migrant | | | | 夏候鳥 Summer Visitor | | 秋季過境遷徙鳥 Autumn Passage Migrant | | 冬候鳥 Winter Visitor | | | |
|---|---|---|---|---|---|---|---|---|---|---|---|
| 1 | 2 | 3 | 4 | 5 | 6 | 7 | 8 | 9 | 10 | 11 | 12 |
| 留鳥 Resident | | | | 迷鳥 Vagrant | | | | 偶見鳥 Occasional Visitor | | | |

常見月份

# 針尾沙錐 (普)zhēn wěi shā zhuī

體長 length：25-27cm

## Pintail Snipe | *Gallinago stenura*

針尾沙錐和大沙錐在野外極難分辨。和扇尾沙錐比較，嘴較短，背部縱紋顏色較淡且較細長。靜止時，翼長不及尾的末端。飛行時次級飛羽沒有白色後緣，翼下有濃密橫紋。（註：圖中鳥未能完全肯定鳥種。）

Pintail and Swinhoe's Snipes are very difficult to distinguish in the field. Compared with Common Snipe, stripes on the back are paler and thinner. At rest, the wings stop short of the tail. In flight, the secondaries show no white edge, and the underwings have dense bars. *(Note: The bird in this photo cannot be definitely identified.)*

[1] Mai Po 米埔：Sep-07, 07 年 9 月；Kinni Ho 何建業

| 春季過境遷徙鳥<br>Spring Passage Migrant | | | 夏候鳥<br>Summer Visitor | | | 秋季過境遷徙鳥<br>Autumn Passage Migrant | | 冬候鳥<br>Winter Visitor | | |
|---|---|---|---|---|---|---|---|---|---|---|
| 常見月份 1 | 2 | 3 | 4 | 5 | 6 | 7 | 8 | 9 | 10 | 11 | 12 |

| 留鳥<br>Resident | 迷鳥<br>Vagrant | 偶見鳥<br>Occasional Visitor |
|---|---|---|

# 大沙錐 <sup>普</sup> dà shā zhuī

體長 length：27-29cm

Swinhoe's Snipe | *Gallinago megala*

針尾沙錐和大沙錐在野外極難分辨。和扇尾沙錐比較，大沙錐背部縱紋顏色較淡且較細長。靜止時，翼長不及尾的末端。飛行時次級飛羽沒有白色後緣，翼下有濃密橫紋。大沙錐的尾羽較針尾沙錐闊，羽毛的白邊較明顯。

Pintail and Swinhoe Snipes are very difficult to distinguish in the field. Compared with Common Snipe, Swinhoe's Snipe has paler and thinner stripes on the back. At rest, the wings stop short of the tail. In flight, the secondaries show no white edge, and the underwings have dense bars. Swinhoe's Snipe has tail feathers which are more broad compared with Pintail Snipe, and the feather edge is more conspicuous.

[1] juvenile 幼鳥；Mai Po 米埔；Sep-08, 08 年 9 月；Cherry Wong 黃卓研
[2] juvenile 幼鳥；Mai Po 米埔；Sep-08, 08 年 9 月；Cherry Wong 黃卓研
[3] juvenile 幼鳥；Mai Po 米埔；Sep-08, 08 年 9 月；Cherry Wong 黃卓研
[4] juvenile 幼鳥；Mai Po 米埔；Sep-08, 08 年 9 月；Cherry Wong 黃卓研

| 春季過境遷徙鳥 Spring Passage Migrant | | | 夏候鳥 Summer Visitor | | | 秋季過境遷徙鳥 Autumn Passage Migrant | | | 冬候鳥 Winter Visitor | | |
|---|---|---|---|---|---|---|---|---|---|---|---|
| 1 | 2 | 3 | 4 | 5 | 6 | 7 | 8 | 9 | 10 | 11 | 12 |

常見月份

| 留鳥 Resident | | | 迷鳥 Vagrant | | | 偶見鳥 Occasional Visitor | | |
|---|---|---|---|---|---|---|---|---|

# 扇尾沙錐

 (普)shàn wěi shā zhuī

體長 length：25-27cm

## Common Snipe | *Gallinago gallinago*

其他名稱 Other names：Fantail Snipe

**嘴** 比其他沙錐長，一般為頭部的2.2倍，背上有鮮明金黃色的粗縱紋。飛行時次級飛羽有明顯白色後緣，翼下偏白。通常躲在水邊挺水植物之間。

Compared with other snipes, it has a longer bill (about 2.2 times the length of the head). Prominent yellowish stripes runs along the back. In flight, the secondaries show white edges, and paler underwings. Unusually hides among emergent aquatic vegetation along waterside.

1 Long Valley 塱原：Nov-07, 07 年 11 月；James Lam 林文華
2 Long Valley 塱原：31-Dec-06, 06 年 12 月 31 日；Matthew Kwan 關朗曦
3 Long Valley 塱原：Oct-04, 04 年 10 月；Henry Lui 呂德恆
4 Long Valley 塱原：Dec-06, 06 年 12 月；Andy Kwok 郭匯昌
5 Long Valley 塱原：Jan-09, 09 年 1 月；Kennic Man 文超凡
6 juvenile 幼鳥：Long Valley 塱原；Owen Chiang 深藍
7 Tsim Bei Tsui 尖鼻咀：Oct-08, 08 年 10 月；Kitty Koo 古愛婉

| | 春季過境遷徙鳥<br>Spring Passage Migrant | | | 夏候鳥<br>Summer Visitor | | 秋季過境遷徙鳥<br>Autumn Passage Migrant | | 冬候鳥<br>Winter Visitor | | | |
|---|---|---|---|---|---|---|---|---|---|---|---|
| 常見月份 | 1 | 2 | 3 | 4 | 5 | 6 | 7 | 8 | 9 | 10 | 11 | 12 |
| | 留鳥<br>Resident | | | | 迷鳥<br>Vagrant | | | 偶見鳥<br>Occasional Visitor | | | |

# 紅頸瓣蹼鷸

(普) hóng jǐng bàn pǔ yù
(粵) 蹼鷸：音樸核

體長 length：18-19cm

## Red-necked Phalarope | *Phalaropus lobatus*

其他名稱 Other names：Northern Phalarope

外 形纖瘦，嘴黑色而細長。飛行時有白色翼帶，尾部兩側有白斑。繁殖期後頸至胸部紅棕色，與白色的喉部對比鮮明。非繁殖期為灰白二色，眼後有黑斑。喜游泳，軀體浮出水面頗多。

Slim body, with thin black bill. In flight, it shows white wing bars and white patches on sides of rump. In breeding plumage, it shows prominent rufous hind neck and breast contrast with white throat. Non-breeding bird is mostly grey and white, with a black patch behind the eyes. A frequent swimmer, stays high out of the water.

1 breeding female with oil stain 沾有油漬的繁殖羽雌鳥；Po Toi 蒲台；May-08, 08 年 5 月；James Lam 林文華
2 juvenile 幼鳥；Long Valley 塱原；Apr-07, 07 年 4 月；Allen Chan 陳志雄
3 breeding female with oil stain 沾有油漬的繁殖羽雌鳥；Po Toi 蒲台；May-08, 08 年 5 月；Eling Lee 李佩玲
4 juvenile 幼鳥；Po Toi 蒲台；Apr-07, 07 年 4 月；Jemi and John Holmes 孔思義、黃亞萍
5 juvenile 幼鳥；Long Valley 塱原；Apr-07, 07 年 4 月；Allen Chan 陳志雄
6 juvenile 幼鳥；Long Valley 塱原；Apr-08, 08 年 4 月；James Lam 林文華
7 Mai Po 米埔；Apr-07, 07 年 4 月；Henry Lui 呂德恆
8 non-breeding adult 非繁殖羽成鳥；Mai Po 米埔；Apr-06, 06 年 4 月；Tam Yip Shing 譚業成

| | 春季過境遷徙鳥<br>Spring Passage Migrant | | | 夏候鳥<br>Summer Visitor | | | 秋季過境遷徙鳥<br>Autumn Passage Migrant | | | 冬候鳥<br>Winter Visitor | | |
|---|---|---|---|---|---|---|---|---|---|---|---|---|
| 常見月份 | 1 | 2 | 3 | 4 | 5 | 6 | 7 | 8 | 9 | 10 | 11 | 12 |
| | 留鳥<br>Resident | | | | 迷鳥<br>Vagrant | | | | 偶見鳥<br>Occasional Visitor | | | |

# 灰瓣蹼鷸

(普) huī bàn pǔ yù
(粵) 蹼鷸：音樸核

體長 length：20-22cm

Red Phalarope | *Phalaropus fulicarius*

其他名稱 Other names：Grey Phalarope

[1]

**體**形比紅頸瓣蹼鷸大，嘴較寬。繁殖期背部黑色，有淡色羽緣，頸及下體栗紅色，頭部深色，臉部有白斑，嘴黃色，嘴尖黑色。雄鳥羽色較淡，頭頂灰黑色，下體有白色橫紋。雌鳥頭部更深色。非繁殖期後枕深灰色，後頸至上體為灰色，下體為白色。嘴深色，嘴基橙黃色，眼後有黑斑。

Larger than Red-necked phalarope with broader bill. Breeding plumage has upperparts black in colour with pale fringes. Neck and underparts chestnut. Blackish head with white patch behind the eye. Bill yellow with dark tip. Male has paler plumages, greyish heads and more whitish strips at underparts. Female has darker heads. Non-breeding plumage is grey in upperparts with dark hind-crown and white in underparts. Black bill with yellowish orange bill base. Dark eye patch.

[1] non-breeding adult, 非繁殖羽成鳥：Nam Chung 南涌：Apr-08, 08 年 4 月：Andy Kwok 郭匯昌
[2] Nam Chung 南涌：Apr-09, 09 年 4 月：Chan Kin Chung Gary 陳建中
[3] Nam Chung 南涌：Owen Chiang 深藍
[4] Nam Chung 南涌：Apr-09, 09 年 4 月：Chan Kin Chung Gary 陳建中
[5] Nam Chung 南涌：Apr-09, 09 年 4 月：Chan Kin Chung Gary 陳建中
[6] Nam Chung 南涌：May-08, 08 年 5 月：Eling Lee 李佩玲
[7] Nam Chung 南涌：Apr-08, 08 年 4 月：Danny Ho 何國海
[8] Nam Chung 南涌：Apr-09, 09 年 4 月：Chan Kin Chung Gary 陳建中

| 春季過境遷徙鳥 Spring Passage Migrant | | | 夏候鳥 Summer Visitor | | | 秋季過境遷徙鳥 Autumn Passage Migrant | | | 冬候鳥 Winter Visitor | | |
|---|---|---|---|---|---|---|---|---|---|---|---|
| 1 | 2 | 3 | 4 | 5 | 6 | 7 | 8 | 9 | 10 | 11 | 12 |

常見月份

| 留鳥 Resident | 迷鳥 Vagrant | 偶見鳥 Occasional Visitor |
|---|---|---|

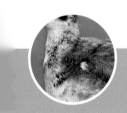

# 翹嘴鷸

(普) qiào zuǐ yù
(粵) 鷸：音核

體長 length：22-25cm

## Terek Sandpiper | *Xenus cinereus*

以 腳短、嘴長而向上彎和其他鷸區別。飛行時次級飛羽白色。繁殖期腳部橙色鮮明，肩膀有一道黑斑。非繁殖鳥和幼鳥羽毛顏色偏褐。常小群出現。

Characterised by short legs and long upturned bill. White secondaries in flight. In breeding plumage, it has obvious orange legs and a dark stripe on scapulars. Non-breeding bird and juvenile are more brownish. Gregarious.

1 breeding 繁殖羽：Mai Po 米埔：May-06, 06 年 5 月：Martin Hale 夏敖天
2 non-breeding adult 非繁殖羽成鳥：Mai Po 米埔：Apr-04, 04 年 4 月：Jemi and John Holmes 孔思義、黃亞萍
3 non-breeding adult 非繁殖羽成鳥：Mai Po 米埔：May-06, 06 年 5 月：Martin Hale 夏敖天
4 non-breeding adult 非繁殖羽成鳥：Mai Po 米埔：Apr-06, 06 年 4 月：Tam Yip Shing 譚業成
5 non-breeding adult 非繁殖羽成鳥：Mai Po 米埔：Apr-04, 04 年 4 月：Jemi and John Holmes 孔思義、黃亞萍

| 春季過境遷徙鳥 Spring Passage Migrant | | | 夏候鳥 Summer Visitor | | 秋季過境遷徙鳥 Autumn Passage Migrant | | | 冬候鳥 Winter Visitor | | | |
|---|---|---|---|---|---|---|---|---|---|---|---|
| 常見月份 | | | | | | | | | | | |
| 1 | 2 | 3 | 4 | 5 | 6 | 7 | 8 | 9 | 10 | 11 | 12 |
| 留鳥 Resident | | | | 迷鳥 Vagrant | | | | 偶見鳥 Occasional Visitor | | | |

# 磯鷸

(普) jī yù
(粵) 鷸：音核

體長 length：19-21cm

Common Sandpiper | *Actitis hypoleucos*

上體褐色，有淡色眼眉，腳黃色。站立時翼前有明顯的彎月形白斑，頭和臀部不斷上下擺動。飛行時貼近水面，有明顯的白色翼帶。

Upperparts brown, with pale supercilium. Legs yellow. At rest, it has prominent white crescents in front of the wings, constantly bobs head and tail. Shows white wing bars in flight. Flies low over water.

1 non-breeding moulting from breeding plumage 繁殖羽轉換非繁殖羽：Sai Kung 西貢；Nov-08, 08 年 11 月；Ken Fung 馮漢城
2 non-breeding adult 非繁殖羽成鳥：Mai Po 米埔；Nov-07, 07 年 11 月；Kitty Koo 古愛婉
3 non-breeding adult 非繁殖羽成鳥：Kam Tin 錦田；Nov-05, 05 年 11 月；Cherry Wong 黃卓研
4 adult 成鳥：Mai Po 米埔；Mar-08, 08 年 3 月；Kami Hui 許淑君
5 breeding 繁殖羽：Tseung Kwan O 將軍澳；23-Aug-07, 07 年 8 月 23 日；Sonia and Kenneth Fung 馮啟文、蕭敏晶
6 breeding 繁殖羽：Mai Po 米埔；Jul-07, 07 年 7 月；James Lam 林文華
7 adult 成鳥：Mai Po 米埔；Mar-20, 20 年 3 月；Henry Lui 呂德恆

| 春季過境遷徙鳥 Spring Passage Migrant | | | 夏候鳥 Summer Visitor | | 秋季過境遷徙鳥 Autumn Passage Migrant | | 冬候鳥 Winter Visitor | | | |
|---|---|---|---|---|---|---|---|---|---|---|
| 1 | 2 | 3 | 4 | 5 | 6 | 7 | 8 | 9 | 10 | 11 | 12 |

常見月份

| 留鳥 Resident | 迷鳥 Vagrant | 偶見鳥 Occasional Visitor |
|---|---|---|

# 灰尾漂鷸

(普) huī wěi piāo yù
(粤) 鷸：音核

體長 length：23-27cm

Grey-tailed Tattler | *Tringa brevipes*

其他名稱 Other names：Grey-tailed Sandpiper, 灰尾鷸

上體呈均勻的灰色，有淡白眼眉，腳黃色。飛行時，上體灰色均勻，無翼斑。繁殖期喉、胸和脇有纖細橫紋，腹部和尾下覆羽純白，對比鮮明。

Upperparts uniformly grey, with pale supercilium and distinct yellow legs. In flight it shows uniform grey on the upperparts and no wing bar. In breeding plumage, the throat, breast and flanks have fine bars, in good contrast with the white belly and undertail coverts.

1 breeding 繁殖羽：Tsim Bei Tsui 尖鼻咀；May-06, 06 年 5 月；Law Kam Man 羅錦文
2 breeding 繁殖羽：Tai A Chau 大鴉洲；May-04, 04 年 5 月；Henry Lui 呂德恒
3 juvenile 幼鳥：Ma On Shan 馬鞍山；Oct-08, 08 年 10 月；Ken Fung 馮漢城
4 juvenile 幼鳥：Ma On Shan 馬鞍山；Oct-08, 08 年 10 月；Martin Hale 夏敖天
5 juvenile 幼鳥：Ma On Shan 馬鞍山；Oct-08, 08 年 10 月；Andy Kwok 郭匯昌
6 juvenile 幼鳥：Sai Kung 西貢；Oct-08, 08 年 10 月；Ng Lin Yau 吳璉宥
7 juvenile 幼鳥：Sai Kung 西貢；Oct-08, 08 年 10 月；Joyce Tang 鄧玉蓮

| 春季過境遷徙鳥 Spring Passage Migrant | | | 夏候鳥 Summer Visitor | | | 秋季過境遷徙鳥 Autumn Passage Migrant | | | 冬候鳥 Winter Visitor | | |
|---|---|---|---|---|---|---|---|---|---|---|---|
| 1 | 2 | 3 | 4 | 5 | 6 | 7 | 8 | 9 | 10 | 11 | 12 |
| 留鳥 Resident | | | | 迷鳥 Vagrant | | | | 偶見鳥 Occasional Visitor | | | |

常見月份

285

# 白腰草鷸

（普）bái yāo cǎo yù
（粵）鷸：音核

體長 length：21-24cm

## Green Sandpiper | *Tringa ochropus*

**中** 等體型水鳥，佇立時似磯鷸。上體深色，和下體對比鮮明，腳綠色。背及翅膀羽毛邊緣有細小淡點。飛行時翼下偏黑，白腰顯眼，與深色的上體對比鮮明。於淡水環境，例如溪流、水道和魚塘邊緣活動。

Medium-sized wader. Resembles Common Sandpiper at rest. Dark upperparts contrasts with white rump and white underparts, green legs. Upperparts and wing coverts have small pale spots lining at the edge. Dark underwings in flight. Lives in freshwater habitats inclnding streams, ditches and fishponds.

1 non-breeding adult 非繁殖羽成鳥：Shek Kong 石崗：Dec-08, 08 年 12 月：Ken Fung 馮漢城
2 breeding 繁殖羽：Long Valley 塱原：Sep-07, 07 年 9 月：Sammy Sam and Winnie Wong 森美與雲妮
3 breeding 繁殖羽：Fung Lok Wai 豐樂圍：Sep-05, 05 年 9 月：Jemi and John Holmes 孔思義、黃亞萍
4 breeding 繁殖羽：Long Valley 塱原：Aug-07, 07 年 8 月：Sammy Sam and Winnie Wong 森美與雲妮
5 juvenile 幼鳥：Mai Po 米埔：Nov-06, 06 年 11 月：Henry Lui 呂德恆

| | 春季過境遷徙鳥<br>Spring Passage Migrant | | | 夏候鳥<br>Summer Visitor | | 秋季過境遷徙鳥<br>Autumn Passage Migrant | | 冬候鳥<br>Winter Visitor | | | |
|---|---|---|---|---|---|---|---|---|---|---|---|
| 常見月份 | 1 | 2 | 3 | 4 | 5 | 6 | 7 | 8 | 9 | 10 | 11 | 12 |
| | 留鳥<br>Resident | | | 迷鳥<br>Vagrant | | | 偶見鳥<br>Occasional Visitor | | | | | |

# 紅腳鷸
(普) hóng jiǎo yù
(粵) 鷸：音核

體長 length：27-29cm

## Common Redshank | *Tringa totanus*

其他名稱 Other names：Redshank

腳 上有鮮明的紅或橙色。佇立時一般比鶴鷸的褐色更濃，嘴直而上下嘴基紅色。飛行時，次級飛羽和腰部明顯白色。繁殖期白色下體有深色濃密縱紋。非繁殖期上體和胸部褐色均勻。幼鳥上體有淺色斑點，嘴基紅色不明顯。

Obvious red or orange legs. At rest, it shows more brownish than Spotted Redshank, straight bill with red bill base. In flight, it shows white secondaries and rump. In breeding plumage, the white underparts have dense dark stripes. In non-breeding plumage, upperparts and breast are brown and plainer. Juvenile has pale spots on upperparts, and dull bill base.

1 breeding 繁殖羽；Mai Po 米埔；Apr-06, 06 年 4 月；Tam Yip Shing 譚業成
2 non-breeding adult 非繁殖羽成鳥；Mai Po 米埔；May-08, 08 年 5 月；Cherry Wong 黃卓研
3 breeding 繁殖羽；Mai Po 米埔；Mar-05, 05 年 3 月；Martin Hale 夏敖天
4 juvenile 幼鳥；Mai Po 米埔；Aug-04, 04 年 8 月；Henry Lui 呂德恆

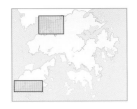

| 春季過境遷徙鳥 Spring Passage Migrant | | | 夏候鳥 Summer Visitor | | | 秋季過境遷徙鳥 Autumn Passage Migrant | | | 冬候鳥 Winter Visitor | | |
|---|---|---|---|---|---|---|---|---|---|---|---|
| 1 | 2 | 3 | 4 | 5 | 6 | 7 | 8 | 9 | 10 | 11 | 12 |

常見月份

| 留鳥 Resident | 迷鳥 Vagrant | 偶見鳥 Occasional Visitor |
|---|---|---|

# 澤鷸

(普) zé yù
(粵) 鷸：音核

體長 length：22-26cm

## Marsh Sandpiper | *Tringa stagnatilis*

中型水鳥，嘴細長，腳長呈黃綠色，上體灰色。飛行時腳伸出尾後，腰部白色延伸至背部，翼均勻深灰褐色。繁殖期上體羽毛有黑色和褐色斑。非繁殖期毛色較淡。

Medium-sized wader, with long thin bill, and long yellowish green legs. Overall greyish upperparts. In flight, legs protrude beyond the tail, and white rump extends to the back. Wings appear dark greyish brown. In breeding plumage, it has black and brown spots at the upperparts. Non-breeding plumage is often paler.

1 breeding 繁殖羽：Mai Po 米埔：Nov-06, 06 年 11 月：Danny Ho 何國海
2 Mai Po 米埔：Oct-07, 07 年 10 月：Isaac Chan 陳家強
3 non-breeding adult 非繁殖羽成鳥：Mai Po 米埔：Nov-03, 03 年 11 月：Jemi and John Holmes 孔思義、黃亞萍
4 juvenile moulting into 1st winter plumage 幼鳥：Mai Po 米埔：Oct-04, 04 年 10 月：Jemi and John Holmes 孔思義、黃亞萍
5 juvenile 幼鳥：Mai Po 米埔：Sep-07, 07 年 9 月：James Lam 林文華
6 breeding 繁殖羽：Mai Po 米埔：Apr-04, 04 年 4 月：Jemi and John Holmes 孔思義、黃亞萍

| 春季過境遷徙鳥 Spring Passage Migrant | | | 夏候鳥 Summer Visitor | | | 秋季過境遷徙鳥 Autumn Passage Migrant | | 冬候鳥 Winter Visitor | | | |
|---|---|---|---|---|---|---|---|---|---|---|---|
| 1 | 2 | 3 | 4 | 5 | 6 | 7 | 8 | 9 | 10 | 11 | 12 |
| | 留鳥 Resident | | | | 迷鳥 Vagrant | | | | 偶見鳥 Occasional Visitor | | |

常見月份

# 林鷸

普 lín yù
粵 鷸：音核

體長 length：19-23cm

## Wood Sandpiper | *Tringa glareola*

上體深褐色而有濃密的淡色斑點；眉紋白色明顯，伸延至眼後；腳黃色。飛行時翼下顏色較淡，腰白色，尾白色及有褐色橫紋。

Dense pale spots on greyish brown upperparts, conspicuous white supercilium extends beyond the eyes, yellow legs. In flight, it shows pale underwings, white rump and white tail with brown bars.

1 non-breeding adult 非繁殖羽成鳥；Kam Tin 錦田；Nov-05, 05 年 11 月；Cherry Wong 黃卓研
2 1st winter 第一年冬天；Fung Lok Wai 豐樂圍；Oct-04, 04 年 10 月；Jemi and John Holmes 孔思義、黃亞萍
3 breeding 繁殖羽；Shek Kong 石崗；Mar-07, 07 年 3 月；James Lam 林文華
4 adult 成鳥；Mai Po 米埔；Feb-20, 20年2月；Henry Lui 呂德恆
5 juvenile 幼鳥；Long Valley 塱原；Oct-04, 04 年 10 月；Henry Lui 呂德恆

| 春季過境遷徙鳥 Spring Passage Migrant | | | 夏候鳥 Summer Visitor | | | 秋季過境遷徙鳥 Autumn Passage Migrant | | | 冬候鳥 Winter Visitor | | |
|---|---|---|---|---|---|---|---|---|---|---|---|
| 1 | 2 | 3 | 4 | 5 | 6 | 7 | 8 | 9 | 10 | 11 | 12 |
| 留鳥 Resident | | | | 迷鳥 Vagrant | | | | 偶見鳥 Occasional Visitor | | | |

常見月份

# 鶴鷸

(普) hè yù
(粵) 鷸：音核

體長 length：29-32cm

Spotted Redshank | *Tringa erythropus*

非 繁殖期和紅腳鷸相似，但嘴部較幼長，嘴端微向下，下嘴基紅色。飛行時次級飛羽深褐色，腰白色。繁殖羽全身炭黑色，上背有白色斑點，腳偏黑色。非繁殖期羽色偏灰，腳紅色，腹部白色。

Resembles Common Redshank, but the black bill is thinner and longer, with a dipped end and red base at lower bill. In flight, it shows dark secondaries and white rump. In breeding plumage, it has black legs, overall black plumage with white spots on upperparts. In non-breeding plumage, it appears grey, with red legs and white belly.

1 breeding 繁殖羽：Mai Po 米埔：May-04, 04 年 5 月；Jemi and John Holmes 孔思義、黃亞萍
2 non-breeding adult 非繁殖羽成鳥；Mai Po 米埔：10-Feb-07, 07 年 2 月 10 日；Michelle and Peter Wong 江敏兒、黃理沛
3 breeding 繁殖羽：1-May-05, 05 年 5 月 1 日；Michelle and Peter Wong 江敏兒、黃理沛
4 non-breeding adult 非繁殖羽成鳥；Mai Po 米埔：10-Feb-07, 07 年 2 月 10 日；Michelle and Peter Wong 江敏兒、黃理沛

| 春季過境遷徙鳥 Spring Passage Migrant | | | 夏候鳥 Summer Visitor | | | 秋季過境遷徙鳥 Autumn Passage Migrant | | 冬候鳥 Winter Visitor | | | |
|---|---|---|---|---|---|---|---|---|---|---|---|
| 1 | 2 | 3 | 4 | 5 | 6 | 7 | 8 | 9 | 10 | 11 | 12 |

常見月份

| 留鳥 Resident | 迷鳥 Vagrant | 偶見鳥 Occasional Visitor |
|---|---|---|

# 青腳鷸

(普) qīng jiǎo yù
(粵) 鷸：音核

體長 length：30-35cm

## Common Greenshank | *Tringa nebularia*

其他名稱 Other names：Greenshank

1

中 型水鳥，嘴粗長而下嘴微向上翹，腳黃綠色。飛行時尾部白色，有淡褐色橫紋，腰部白色伸延到上背，翼均勻深灰褐色。非繁殖期上體灰色，腹部純白無斑點。

Medium-sized wader, with thick and slightly up-curved lower bill, legs yellowish green. In flight, it shows white tail with pale brown bars, white rump which extends to the mantle, and dark greyish brown wings. In non-breeding plumage, it shows greyish upperparts and white belly.

[1] Mai Po 米埔；Oct-04, 04 年 10 月；Jemi and John Holmes 孔思義，黃亞萍
[2] Mai Po 米埔；Mar-03, 03 年 3 月；Marcus Ho 何萬邦
[3] breeding 繁殖羽；14-May-05, 05 年 5 月 14 日；Mai Po 米埔；Michelle and Peter Wong 江敏兒，黃理沛
[4] Mai Po 米埔；May-03, 03 年 5 月；Marcus Ho 何萬邦

| | 春季過境遷徙鳥<br>Spring Passage Migrant | | | 夏候鳥<br>Summer Visitor | | 秋季過境遷徙鳥<br>Autumn Passage Migrant | | 冬候鳥<br>Winter Visitor | | |
|---|---|---|---|---|---|---|---|---|---|---|
| 常見月份 | 1 | 2 | 3 | 4 | 5 | 6 | 7 | 8 | 9 | 10 | 11 | 12 |

| 留鳥<br>Resident | 迷鳥<br>Vagrant | 偶見鳥<br>Occasional Visitor |
|---|---|---|

# 小青腳鷸

(普) xiǎo qīng jiǎo yù
(粵) 鷸：音核

體長 length：29-32cm

## Nordmann's Greenshank | *Tringa guttifer*

與青腳鷸相似，但體型較為小巧，黃色腳亦較短，尤其是脛部；嘴較短及直，下嘴端微微向上彎，嘴基有時沾淡黃色。繁殖期胸部白色及有深色斑點，背部羽緣上的白斑較青腳鷸大。非繁殖期頭、頸和上體顏色較淡。飛行時，翼底白色，腳後伸僅及尾端。

Resembles Common Greenshank, but appears more compact. Yellow legs are short, especially the tibia. Bill straight and short, tip of lower bill upturned, bill base sometimes yellow in colour. In breeding plumage, white breast are with dark spots, and the whitish spots lining at feather edge at the back are larger than Common Greenshank. In non-breeding plumage, the colour of head, neck and upperparts are paler. In flight, it shows white underwing, legs do not extend beyond the tail.

[1] breeding 繁殖羽；Mai Po 米埔；19-Apr-04, 04 年 4 月 19 日；Jemi and John Holmes 孔思義、黃亞萍
[2] non-breeding adult moulting into breeding plumage 非繁殖羽成鳥轉換繁殖羽；Mai Po 米埔；Apr-08, 08 年 4 月；Andy Kwok 郭匯昌
[3] non-breeding adult 非繁殖羽成鳥；Mai Po 米埔；24-Apr-04, 04 年 4 月 24 日；Lo Kar Man 盧嘉孟
[4] left breeding, right non-breeding 左繁殖羽，右非繁殖羽；Mai Po 米埔；28-Mar-05, 05 年 3 月 28 日；Michelle and Peter Wong 江敏兒、黃理沛
[5] non-breeding adult moulting into breeding plumage 非繁殖羽成鳥轉換繁殖羽；Mai Po 米埔；Apr-08, 08 年 4 月；Andy Kwok 郭匯昌
[6] non-breeding adult moulting into breeding plumage 非繁殖羽成鳥轉換繁殖羽；Mai Po 米埔；Apr-08, 08 年 4 月；Andy Kwok 郭匯昌

| | 春季過境遷徙鳥 Spring Passage Migrant | | 夏候鳥 Summer Visitor | | 秋季過境遷徙鳥 Autumn Passage Migrant | | 冬候鳥 Winter Visitor | |
|---|---|---|---|---|---|---|---|---|
| 常見月份 | 1 | 2 | 3 | 4 | 5 | 6 | 7 | 8 | 9 | 10 | 11 | 12 |

| 留鳥 Resident | 迷鳥 Vagrant | 偶見鳥 Occasional Visitor |
|---|---|---|

燕鴴科
**Glareolidae**

# 普通燕鴴

(普) pǔ tōng yàn héng
(粵) 鴴：音恆

體長 length：23-25cm

## Oriental Pratincole | *Glareola maldivarum*

飛行時像大型燕子，能在空中捕捉昆蟲。嘴闊而短，靜立時可見翼比尾長，尾端黑色開叉。成鳥上體灰褐色，喉淡黃色及有明顯黑線圍繞。幼鳥全身偏褐，背部、頸側及翼上均有斑點。飛行時腰部明顯白色，翼底紅褐色。

Resembles a big swallow in flight. Preys on flying insects. Bill broad and short. Wings extend beyond the tail at rest. Forked, black-tipped tail. Adult is uniformly grey brown, pale yellow throat bordered by obvious black line. Juvenile is browner, with scaly marks on mantle, both sides of the neck and wing coverts. Clear white rump and reddish brown underwing in flight.

[1] breeding 繁殖羽；Long Valley 塱原；Mar-07, 07 年 3 月；Allen Chan 陳志雄
[2] breeding 繁殖羽；Mai Po 米埔；Apr-07, 07 年 4 月；James Lam 林文華
[3] breeding 繁殖羽；Long Valley 塱原；Apr-08, 08 年 4 月；Joyce Tang 鄧玉蓮
[4] breeding 繁殖羽；Mai Po 米埔；Apr-07, 07 年 4 月；Martin Hale 夏敖天
[5] breeding 繁殖羽；Mai Po 米埔；Apr-07, 07 年 4 月；Eling Lee 李佩玲
[6] breeding 繁殖羽；Mai Po 米埔；Apr-07, 07 年 4 月；Kinni Ho 何建業
[7] juvenile 幼鳥；Long Valley 塱原；Cheung Ho Fai 張浩輝
[8] breeding 繁殖羽；Mai Po 米埔；Apr-07, 07 年 4 月；Winnie Wong and Sammy Sam 森美與雲妮

| | 春季過境遷徙鳥 Spring Passage Migrant | | | 夏候鳥 Summer Visitor | | | 秋季過境遷徙鳥 Autumn Passage Migrant | | | 冬候鳥 Winter Visitor | | |
|---|---|---|---|---|---|---|---|---|---|---|---|---|
| 常見月份 | 1 | 2 | 3 | 4 | 5 | 6 | 7 | 8 | 9 | 10 | 11 | 12 |
| | 留鳥 Resident | | | | 迷鳥 Vagrant | | | | 偶見鳥 Occasional Visitor | | | |

燕鴴科
**Glareolidae**

# 灰燕鴴

(普) huī yàn héng
(粵) 鴴：音恆

體長 length：15.5-19cm

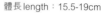

## Small Pratincole | *Glareola lactea*

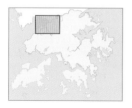

小 型燕鴴。全身大致淡灰褐色，腹部偏白，嘴和腳黑色。翼尖黑色，飛行時可見翼底大範圍黑色，方形的尾上白色並有黑斑。

Small pratincole. Mainly pale greyish-brown body with whitish belly. Black bill and legs. Black wing tips. In flight, underwings mainly black. Square tail with white upper tail and black spots.

1 adult 成鳥：San Tin 新田：Apr-17, 17 年 4 月：Oldcar Lee 李啟康
2 adult 成鳥：San Tin 新田：Apr-17, 17 年 4 月：Oldcar Lee 李啟康
3 adult 成鳥：San Tin 新田：Apr-17, 17 年 4 月：Oldcar Lee 李啟康
4 adult 成鳥：San Tin 新田：Apr-17, 17 年 4 月：Kwok Tsz Ki 郭子祈

| 春季過境遷徙鳥 Spring Passage Migrant | | | 夏候鳥 Summer Visitor | | | 秋季過境遷徙鳥 Autumn Passage Migrant | | | 冬候鳥 Winter Visitor | | |
|---|---|---|---|---|---|---|---|---|---|---|---|
| 1 | 2 | 3 | 4 | 5 | 6 | 7 | 8 | 9 | 10 | 11 | 12 |

常見月份

| 留鳥 Resident | | | 迷鳥 Vagrant | | | 偶見鳥 Occasional Visitor | | |
|---|---|---|---|---|---|---|---|---|

# 玄燕鷗

(普) xuán yàn ōu

體長 length：35-39cm

## Black Noddy | *Anous minutus*

全身大致黑色，嘴和腳黑色，與頭頂灰白色形成強烈對比。飛行時可見翼底亦是近全黑色。

Mainly black body with black bill and legs, which contrast with greyish-white crown. In flight, underwings almost entirely black.

1 adult 成鳥：Sai Kung 西貢；Jun-17, 17 年 6 月：Jemi and John Holmes 孔思義、黃亞萍
2 adult 成鳥：Sai Kung 西貢；Jun-17, 17 年 6 月：Jemi and John Holmes 孔思義、黃亞萍

| 春季過境遷徙鳥 Spring Passage Migrant | | | 夏候鳥 Summer Visitor | | | 秋季過境遷徙鳥 Autumn Passage Migrant | | | 冬候鳥 Winter Visitor | | |
|---|---|---|---|---|---|---|---|---|---|---|---|
| 1 | 2 | 3 | 4 | 5 | 6 | 7 | 8 | 9 | 10 | 11 | 12 |
| 留鳥 Resident | | | | 迷鳥 Vagrant | | | | 偶見鳥 Occasional Visitor | | | |

常見月份

# 三趾鷗 普 sān zhǐ ōu

體長 length：38-40cm

## Black-legged Kittiwake | *Rissa tridactyla*

體型細小的鷗類。在香港出沒的通常都是第一年冬羽。嘴黑色，耳後及頸部有黑色紋。飛行時，翅膀上的黑色部分組成一個M字。成鳥的嘴黃色，翅膀只有尖端是黑色，但在香港較難見到成鳥。

Small-sized gull. Those occurring in Hong Kong are usually first winter. Dark bill and dark bars behind ear and neck. A prominent pattern "M" on the upper wing in flight. Adult bird has yellow bill and black wing tip. Adults are seldom seen in Hong Kong.

[1] immature 未成年鳥：HK Southern waters 香港南面水域：Apr-08, 08 年 4 月：Pippen Ho 何志剛
[2] immature 未成年鳥：HK Southern waters 香港南面水域：Apr-08, 08 年 4 月：Michelle and Peter Wong 江敏兒、黃理沛
[3] immature 未成年鳥：HK Southern waters 香港南面水域：Apr-08, 08 年 4 月：Andy Kwok 郭匯昌

| 春季過境遷徙鳥<br>Spring Passage Migrant | | | 夏候鳥<br>Summer Visitor | | 秋季過境遷徙鳥<br>Autumn Passage Migrant | | | 冬候鳥<br>Winter Visitor | | |
|---|---|---|---|---|---|---|---|---|---|---|

| 常見月份 | 1 | 2 | 3 | 4 | 5 | 6 | 7 | 8 | 9 | 10 | 11 | 12 |
|---|---|---|---|---|---|---|---|---|---|---|---|---|

| 留鳥<br>Resident | 迷鳥<br>Vagrant | 偶見鳥<br>Occasional Visitor |
|---|---|---|

# 細嘴鷗

(普) xì zuǐ ōu

體長 length：42-44cm

## Slender-billed Gull | *Chroicocephalus genei*

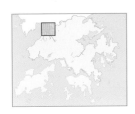

嘴及腳紅色，身體白色而帶點粉紅的感覺。背及覆羽淺灰色。大部分於香港的記錄都是首次度冬鳥，牠們跟紅嘴鷗分別在於體型略大、頸較長、嘴較幼及淡色、腳淡色。虹膜白色，眼後的黑點較不明顯。

Breeding bird shows white plumage with pink wash, red bill and legs. Pale mantle and wing coverts. Most records in Hong Kong were first winter birds. Different from Black-headed Gull by larger size, long neck, slender and pale bill, and pale legs. White iris and pale marks behind eyes.

1 immature 未成年鳥；Mai Po 米埔；Mar-08, 08 年 3 月：Lee Kai Hong 李啟康
2 immature 未成年鳥；Mai Po 米埔；1-Mar-08, 08 年 3 月 1 日：Owen Chiang 深藍
3 immature 未成年鳥；Mai Po 米埔；1-Mar-08, 08 年 3 月 1 日：Michelle and Peter Wong 江敏兒、黃理沛
4 breeding 繁殖羽；Mai Po 米埔；Mar-10, 10 年 3 月：Michelle and Peter Wong 江敏兒、黃理沛
5 breeding 繁殖羽；Mai Po 米埔；Mar-10, 10 年 3 月：Michelle and Peter Wong 江敏兒、黃理沛
6 immature 未成年鳥；Mai Po 米埔；2-Mar-08, 08 年 3 月 2 日：Thomas Chan 陳土飛

| 春季過境遷徙鳥 Spring Passage Migrant | | | 夏候鳥 Summer Visitor | | | 秋季過境遷徙鳥 Autumn Passage Migrant | | | 冬候鳥 Winter Visitor | | |
|---|---|---|---|---|---|---|---|---|---|---|---|
| 1 | 2 | 3 | 4 | 5 | 6 | 7 | 8 | 9 | 10 | 11 | 12 |

常見月份

| 留鳥 Resident | 迷鳥 Vagrant | 偶見鳥 Occasional Visitor |
|---|---|---|

# 紅嘴鷗 (普)hóng zuǐ ōu

## Black-headed Gull | *Chroicocephalus ridibundus*

體長 length：37-43cm

① 

香 港最常見的海鷗，體型細小，上體淡灰色，嘴及腳深紅色。非繁殖羽時頭部白色，耳羽有黑斑。繁殖羽頭部深褐色，有明顯的白眼圈。幼鳥嘴和腳顏色較淡，尾端有一黑色橫斑。

Commonest small-sized gull in Hong Kong. Upperparts pale grey, bill and legs deep red. In non-breeding plumage, it has white head and black mark on ear coverts. In breeding plumage, it shows dark brown head with prominent white eyerings. Juvenile has paler bill and legs, and a black band on tail.

① breeding 繁殖羽；Mai Po 米埔；Mar-07, 07 年 3 月：Jemi and John Holmes 孔思義、黃亞萍
② immature 未成年鳥；Mai Po 米埔；Feb-07, 07 年 2 月：Michelle and Peter Wong 江敏兒、黃理沛
③ immature 未成年鳥；Nam Sang Wai 南生圍；Jan-08, 08 年 1 月：Horman Ip 葉紀江
④ immature 未成年鳥；Mai Po 米埔；Mar-05, 05 年 3 月：Cherry Wong 黃卓研
⑤ immature 未成年鳥；Nam Sang Wai 南生圍；Feb-05, 05 年 2 月：Cherry Wong 黃卓研
⑥ immature 未成年鳥；Mai Po 米埔；Feb-07, 07 年 2 月：Michelle and Peter Wong 江敏兒、黃理沛
⑦ immature 未成年鳥；Mai Po 米埔；Mar-05, 05 年 3 月：Michelle and Peter Wong 江敏兒、黃理沛

| | 春季過境遷徙鳥<br>Spring Passage Migrant | | | 夏候鳥<br>Summer Visitor | | 秋季過境遷徙鳥<br>Autumn Passage Migrant | | | 冬候鳥<br>Winter Visitor | | |
|---|---|---|---|---|---|---|---|---|---|---|---|---|
| 常見月份 | 1 | 2 | 3 | 4 | 5 | 6 | 7 | 8 | 9 | 10 | 11 | 12 |
| | 留鳥<br>Resident | | | | 迷鳥<br>Vagrant | | | | 偶見鳥<br>Occasional Visitor | | | |

# 黑嘴鷗 ⟨普⟩hēi zuǐ ōu

體長 length : 29-32cm

## Saunders's Gull | *Chroicocephalus saundersi*

1

**體**型比紅嘴鷗小，嘴黑色而短厚，腳部暗紅色，站立時翼尖黑而有白點。繁殖羽整個頭部黑色，有明顯的白眼圈，看似在眼部前段斷開。非繁殖羽頭部白色，耳羽有黑斑。幼鳥身體有褐色斑，頭頂及後枕有深色闊橫紋。

Smaller than Black-headed Gull, with thick black bill and dark-red legs. At rest, it shows white spots on black wing tips. In breeding plumage, it has black head and prominent white eyerings which appear broken in front of the eyes. In non-breeding plumage, the head is white with a dark spot on ear coverts. Juvenile body has brown markings, with dark broad bands on crown and nape.

1 breeding 繁殖羽：Mai Po 米埔：Mar-07, 07 年 3 月：Jemi and John Holmes 孔思義，黃亞萍
2 brooding 繁殖羽：Mai Po 米埔：Mar-07, 07 年 3 月：Jemi and John Holmes 孔思義，黃亞萍
3 breeding 繁殖羽：Mai Po 米埔：Mar-05, 05 年 3 月：Martin Hale 夏敖天
4 breeding 繁殖羽：Mai Po 米埔：Mar-07, 07 年 3 月：Henry Lui 呂德恆
5 breeding 繁殖羽：Mai Po 米埔：Mar-07, 07 年 3 月：Bill Man 文權溢
6 immature 未成年鳥：Mai Po 米埔：Mar-07, 07 年 3 月：James Lam 林文華
7 breeding 繁殖羽：Mai Po 米埔：Feb-07, 07 年 2 月：Jemi and John Holmes 孔思義，黃亞萍
8 immature 未成年鳥：Mai Po 米埔：Mar-07, 07 年 3 月：Jemi and John Holmes 孔思義，黃亞萍

| 春季過境遷徙鳥 Spring Passage Migrant | | | 夏候鳥 Summer Visitor | | | 秋季過境遷徙鳥 Autumn Passage Migrant | | 冬候鳥 Winter Visitor | | | |
|---|---|---|---|---|---|---|---|---|---|---|---|
| 1 | 2 | 3 | 4 | 5 | 6 | 7 | 8 | 9 | 10 | 11 | 12 |

常見月份

| 留鳥 Resident | 迷鳥 Vagrant | 偶見鳥 Occasional Visitor |
|---|---|---|

# 棕頭鷗

（普）zōng tóu ōu

體長 length：41-45cm

## Brown-headed Gull | *Chroicocephalus brunnicephalus*

與紅嘴鷗相似，不過體型較大，嘴和腳較粗。虹膜淡色。上體淡灰色，嘴及腳深紅色。非繁殖羽時頭部白色，耳羽有黑斑。繁殖羽頭部深褐色，有明顯缺口的白眼圈。飛行時，於初級飛羽近羽尖位置有顯眼的白斑。

Resemble but bigger than Black-headed Gull with a thicker bill and legs. Pale iris. Upperparts pale grey, bill and legs deep red. In non-breeding plumage, it has white head and black marks on ear coverts. In breeding plumage, it shows dark brown head with prominent broken white eyering. In flight, it shows white patch near wing tip on primaries.

[1] non-breeding adult 非繁殖羽成鳥：Nam Sang Wai 南生圍：Jan-06, 06 年 1 月：Jemi and John Holmes 孔思義、黃亞萍
[2] non-breeding adult 非繁殖羽成鳥：Mai Po 米埔：22-Apr-07, 07 年 4 月 22 日：Helen Chan 陳燕芳
[3] non-breeding adult 非繁殖羽成鳥，Mai Po 米埔：Cheung Mok Jose Alberto 張振國

| 春季過境遷徙鳥<br>Spring Passage Migrant | | | 夏候鳥<br>Summer Visitor | | 秋季過境遷徙鳥<br>Autumn Passage Migrant | | | 冬候鳥<br>Winter Visitor | | |
|---|---|---|---|---|---|---|---|---|---|---|
| 常見月份 1 | 2 | 3 | 4 | 5 | 6 | 7 | 8 | 9 | 10 | 11 | 12 |

| 留鳥<br>Resident | 迷鳥<br>Vagrant | 偶見鳥<br>Occasional Visitor |
|---|---|---|

# 弗氏鷗

(普) fú shì ōu

Franklin's Gull | *Leucophaeus pipixcan*

體長 length：32-38cm

中 小型海鷗。頭至下體白色，翼灰色，眼後大片黑斑過渡至枕部黑白斑駁，腳黑色，嘴黑色而嘴端紅色，繁殖期頭部黑色而嘴紅色。飛行時可見翼上灰色，後緣有白色邊，翼尖黑色而初級飛羽末端白色，翼下偏白而翼尖黑色。

Medium small-sized gull. White from head to underparts. Grey wings. Black hood behind eyes becomes patchy on nape. Black legs. Black bill with red tip. Black head with red bill in breeding plumage. In flight, upper wings are grey with white hind edges. Black wing tips with white tips at the end of the primaries. Whitish underwings with black wing tips.

[1] immature 未成年鳥；Mai Po 米埔；Dec-15, 15 年 12 月；Kinni Ho 何建業
[2] immature 未成年鳥；Mai Po 米埔；Dec-15, 15 年 12 月；Kinni Ho 何建業

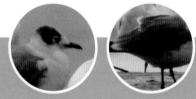

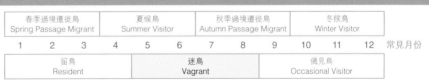

| 春季過境遷徙鳥 Spring Passage Migrant | | | 夏候鳥 Summer Visitor | | | 秋季過境遷徙鳥 Autumn Passage Migrant | | | 冬候鳥 Winter Visitor | | |
|---|---|---|---|---|---|---|---|---|---|---|---|
| 1 | 2 | 3 | 4 | 5 | 6 | 7 | 8 | 9 | 10 | 11 | 12 |

常見月份

| 留鳥 Resident | 迷鳥 Vagrant | 偶見鳥 Occasional Visitor |
|---|---|---|

鷗科
Laridae

# 遺鷗 <sup>普</sup>yí ōu

Relict Gull | *Ichthyaetus relictus*

體長 length：44cm

成 鳥的嘴及腳紅色，身體白色，背及覆羽淺灰色。大部分本港記錄為第一年冬羽，辨識特徵為白而粗的頸有褐色斑。嘴淺黃，嘴端黑色。跟紅嘴鷗的分別在於牠體型較大，及第一條初級飛羽全黑。

Breeding bird shows red bill and legs, white body, pale grey mantle and coverts. Most Hong Kong records were first winters with brown streaks on thick white neck, pale yellow bill with black tip. Separated from the Black-headed Gull by its larger size and completehy black outer primaries.

1 immature 未成年鳥：Mai Po 米埔：19-Jan-08, 08 年 1 月 19 日：Kinni Ho 何建業
2 immature 未成年鳥：Mai Po 米埔：20-Jan-08, 08 年 1 月 20 日：Thomas Chan 陳土飛
3 immature 未成年鳥：Mai Po 米埔：Jan-08, 08 年 1 月：Kinni Ho 何建業
4 immature 未成年鳥：Mai Po 米埔：Mar-08, 08 年 3 月：Lee Kai Hong 李啟康

| 春季過境遷徙鳥 Spring Passage Migrant | | | | 夏候鳥 Summer Visitor | | 秋季過境遷徙鳥 Autumn Passage Migrant | | | 冬候鳥 Winter Visitor | | |
|---|---|---|---|---|---|---|---|---|---|---|---|
| 常見月份 | 1 | 2 | 3 | 4 | 5 | 6 | 7 | 8 | 9 | 10 | 11 | 12 |
| 留鳥 Resident | | | | 迷鳥 Vagrant | | | 偶見鳥 Occasional Visitor | | | | |

# 漁鷗 <sup>普</sup> yú ōu

## Pallas's Gull | *Ichthyaetus ichthyaetus*

體長 length：60-72cm

黑色頭的大型海鷗。前額斜度較緩，嘴大而粗壯，眼皮白色。繁殖期頭部全黑，非繁殖期及幼鳥額及喉部沾白，比其他大型海鷗淺色，眼後及頸部羽色較深。

Easily distinguished by its black head and huge size. Flat and long forehead, long and thick bill, white eyelids. In breeding plumage, it shows conspicuous black head. Non-breeding bird and juvenile have white forehead and throat, paler than other big gulls, and darker patches behind eyes and neck.

1 breeding 繁殖羽：Mai Po 米埔；25-Mar-09, 09 年 3 月 25 日：Martin Hale 夏敖天
2 immature 未成年鳥：Mai Po 米埔；28-Feb-09, 09 年 2 月 28 日：Owen Chiang 深藍
3 breeding 繁殖羽：Mai Po 米埔；12-Mar-09, 09 年 3 月 12 日：Thomas Chan 陳土飛

| 春季過境遷徙鳥<br>Spring Passage Migrant | | | | 夏候鳥<br>Summer Visitor | | 秋季過境遷徙鳥<br>Autumn Passage Migrant | | | 冬候鳥<br>Winter Visitor | | |
|---|---|---|---|---|---|---|---|---|---|---|---|
| 1 | 2 | 3 | 4 | 5 | 6 | 7 | 8 | 9 | 10 | 11 | 12 |

常見月份

| 留鳥<br>Resident | 迷鳥<br>Vagrant | 偶見鳥<br>Occasional Visitor |
|---|---|---|

# 黑尾鷗 <sup>普</sup> hēi wěi ōu

體長 length：43-51cm

## Black-tailed Gull | *Larus crassirostris*

中 型海鷗，在香港出現的多是幼鳥。嘴部較其他海鷗長，翼尖黑色，腳黃綠色。成鳥背及翼均呈深灰色；嘴部黃色，末端紅色而有黑環；尾部有粗黑橫帶。幼鳥大致褐色，面部淡色，嘴部淡粉紅色而末端深色。

Medium-sized gull. Most of the records in Hong Kong were young birds. Bill is relatively longer than other gulls. Black wing tips, greenish yellow legs. Adult has dark grey mantle and wings. Bill is yellow and the tip is red with black ring. It has a broad black band on tail. Immature birds are mainly brown, with prominent pale face, pink bill with dark tip.

1 breeding 繁殖羽；Mirs Bay 大鵬灣；Jun-08, 08 年 6 月；Isaac Chan 陳家強
2 breeding 繁殖羽；Mirs Bay 大鵬灣；Jun-08, 08 年 6 月；Isaac Chan 陳家強
3 immature 未成年鳥；HK Southern waters 香港南面水域；Owen Chiang 深藍
4 immature 未成年鳥；HK Southern waters 香港南面水域；Owen Chiang 深藍
5 immature 未成年鳥；Mai Po 米埔；25-Mar-09, 09 年 3 月 25 日；Martin Hale 夏敖天
6 immature 未成年鳥；Mai Po 米埔；Aug-03, 03 年 8 月；Cherry Wong 黃卓研
7 immature 未成年鳥；Mai Po 米埔；Oct-03, 03 年 10 月；Jemi and John Holmes 孔思義、黃亞萍

| 春季過境遷徙鳥<br>Spring Passage Migrant | | | 夏候鳥<br>Summer Visitor | | | 秋季過境遷徙鳥<br>Autumn Passage Migrant | | 冬候鳥<br>Winter Visitor | | | |
|---|---|---|---|---|---|---|---|---|---|---|---|
| 1 | 2 | 3 | 4 | 5 | 6 | 7 | 8 | 9 | 10 | 11 | 12 |
| 留鳥<br>Resident | | | | 迷鳥<br>Vagrant | | | | 偶見鳥<br>Occasional Visitor | | | |

常見月份

# 海鷗 ⑧hǎi ōu

體長 length：40-46cm

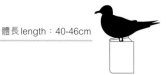

## Mew Gull | *Larus canus*

其他名稱 Other names：Common Gull

中 等大小的海鷗，體形比休氏銀鷗小。翼和上背灰色，翼尖深色有白斑。嘴黃色，腳綠黃色。站立時，肩上有新月形淺色斑。非繁殖羽，頭及後頸有縱紋，嘴尖附近深色。

Medium-sized gull, smaller than Lesser Black-backed Gull. Grey wings and mantle with dark wing tips that have white mirrors. Yellow bill and greenish yellow legs. At rest, it shows scapular crescent. Non-breeding bird has streaking on the head and hindneck and a dark subterminal mark on the bill.

1 immature 未成年鳥；Mai Po 米埔；Mar-10, 10 年 3 月；Michelle and Peter Wong 江敏兒・黃理沛
2 immature 未成年鳥；Mai Po 米埔；Mar-10, 10 年 3 月；Michelle and Peter Wong 江敏兒・黃理沛
3 immature 未成年鳥；Mai Po 米埔；Mar-10, 01 年 3 月：Allen Chan 陳志雄
4 immature 未成年鳥；Mai Po 米埔；15-Mar-09, 09 年 3 月 15 日；Owen Chiang 深藍
5 immature 未成年鳥；Mai Po 米埔；15-Mar-09, 09 年 3 月 15 日；Thomas Chan 陳土飛

| 春季過境遷徙鳥 Spring Passage Migrant | | | | 夏候鳥 Summer Visitor | | | 秋季過境遷徙鳥 Autumn Passage Migrant | | 冬候鳥 Winter Visitor | | |
|---|---|---|---|---|---|---|---|---|---|---|---|
| 常見月份 1 | 2 | 3 | 4 | 5 | 6 | 7 | 8 | 9 | 10 | 11 | 12 |
| 留鳥 Resident | | | | 迷鳥 Vagrant | | | | 偶見鳥 Occasional Visitor | | | |

# 灰翅鷗 （普）huī chì ōu

體長 length：61-68cm

## Glaucous-winged Gull | *Larus glaucescens*

**體**型大，越冬成鳥頭後和頸背略帶褐斑，上背灰色，腳粉紅色，嘴黃，灰白的初級飛羽。第一年度冬鳥，全身佈滿淺褐色的斑點，後頸較淡色，嘴黑而豐厚。

Large-sized gull. Wintering adult with tinged brown spots on back of head and nape, upperparts pale grey, pinkish legs, yellow bill, pale grey primaries. First year wintering bird has brown spots cover all body, pale nape, black and thick bill.

1 immature 未成年鳥；Mai Po 米埔；14-Apr-09, 09 年 4 月 14 日；James Lam 林文華
2 immature 未成年鳥；Mai Po 米埔；16-Apr-09, 09 年 4 月 16 日；Yue Pak Wai 余柏維
3 immature 未成年鳥；Mai Po 米埔；16-Apr-09, 09 年 4 月 16 日；Yue Pak Wai 余柏維

| 春季過境遷徙鳥 Spring Passage Migrant | | | 夏候鳥 Summer Visitor | | | 秋季過境遷徙鳥 Autumn Passage Migrant | | | 冬候鳥 Winter Visitor | | |
|---|---|---|---|---|---|---|---|---|---|---|---|
| 1 | 2 | 3 | 4 | 5 | 6 | 7 | 8 | 9 | 10 | 11 | 12 |

常見月份

| 留鳥 Resident | 迷鳥 Vagrant | 偶見鳥 Occasional Visitor |
|---|---|---|

# 北極鷗 <sub>普</sub>běi jí ōu

體長 length：64-77cm

## Glaucous Gull | *Larus hyperboreus*

大型淺色的海鷗，體型僅略大於烏灰銀鷗。在香港出現的多是幼鳥。初級飛羽末端白色，腳部粉紅色。成鳥上背淺灰色，頭部白色而有灰色縱紋；嘴黃色，下嘴末端有紅色斑點。未成年鳥身上有淺褐色橫紋和斑點，嘴部粉紅色而嘴尖黑色。

Large-sized and pale in colour. Noticably bigger than Lesser Black-backed Gull. Most of the records in Hong Kong were young birds. Whitish primaries and pink legs. Adult has pale grey mantle, whitish head with some greyish streaks. Bill yellow with red spot on tip of lower bill. Immature bird shows pale brownish bars and spots on the body. Pink bill with black tip.

1 immature 未成年鳥；Mai Po 米埔；Mar-01, 01 年 3 月；Jemi and John Holmes 孔思義、黃亞萍
2 immature 未成年鳥；Mai Po 米埔；13-Jan-08, 08 年 1 月 13 日；Kinni Ho 何建業

| 春季過境遷徙鳥<br>Spring Passage Migrant | | | | 夏候鳥<br>Summer Visitor | | | 秋季過境遷徙鳥<br>Autumn Passage Migrant | | | 冬候鳥<br>Winter Visitor | | |
|---|---|---|---|---|---|---|---|---|---|---|---|---|
| 1 | 2 | 3 | 4 | 5 | 6 | 7 | 8 | 9 | 10 | 11 | 12 | 常見月份 |
| | 留鳥<br>Resident | | | | 迷鳥<br>Vagrant | | | | 偶見鳥<br>Occasional Visitor | | | |

# 織女銀鷗

(普) zhī nǔ yín ōu

體長 length：55-67cm

Vega Gull | *Larus vegae*

其他名稱 Other names：紅腳銀鷗

大型鷗類，上背灰色，翼端黑色，下體白色。白色的頭、頸和上胸雜有褐色。嘴黃色，下嘴前端有紅點，腳粉紅色。首次度冬的未成年鳥全身夾雜褐色斑塊，嘴黑色。

Large-sized gull. Mantle grey with black wing tips. Underparts white. White head, neck and upper breast scattered with brown. Yellow bill with a red spot at the front of lower bill. Pink legs. First winter immature bird overall scattered with brown patches, black bill.

1 non-breeding adult, race vegae 非繁殖羽成鳥；vegae亞種；Mai Po 米埔；29-Mar-09, 09 年 3 月 29 日；Gary Chow 周家禮
2 immature, race vegae 未成年鳥；vegae亞種；Mai Po 米埔；2-Feb-08, 08 年 2 月 2 日；James Lam 林文華
3 immature, race vegae 未成年鳥；vegae亞種；Mai Po 米埔；2-Feb-09, 09 年 2 月 2 日；Kinni Ho 何建業
4 immature, race vegae 未成年鳥；vegae亞種；Mai Po 米埔；29-Mar-09, 09 年 3 月 29 日；Gary Chow 周家禮
5 immature, race vegae 未成年鳥；vegae亞種；Mai Po 米埔；2-Feb-09, 09 年 2 月 2 日；Kinni Ho 何建業
6 immature, race vegae 未成年鳥；vegae亞種；Mai Po 米埔；29-Mar-09, 09 年 3 月 29 日；Christina Chan 陳燕明
7 immature, race vegae 未成年鳥；vegae亞種；Mai Po 米埔；24-Mar-08, 08 年 3 月 24 日；Thomas Chan 陳土飛
8 breeding, race mongolicus 繁殖羽；mongolicus亞種；Mai Po 米埔；Owen Chiang 深藍
9 non-breeding adult, race mongolicus 非繁殖羽成鳥；mongolicus亞種；Nam Sang Wai 南生圍；Jan-06, 06 年 1 月；Jemi and John Holmes 孔思義、黃亞萍
10 non-breeding adult, race mongolicus 非繁殖羽成鳥；mongolicus亞種；Nam Sang Wai 南生圍；Jan-06, 06 年 1 月；Jemi and John Holmes 孔思義、黃亞萍
11 immature, race vegae 未成年鳥；vegae亞種；Mai Po 米埔；2-Feb-09, 09 年 2 月 2 日；Kinni Ho 何建業

| | 春季過境遷徙鳥<br>Spring Passage Migrant | | | 夏候鳥<br>Summer Visitor | | 秋季過境遷徙鳥<br>Autumn Passage Migrant | | | 冬候鳥<br>Winter Visitor | | |
|---|---|---|---|---|---|---|---|---|---|---|---|
| 常見月份 | 1 | 2 | 3 | 4 | 5 | 6 | 7 | 8 | 9 | 10 | 11 | 12 |

| 留鳥<br>Resident | 迷鳥<br>Vagrant | 偶見鳥<br>Occasional Visitor |
|---|---|---|

# 灰背鷗 <small>普 huī bèi ōu</small>

體長 length：55-67cm

## Slaty-backed Gull | *Larus schistisagus*

**體**型上較同類的鷗類較壯而胖，翼腳及頸較短，嘴較厚、腳粉紅色。第一年度冬全身褐色，下腹較深色，尾端有褐至黑色粗橫紋。

Sturdier than Lesser Black-backed Gull. Shorter wings, legs and neck; thicker bill. Pink legs. First winter plumage is overall brown in colour with darker belly. Dark band on tail.

1 immature 未成年鳥；HK Southwestern waters 香港西南水域：Owen Chiang 深藍
2 immature 未成年鳥；HK Southern waters 香港南面水域：Owen Chiang 深藍
3 immature 未成年鳥；Nam Sang Wai 南生圍：17-Feb-09, 09 年 2 月 17 日：Jemi and John Holmes 孔思義、黃亞萍（seperate from Lesser Black-backed Gull difficult, 不易與烏灰銀鷗分辨）
4 immature 未成年鳥；Nam Sang Wai 南生圍：17-Feb-09, 09 年 2 月 17 日：Jemi and John Holmes 孔思義、黃亞萍

| 春季過境遷徙鳥<br>Spring Passage Migrant | | | | 夏候鳥<br>Summer Visitor | | 秋季過境遷徙鳥<br>Autumn Passage Migrant | | | 冬候鳥<br>Winter Visitor | | |
|---|---|---|---|---|---|---|---|---|---|---|---|
| 1 | 2 | 3 | 4 | 5 | 6 | 7 | 8 | 9 | 10 | 11 | 12 |

常見月份

| 留鳥<br>Resident | 迷鳥<br>Vagrant | 偶見鳥<br>Occasional Visitor |
|---|---|---|

# 烏灰銀鷗

(普) wū huī yín ōu

體長 length：51-61cm

## Lesser Black-backed Gull | *Larus fuscus*

其他名稱 Other names：休氏銀鷗

大 型海鷗，為本港最常見的銀鷗。任何年紀腳部皆為黃色。成鳥背部深灰色，較黃腳銀鷗深色。翼尖黑色帶有白點。成鳥嘴部黃色，下嘴末端有紅色斑點。頭部污白色而有較多細小褐色縱紋。未成年鳥軀體沾褐色，嘴部深色。

Large-sized and the commonest herring gull in Hong Kong. Legs are yellow in all ages. Adult has dark grey mantle which is comparatively darker than Yellow-legged Gull. Black wing tip with white spots near feather tips. Bill yellow with red spots on tip of lower bill. Heavy brownish streaks on dirty white head. Immature shows overall brownish body and dark bill.

[1] breeding 繁殖羽；Mai Po 米埔；Mar-07, 07 年 3 月；Henry Lui 呂德恆
[2] breeding 繁殖羽；Mar-07, 07 年 3 月；Cherry Wong 黃卓研
[3] non-breeding adult 非繁殖羽成鳥；Mai Po 米埔；May-08, 08 年 5 月；Cherry Wong 黃卓研
[4] immature 未成年鳥；Mai Po 米埔；28-Mar-05, 05 年 3 月 28 日；Michelle and Peter Wong 江敏兒、黃理沛

| 春季過境遷徙鳥<br>Spring Passage Migrant | | | 夏候鳥<br>Summer Visitor | | 秋季過境遷徙鳥<br>Autumn Passage Migrant | | 冬候鳥<br>Winter Visitor | | |
|---|---|---|---|---|---|---|---|---|---|
| 1 | 2 | 3 | 4 | 5 | 6 | 7 | 8 | 9 | 10 | 11 | 12 | 常見月份 |
| 留鳥<br>Resident | | | 迷鳥<br>Vagrant | | | 偶見鳥<br>Occasional Visitor | | | |

# 鷗嘴噪鷗

(普) ōu zuǐ zào ōu

體長 length：33-43cm

## Gull-billed Tern | *Gelochelidon nilotica*

大型燕鷗，嘴黑色而粗壯，上體淡灰色，下體白色，尾稍開叉，翼尖偏黑。繁殖羽頭上半部全黑。非繁羽及幼鳥頭部白色，眼後有黑斑。

Large-sized tern. Black bill thick and strong, upperparts pale grey, underparts white in colour, slightly forked tail and black wing tips. In breeding plumage, the top half of the head completely black in colour. Non-breeding plumage and juvenile birds are white-headed, with a dark patch behind the eye.

1 breeding 繁殖羽；Mai Po 米埔；Apr-05, 05 年 4 月；Cherry Wong 黃卓研
2 breeding 繁殖羽；Mai Po 米埔；Apr-03, 03 年 4 月；Henry Lui 呂德恆
3 breeding 繁殖羽；Mai Po 米埔；Apr-08, 08 年 4 月；Andy Kwok 郭匯昌
4 breeding 繁殖羽；Mai Po 米埔；Apr-07, 07 年 4 月；Martin Hale 夏敦天
5 breeding 繁殖羽；HK Southern waters 香港南面水域；May-07, 07 年 5 月；Jemi and John Holmes 孔思義、黃亞萍

| | 春季過境遷徙鳥<br>Spring Passage Migrant | | 夏候鳥<br>Summer Visitor | | 秋季過境遷徙鳥<br>Autumn Passage Migrant | | 冬候鳥<br>Winter Visitor | | |
|---|---|---|---|---|---|---|---|---|---|
| 常見月份 | 1 | 2 | 3 | 4 | 5 | 6 | 7 | 8 | 9 | 10 | 11 | 12 |

| 留鳥<br>Resident | 迷鳥<br>Vagrant | 偶見鳥<br>Occasional Visitor |
|---|---|---|

# 紅嘴巨鷗

(普)hóng zuǐ jù ōu

體長 length：48-56cm

## Caspian Tern | *Hydroprogne caspia*

香 港體型最大的燕鷗，鮮紅色大嘴十分顯眼。上體淡灰色，下體白色，翼尖偏黑。繁殖羽頭上半部全黑。非繁羽及幼鳥頭部的黑色羽毛沾白。

The largest tern in Hong Kong. Prominent bright red bill. Upperparts pale grey, underparts white colour with darker wing tip. In breeding plumage, top half of the head is completely black. Non-breeding and juvenile birds have forehead black mottled with white.

1 breeding 繁殖羽；Mai Po 米埔；28-Mar-05, 05 年 3 月 28 日；Michelle and Peter Wong 江敏兒、黃理沛
2 breeding 繁殖羽；Mai Po 米埔；28-Mar-05, 05 年 3 月 28 日；Michelle and Peter Wong 江敏兒、黃理沛
3 non-breeding 非繁殖羽；Mai Po 米埔；Mar-07, 07 年 3 月；Jemi and John Holmes 孔思義、黃亞萍
4 breeding 繁殖羽；Mai Po 米埔；Aug-04, 04 年 8 月；Henry Lui 呂德恆
5 non-breeding adult 非繁殖羽成鳥；Mai Po 米埔；Mar-08, 08 年 3 月；James Lam 林文華
6 non-breeding adult moulting into breeding plumage 非繁殖羽成鳥轉換繁殖羽；Mai Po 米埔；Apr-07, 07 年 4 月；Henry Lui 呂德恆

| 春季過境遷徙鳥<br>Spring Passage Migrant | | | | 夏候鳥<br>Summer Visitor | | | 秋季過境遷徙鳥<br>Autumn Passage Migrant | | | 冬候鳥<br>Winter Visitor | | |
|---|---|---|---|---|---|---|---|---|---|---|---|---|
| 1 | 2 | 3 | 4 | 5 | 6 | 7 | 8 | 9 | 10 | 11 | 12 | 常見月份 |

| 留鳥<br>Resident | 迷鳥<br>Vagrant | 偶見鳥<br>Occasional Visitor |
|---|---|---|

# 大鳳頭燕鷗

（普）dà fèng tóu yàn ōu

體長 length：43-53cm

## Greater Crested Tern | *Thalasseus bergii*

大型燕鷗，嘴粗大而黃色，有一簇黑色的長冠羽，前額、面頰及尾下白色，背部和翼上灰色。飛行時尾部深叉，翼底大致白色。非繁殖及幼鳥頭頂黑色羽毛沾白。幼鳥上體白色褐色斑駁。

Large-sized tern. Large, thick yellow bill. Black tuft of crown feathers. Forehead, cheeks and under tail white. Mantle and upperwing grey. In flight, it shows deeply forked tail and mainly white underwing. Non-breeding and juvenile birds have more white on forecrown. Upperparts of juvenile marked with brown and white.

1 breeding 繁殖羽：HK Southeastern waters 香港東南水域；May-08, 08 年 5 月；Isaac Chan 陳家強
2 breeding 繁殖羽：HK Southern waters 香港南面水域；May-08, 08 年 5 月；Michelle and Peter Wong 江敏兒、黃理沛
3 juvenile 幼鳥：HK Southern waters 香港南面水域；Jul-07, 07 年 7 月；Kinni Ho 何建業
4 juvenile 幼鳥：HK Southern waters 香港南面水域；Jul-07, 07 年 7 月；Kinni Ho 何建業
5 breeding 繁殖羽：HK Southern waters 香港南面水域；Apr-08, 08 年 4 月；Cherry Wong 黃卓研
6 breeding 繁殖羽：HK Southern waters 香港南面水域；Apr-07, 07 年 4 月；Matthew and TH Kwan 關朗曦、關子凱
7 breeding 繁殖羽：HK Southern waters 香港南面水域；Apr-07, 07 年 4 月；Matthew and TH Kwan 關朗曦、關子凱
8 non-breeding adult 非繁殖羽成鳥：HK Southern waters 香港南面水域；Jul-07, 07 年 7 月；Kinni Ho 何建業

| 春季過境遷徙鳥<br>Spring Passage Migrant | | | 夏候鳥<br>Summer Visitor | | 秋季過境遷徙鳥<br>Autumn Passage Migrant | | | 冬候鳥<br>Winter Visitor | | |
|---|---|---|---|---|---|---|---|---|---|---|
| 常見月份 1 | 2 | 3 | 4 | 5 | 6 | 7 | 8 | 9 | 10 | 11 | 12 |
| 留鳥<br>Resident | | | 迷鳥<br>Vagrant | | | 偶見鳥<br>Occasional Visitor | | | | | |

# 白腰燕鷗

(普) bái yāo yàn ōu

體長 length：32-34cm

Aleutian Tern | *Onychoprion aleuticus*

[1]

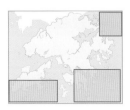

中型燕鷗，上體淡灰色，頭上半部黑色而前額白色，下體淺灰色，腰和尾白色。翼尖和翼底次級飛羽後緣有深色帶。站立時，翼尖向上翹。幼鳥上體偏褐。

Medium-sized tern. Upperparts are pale grey, while top half of the head black and forehead are white. Underparts light grey, rump and tail white. Wing tips and the edge of secondaries underwing marked with dark band. At rest, the wing tip points upwards. Juvenile has brownish upperparts.

1 adult breeding 繁殖羽成鳥：Lamma Channel 博寮海峽；Sep-08, 08 年 9 月；Cherry Wong 黃卓研
2 adult breeding 繁殖羽成鳥：HK Southern waters 香港南面水域；Sep-08, 08 年 9 月；Michelle and Peter Wong 江敏兒、黃理沛
3 breeding 繁殖羽：Po Toi 蒲台；May-08, 08 年 5 月；Pippen Ho 何志剛
4 breeding 繁殖羽：HK Southern waters 香港南面水域；May-06, 06 年 5 月；Jemi and John Holmes 孔思義、黃亞萍
5 adult breeding and juvenile 繁殖羽成鳥和幼鳥：HK Southern waters 香港南面水域；Sep-08, 08 年 9 月；Michelle and Peter Wong 江敏兒、黃理沛
6 breeding 繁殖羽：HK Southern waters 香港南面水域；May-06, 06 年 5 月；Martin Hale 夏敖天
7 juvenile 幼鳥：Lamma Channel 博寮海峽；May-08, 08 年 5 月；Cherry Wong 黃卓研
8 non-breeding adult 非繁殖羽成鳥：HK Southern waters 香港南面水域；Sep-08, 08 年 9 月；Michelle and Peter Wong 江敏兒、黃理沛

| 春季過境遷徙鳥<br>Spring Passage Migrant | | | 夏候鳥<br>Summer Visitor | | | 秋季過境遷徙鳥<br>Autumn Passage Migrant | | | 冬候鳥<br>Winter Visitor | | |
|---|---|---|---|---|---|---|---|---|---|---|---|
| 1 | 2 | 3 | 4 | 5 | 6 | 7 | 8 | 9 | 10 | 11 | 12 |

常見月份

| 留鳥<br>Resident | 迷鳥<br>Vagrant | 偶見鳥<br>Occasional Visitor |
|---|---|---|

# 褐翅燕鷗

(普)hè chì yàn ōu

體長length：35-38cm

Bridled Tern | *Onychoprion anaethetus*

1

中型燕鷗。頭上半部黑色而前額白色，上背、腰、尾及翼上褐色，尾羽外側白色。嘴和腳黑色，下體和翼底白色，翼下的飛羽後緣深色。幼鳥頭部和上背褐色斑駁，下體有時沾灰色。

Medium-sized tern. Top half of the head black and forehead is white. Mantle, rump, tail and upperwing dark brown. Outer tail feathers white. Black bill and legs. Underparts and underwing white, the edge of flight feathers at underwing marked with dark bands. Head and upperparts of juvenile have brownish markings, underparts sometimes grey.

[1] breeding 繁殖羽：HK Southern waters 香港南面水域；Sep-08, 08 年 9 月；Michelle and Peter Wong 江敏兒、黃理沛
[2] breeding 繁殖羽：Mirs Bay 大鵬灣；Aug-03, 03 年 8 月；Cherry Wong 黃卓研
[3] breeding 繁殖羽：Mirs Bay 大鵬灣；Jul-05, 05 年 7 月；Isaac Chan 陳家強
[4] breeding 繁殖羽：HK Southern waters 香港南面水域；Sep-08, 08 年 9 月；Michelle and Peter Wong 江敏兒、黃理沛
[5] breeding 繁殖羽：Mirs Bay 大鵬灣；May-05, 05 年 5 月；Pippen Ho 何志剛
[6] breeding 繁殖羽：Mirs Bay 大鵬灣；Jul-06, 06 年 7 月；Isaac Chan 陳家強
[7] breeding 繁殖羽：HK Southern waters 香港南面水域；Aug-07, 07 年 8 月；Winnie Wong and Sammy Sam 森美與雲妮

| | 春季過境遷徙鳥<br>Spring Passage Migrant | | | 夏候鳥<br>Summer Visitor | | | 秋季過境遷徙鳥<br>Autumn Passage Migrant | | | 冬候鳥<br>Winter Visitor | | |
|---|---|---|---|---|---|---|---|---|---|---|---|---|
| 常見月份 | 1 | 2 | 3 | 4 | 5 | 6 | 7 | 8 | 9 | 10 | 11 | 12 |
| | 留鳥<br>Resident | | | 迷鳥<br>Vagrant | | | 偶見鳥<br>Occasional Visitor | | | | | |

# 白額燕鷗

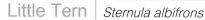 普 bái é yàn ōu

體長 length：22-28cm

Little Tern | *Sternula albifrons*

小 型燕鷗。頭上半部黑色而前額白色，背部及翼上灰色，翼前緣偏黑，下體、腰、尾和翼底白色。繁殖時，嘴黃而嘴尖黑色，腳橙色。非繁殖羽及幼鳥嘴黑色，腳部淡色，頭頂黑色羽毛沾白。

Small-sized tern. Top half of the head black and forehead are white in colour. Mantle and upperwings grey. Front edge of the wing is darker. Underparts, rump, tail and underwing white. In breeding plumage, it shows yellow bill with dark tip. Reddish orange legs. Non-breeding plumage and juvenile have black bill, pale legs, and some white on black crown tinted with white colour.

[1] breeding 繁殖羽；HK Southern waters 香港南面水域；Apr-08, 08 年 4 月；Michelle and Peter Wong 江敏兒、黃理沛
[2] breeding 繁殖羽；Mai Po 米埔；Apr-06, 06 年 4 月；Liu Jianzhong 劉健忠
[3] breeding 繁殖羽；HK Southern waters 香港南面水域；14-Apr-07, 07 年 4 月 14 日；Christina Chan 陳燕明
[4] breeding 繁殖羽；HK Southern waters 香港南面水域；Apr-08, 08 年 4 月；Cherry Wong 黃卓研
[5] breeding 繁殖羽；Mai Po 米埔；Apr-06, 06 年 4 月；Owen Chiang 深藍

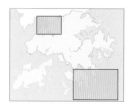

| 春季過境遷徙鳥<br>Spring Passage Migrant | | | 夏候鳥<br>Summer Visitor | | 秋季過境遷徙鳥<br>Autumn Passage Migrant | | | 冬候鳥<br>Winter Visitor | | |
|---|---|---|---|---|---|---|---|---|---|---|
| 常見月份 1 | 2 | 3 | 4 | 5 | 6 | 7 | 8 | 9 | 10 | 11 | 12 |
| 留鳥<br>Resident | | | | 迷鳥<br>Vagrant | | | | 偶見鳥<br>Occasional Visitor | | |

# 烏燕鷗

（普）wū yàn ōu

體長 length：36-45cm

Sooty Tern | *Onychoprion fuscatus*

於海洋活動的燕鷗，上體黑色，下體呈白色。成鳥很容易與褐翅燕鷗混淆，但可從較明顯而不延伸至眼後的白色額部去區分。幼鳥全身黑色，滿佈白色斑點。

An oceanic tern mainly dark in color with white underparts. Adult could be confused with Bridled Tern but has a clear white forehead line not extending behind eye. Juvenile is distinctively blackish, peppered white.

1 juvenile together with Bridled Tern on the right 幼鳥，褐翅燕鷗在右面；Po Toi 蒲台；13-Sep-06, 06 年 9 月 13 日；Geoff Welch
2 HK Southern waters 香港南面水域；5-Apr-09, 09 年 4 月 5 日；Yu Yat Tung 余日東
3 HK Southern waters 香港南面水域；5-Apr-09, 09 年 4 月 5 日；Yu Yat Tung 余日東
4 juvenile 幼鳥；Po Toi 蒲台；13-Sep-06, 06 年 9 月 13 日；Geoff Welch
5 juvenile 幼鳥；HK Southern waters 香港南面水域；May-06, 06 年 5 月；Jemi and John Holmes 孔思義、黃亞萍

5

| 春季過境遷徙鳥<br>Spring Passage Migrant | | | 夏候鳥<br>Summer Visitor | | | 秋季過境遷徙鳥<br>Autumn Passage Migrant | | | 冬候鳥<br>Winter Visitor | | |
|---|---|---|---|---|---|---|---|---|---|---|---|
| 1 | 2 | 3 | 4 | 5 | 6 | 7 | 8 | 9 | 10 | 11 | 12 |

常見月份

| 留鳥<br>Resident | 迷鳥<br>Vagrant | 偶見鳥<br>Occasional Visitor |
|---|---|---|

# 粉紅燕鷗

(普) fěn hóng yàn ōu

**Roseate Tern** | *Sterna dougallii*

體長 length：33-43cm

中 型燕鷗。上體淡灰，下體白色，有時有淡淡的粉紅色，尾長開叉成深V形。繁殖羽嘴鮮明橙紅色，嘴端有時沾黑，頭頂上半部黑色，腳紅色。非繁殖羽前額至冠部白色，腳淡紅色。幼鳥上體有褐色斑，嘴全黑。

Medium-sized tern. Upperparts pale grey, underparts white and sometimes stained with pale pink. Tail is deeply forked forming an "V" shape. In breeding plumage, it shows bright orange red bill, with tip sometimes tinted black. Top half of head is black. Red legs. Non-breeding plumage has white forehead and crown, pale red legs, while juvenile has brownish patches and completely black bill.

1 breeding 繁殖羽；Tap Mun 塔門；Jul-08, 08 年 7 月；Michelle and Peter Wong 江敏兒、黃理沛
2 breeding male 繁殖羽雄鳥；Po Toi 蒲台；Aug-07, 07 年 8 月；James Lam 林文華
3 breeding female 繁殖羽雌鳥；HK Southern waters 香港南面水域；Jul-08, 08 年 7 月；Kinni Ho 何建業
4 non-breeding adult 非繁殖羽成鳥；HK Southwestern waters 香港西南水域；Jun-08, 08 年 6 月；Isaac Chan 陳家強
5 breeding male 繁殖羽雄鳥；Sai Kung Sea 西貢海；Jul-07, 07 年 7 月；Isaac Chan 陳家強
6 breeding male 繁殖羽雄鳥；Sai Kung Sea 西貢海；Jun-07, 07 年 6 月；Isaac Chan 陳家強
7 breeding female 繁殖羽雌鳥；HK Southern waters 香港南面水域；Aug-07, 07 年 8 月；Winnie Wong and Sammy Sam 森美與雲妮

| | 春季過境遷徙鳥<br>Spring Passage Migrant | | | 夏候鳥<br>Summer Visitor | | | 秋季過境遷徙鳥<br>Autumn Passage Migrant | | | 冬候鳥<br>Winter Visitor | | |
|---|---|---|---|---|---|---|---|---|---|---|---|---|
| 常見月份 | 1 | 2 | 3 | 4 | 5 | 6 | 7 | 8 | 9 | 10 | 11 | 12 |
| | 留鳥<br>Resident | | | | 迷鳥<br>Vagrant | | | | 偶見鳥<br>Occasional Visitor | | | |

# 黑枕燕鷗
(普)hēi zhěn yàn ōu

體長 length：34-35cm

Black-naped Tern | *Sterna sumatrana*

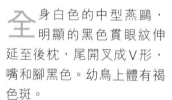

全身白色的中型燕鷗，明顯的黑色貫眼紋伸延至後枕，尾開叉成V形，嘴和腳黑色。幼鳥上體有褐色斑。

Medium-sized white tern. Prominent eyestripe extends to nape. Deeply forked tail, bill and legs black. The upperparts of juvenile are heavily streaked brown.

[1] breeding 繁殖羽；Po Toi 蒲台；Jul-07, 07 年 7 月；Andy Cheung 張玉良
[2] breeding 繁殖羽；Mirs Bay 大鵬灣；May-05, 05 年 5 月；Pippen Ho 何志剛
[3] breeding pair in courtship 求偶的繁殖鳥；Po Toi 蒲台；May-08, 08 年 5 月；Joyce Tang 鄧玉蓮
[4] breeding pair in courtship 求偶的繁殖鳥；Po Toi 蒲台；May-08, 08 年 5 月；James Lam 林文華
[5] breeding pair attempt to mate 意圖交配的繁殖鳥；Po Toi 蒲台；May-08, 08 年 5 月；James Lam 林文華
[6] juvenile 幼鳥（中Middle）；Tap Mun 塔門；Jul-08, 08 年 7 月；Michelle and Peter Wong 江敏兒・黃理沛
[7] juvenile 幼鳥；HK Southern waters 香港南面水域；Aug-07, 07 年 8 月；Sammy Sam and Winnie Wong 森美與雲妮

| | 春季過境遷徙鳥<br>Spring Passage Migrant | | | 夏候鳥<br>Summer Visitor | | | 秋季過境遷徙鳥<br>Autumn Passage Migrant | | | 冬候鳥<br>Winter Visitor | |
|---|---|---|---|---|---|---|---|---|---|---|---|
| 常見月份 | 1 | 2 | 3 | 4 | 5 | 6 | 7 | 8 | 9 | 10 | 11 | 12 |
| | 留鳥<br>Resident | | | | 迷鳥<br>Vagrant | | | | 偶見鳥<br>Occasional Visitor | | |

# 普通燕鷗

(普) pǔ tōng yàn ōu

體長 length：32-39cm

## Common Tern | *Sterna hirundo*

中型燕鷗，在香港錄得的多是 *longipennis* 亞種。嘴和腳黑色，上體淡灰色，下體白色，飛行時可見翼外側邊緣偏灰。繁殖羽頭上半部全黑。非繁殖羽及幼鳥前額白色，而幼鳥中覆羽有深色橫斑。

Medium-sized tern. Subspecies *longipennis* in Hong Kong has black legs and bill. Upperparts pale grey, underparts white in colour. In flight, it shows prominent grey on outer wing. In breeding plumage, top half of the head is completely black. Non-breeding and juvenile birds have white forehead. Juvenile also has dark band on median coverts.

[1] breeding, race *longipennis* 繁殖羽, *longipennis* 亞種；HK Southern waters 香港南面水域；May-00, 00 年 5 月；Martin Hale 夏敖天
[2] breeding, race *longipennis* 繁殖羽, *longipennis* 亞種；HK Southern waters 香港南面水域；May-08, 08 年 5 月；Cherry Wong 黃卓研
[3] breeding, race *minussensis* 繁殖羽, *minussensis* 亞種；HK Southern waters 香港南面水域；May-06, 06 年 5 月；Jemi and John Holmes 孔思義、黃亞萍
[4] breeding, race *longipennis* 繁殖羽, *longipennis* 亞種；HK Southern waters 香港南面水域；Apr-06, 06 年 4 月；Tam Yip Shing 譚業成
[5] breeding, race *longipennis* 繁殖羽, *longipennis* 亞種；HK Southeastern waters 香港東南水域；May-08, 08 年 5 月；Isaac Chan 陳家強
[6] breeding, race *minussensis* 繁殖羽, *minussensis* 亞種；HK Southern waters 香港南面水域；Apr-08, 08 年 4 月；Winnie Wong and Sammy Sam 森美與雲妮
[7] breeding, race *minussensis* 繁殖羽, *minussensis* 亞種；HK Southern waters 香港南面水域；May-06, 06 年 5 月；Jemi and John Holmes 孔思義、黃亞萍

| | 春季過境遷徙鳥<br>Spring Passage Migrant | | | 夏候鳥<br>Summer Visitor | | | 秋季過境遷徙鳥<br>Autumn Passage Migrant | | | 冬候鳥<br>Winter Visitor | | |
|---|---|---|---|---|---|---|---|---|---|---|---|---|
| 常見月份 | 1 | 2 | 3 | 4 | 5 | 6 | 7 | 8 | 9 | 10 | 11 | 12 |
| | 留鳥<br>Resident | | | | 迷鳥<br>Vagrant | | | | 偶見鳥<br>Occasional Visitor | | | |

鷗科
**Laridae**

# 鬚浮鷗 <sup>普</sup>xū fú ōu

體長 length：23-29cm

## Whiskered Tern | *Chlidonias hybrida*

**中**型燕鷗。繁殖羽頭上半部黑色，面頰白色，胸及腹深灰色，嘴和腳紅色。非繁殖期與白翅浮鷗非常相似，但耳羽黑斑伸延至後枕。幼鳥身體沾深褐色斑。

Medium-sized tern. In breeding plumage, its head is black on the top half with white cheeks. It also has dark grey breast and belly, red bill and legs. In non-breeding plumage, it resembles White-winged Tern, but the black mark on ear coverts extends to nape. Juvenile bird has dark brown markings on body.

[1] breeding 繁殖羽；Mai Po 米埔；Sep-05, 05 年 9 月；Martin Hale 夏敦天
[2] non-breeding 非繁殖羽；Mai Po 米埔；Oct-07, 07 年 10 月；Killy Kuo 吉婓婥
[3] breeding 繁殖羽；Mai Po 米埔；May-06, 06 年 5 月；Jemi and John Holmes 孔思義、黃亞萍
[4] 1st winter 第一年冬天；Mai Po 米埔；Oct-07, 07 年 10 月；Yue Pak Wai 余柏維
[5] non-breeding adult 非繁殖羽成鳥；Tai Sang Wai 大生圍；Sep-07, 07 年 9 月；Chan Kai Wai 陳佳瑋
[6] 1st winter 第一年冬天；Lok Ma Chau 落馬洲；Sep-05, 05 年 9 月；Martin Hale 夏敦天
[7] breeding 繁殖羽；Lok Ma Chau 落馬洲；Sep-05, 05 年 9 月；Martin Hale 夏敦天
[8] non-breeding 非繁殖羽；Mai Po 米埔；Sep-05, 05 年 9 月；Martin Hale 夏敦天

| 春季過境遷徙鳥 Spring Passage Migrant | | | 夏候鳥 Summer Visitor | | 秋季過境遷徙鳥 Autumn Passage Migrant | | 冬候鳥 Winter Visitor | | |
|---|---|---|---|---|---|---|---|---|---|

常見月份

| 1 | 2 | 3 | 4 | 5 | 6 | 7 | 8 | 9 | 10 | 11 | 12 |
|---|---|---|---|---|---|---|---|---|---|---|---|

| 留鳥 Resident | 迷鳥 Vagrant | 偶見鳥 Occasional Visitor |
|---|---|---|

# 白翅浮鷗
(普) bái chì fú ōu

體長 length：23-27cm

## White-winged Tern | *Chlidonias leucopterus*

其他名稱 Other names：White-winged Black Tern

[1]

中 型燕鷗，嘴黑色，腰白色，白色尾部有時偏灰。繁殖羽頭至腹部全黑，上體灰色，翼偏白。非繁殖羽與鬚浮鷗非常相似，但耳羽黑斑與頭頂濃密黑點相連。幼鳥身體沾深褐色斑。

Medium-sized tern. Black bill, white rump, white tail sometimes greyish. In breeding plumage, head to belly completely black, upperparts grey, with whitish wings. In non-breeding plumage, resembles Whiskered Tern, but the black mark on ear coverts extends to dense black spots on crown. Juvenile bird has dark brown markings on body.

[1] non-breeding adult 非繁殖羽成鳥；Lok Ma Chau 落馬洲；Sep-05, 05 年 9 月；Martin Hale 夏敖天
[2] breeding 繁殖羽；Mai Po 米埔；May-07, 07 年 5 月；Cherry Wong 黃卓研
[3] 1st summer 第一年夏天；HK Southeastern waters 香港東南水域；May-08, 08 年 5 月；Isaac Chan 陳家強
[4] breeding 繁殖羽；Mai Po 米埔；May-06, 06 年 5 月；Jemi and John Holmes 孔思義、黃亞萍
[5] breeding 繁殖羽；Mai Po 米埔；May-06, 06 年 5 月；Jemi and John Holmes 孔思義、黃亞萍
[6] juvenile moulting into 1st winter plumage 幼鳥；HK Southern waters 香港南面水域；Sep-08, 08 年 9 月；Michelle and Peter Wong 江敏兒、黃理沛
[7] breeding and non-breeding 繁殖羽和非繁殖羽；HK Southeastern waters 香港東南水域；May-08, 08 年 5 月；Isaac Chan 陳家強
[8] 1st winter 第一年冬天；Mai Po 米埔；May-06, 06 年 5 月；Jemi and John Holmes 孔思義、黃亞萍

| | 春季過境遷徙鳥<br>Spring Passage Migrant | | | 夏候鳥<br>Summer Visitor | | | 秋季過境遷徙鳥<br>Autumn Passage Migrant | | | 冬候鳥<br>Winter Visitor | | |
|---|---|---|---|---|---|---|---|---|---|---|---|---|
| 常見月份 | 1 | 2 | 3 | 4 | 5 | 6 | 7 | 8 | 9 | 10 | 11 | 12 |
| | 留鳥<br>Resident | | | | 迷鳥<br>Vagrant | | | | 偶見鳥<br>Occasional Visitor | | | |

# 中賊鷗 zhōng zéi ōu

體長 length：46-51cm

## Pomarine Jaeger | *Stercorarius pomarinus*

香港最大的賊鷗，強壯，胸部健碩，上體深褐色。有顯眼白色翼鏡，尾部楔形，繁殖羽末端有扭曲的中央尾羽。分淡色型和深色型，淡色型下體大致白色夾雜灰褐色斑，面頰淡黃，上胸至背部有深色胸帶。較少見的深色型，除翼上有淺色斑外，全身大致深褐色。

Largest jaeger in Hong Kong. Looks powerful and deep-chested, upperparts dark brown, obvious white wing panels, wedge-shaped tail. Breeding bird has projecting central feathers twisted at the tip. Two morphs. Pale morph with lighter belly and yellowish chin and dark breast band. Uncommon dark morph is all dark apart from lighter wing flashes.

1 HK Southwestern waters 香港西南水域；Apr-09, 09 年 4 月；Pang Chun Chiu 彭俊超
2 HK Southwestern waters 香港西南水域；5-Apr-09, 09 年 4 月 5 日；Tony Hung 洪敦熹
3 non-breeding adult moulting into breeding plumage 非繁殖羽成鳥轉換繁殖羽；HK Southern waters 香港南面水域；12-Mar-06, 06 年 3 月 12 日；Michelle and Peter Wong 江敏兒、黃理沛
4 non-breeding adult moulting into breeding plumage 非繁殖羽成鳥轉換繁殖羽；HK Southern waters 香港南面水域；12-Mar-06, 06 年 3 月 12 日；Michelle and Peter Wong 江敏兒、黃理沛

| 春季過境遷徙鳥<br>Spring Passage Migrant | | | | 夏候鳥<br>Summer Visitor | | 秋季過境遷徙鳥<br>Autumn Passage Migrant | | 冬候鳥<br>Winter Visitor | | | |
|---|---|---|---|---|---|---|---|---|---|---|---|
| 常見月份 1 | 2 | 3 | 4 | 5 | 6 | 7 | 8 | 9 | 10 | 11 | 12 |
| 留鳥<br>Resident | | | | 迷鳥<br>Vagrant | | | | 偶見鳥<br>Occasional Visitor | | | |

# 短尾賊鷗

（普）duǎn wěi zéi ōu

體長 length：41-46cm

## Parasitic Jaeger | *Stercorarius parasiticus*

其他名稱 Other names：賊鷗 Parasitic Skua, Arctic Skua, Parasitic Jaeger

繁 殖羽頭頂和背部暗褐色，耳羽附近淡黃色，下體淡色，翼有白色斑塊，尾部楔形，中央尾羽較長。非繁殖羽上體深色，下體淺色，翼有白色斑塊。同樣有深色型，全身深褐色，翼也有白色斑塊。

Breeding plumage has dark cap and yellow ear covert. Upperparts dark brown, underparts paler. White wing panels and wedge-shaped tail with pointed tail projection. Non-breeding plumage has dark upperparts, pale colour underparts, and white wing panels. Dark morph is uniformly dark brown in colour, also with white wing panels.

1 breeding 繁殖羽：HK Southwestern waters 香港西南水域：17-Apr-05, 05 年 4 月 17 日；Michelle and Peter Wong 江敏兒、黃理沛
2 breeding 繁殖羽：HK Southwestern waters 香港西南水域：17-Apr-05, 05 年 4 月 17 日；Michelle and Peter Wong 江敏兒、黃理沛
3 breeding 繁殖羽：HK Southwestern waters 香港西南水域：17-Apr-05, 05 年 4 月 17 日；Michelle and Peter Wong 江敏兒、黃理沛
4 breeding 繁殖羽：HK Southwestern waters 香港西南水域：Apr-05, 05 年 4 月；Cherry Wong 黃卓研

| 春季過境遷徙鳥<br>Spring Passage Migrant | | 夏候鳥<br>Summer Visitor | | 秋季過境遷徙鳥<br>Autumn Passage Migrant | | 冬候鳥<br>Winter Visitor | |
|---|---|---|---|---|---|---|---|
| 1 | 2 | 3 | 4 | 5 | 6 | 7 | 8 | 9 | 10 | 11 | 12 | 常見月份 |

| 留鳥<br>Resident | 迷鳥<br>Vagrant | 偶見鳥<br>Occasional Visitor |
|---|---|---|

# 長尾賊鷗

(普) cháng wěi zéi ōu

體長 length：48-53cm

## Long-tailed Jaeger | *Stercorarius longicaudus*

其他名稱 Other names：Long-tailed Skua

[1]

小型的賊鷗。繁殖期頭頂黑色和背部灰褐色，耳羽附近淡黃色，下體淡色，有極長中央尾羽。非繁殖期上體深灰色，下體淺色。全身深灰褐色的深色型極為罕見。

Small and slim. Breeding plumage has black cap and yellow ear-coverts. Upperparts grey in colour. Underparts paler. Extremely long central tail feathers. Non-breeding plumage has dark grey upperparts, underparts paler. Overall brownish grey dark morph is extremely rare.

[1] breeding 繁殖羽：HK Southern waters 香港南面水域；17-Apr-05, 05 年 4 月 17 日；Michelle and Peter Wong 汪敏兒、黃理沛
[2] breeding 繁殖羽：HK Southern waters 香港南面水域；Apr-06, 06 年 4 月；Matthew and TH Kwan 關朗曦、關子凱
[3] breeding 繁殖羽：HK Southern waters 香港南面水域；17-Apr-05, 05 年 4 月 17 日；Michelle and Peter Wong 江敏兒、黃理沛
[4] breeding 繁殖羽：HK Southern waters 香港南面水域；17-Apr-05, 05 年 4 月 17 日；Michelle and Peter Wong 江敏兒、黃理沛
[5] breeding 繁殖羽：HK Southern waters 香港南面水域；Apr-08, 08 年 4 月；Andy Kwok 郭匯昌
[6] breeding 繁殖羽：HK Southern waters 香港南面水域；17-Apr-05, 05 年 4 月 17 日；Michelle and Peter Wong 江敏兒、黃理沛
[7] breeding 繁殖羽：HK Southern waters 香港南面水域；Apr-05, 05 年 4 月；Cherry Wong 黃卓研

| 春季過境遷徙鳥 Spring Passage Migrant | | | 夏候鳥 Summer Visitor | | | 秋季過境遷徙鳥 Autumn Passage Migrant | | | 冬候鳥 Winter Visitor | | |
|---|---|---|---|---|---|---|---|---|---|---|---|
| 1 | 2 | 3 | 4 | 5 | 6 | 7 | 8 | 9 | 10 | 11 | 12 |

常見月份

| 留鳥 Resident | 迷鳥 Vagrant | 偶見鳥 Occasional Visitor |
|---|---|---|

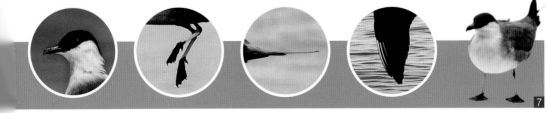

# 扁嘴海雀

(普)biǎn zuǐ hǎi què

體長 length：24-27cm

Ancient Murrelet | *Synthliboramphus antiquus*

其他名稱 Other names：Ancient Auk

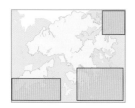

身型短小的小海鳥。上體灰色，頭部黑色，嘴小而淡粉紅色。下體白色。常潛入水中，能在水下長距離潛泳。

Small compact seabird. Grey upperparts with black head. Small, pale pink bill. White underparts. Dives continuously and may swim underwater for considerable distance.

1 non-breeding 非繁殖羽：Po Toi 蒲台：4-May-06, 06 年 5 月 4 日：Geoff Welch
2 non-breeding 非繁殖羽：Po Toi 蒲台：14-Apr-02, 02 年 4 月 14 日：Michelle and Peter Wong 江敏兒、黃理沛
3 non-breeding 非繁殖羽：HK Southern Waters, 香港南部水域：13 Apr-18, 18 年 4 月 13 日：Beetle Cheng 鄭諾銘
4 breeding 繁殖羽：Tap Mun 塔門：Yu Yat Tung 余日東
5 non-breeding 非繁殖羽：HK Southern Waters, 香港南部水域：13 Apr-18, 18 年 4 月 13 日：Beetle Cheng 鄭諾銘
6 Po Toi 蒲台：Apr-25-10, 10 年 4 月 27 日：Geoff Carey 賈知行

| 春季過境遷徙鳥<br>Spring Passage Migrant | | 夏候鳥<br>Summer Visitor | | 秋季過境遷徙鳥<br>Autumn Passage Migrant | | 冬候鳥<br>Winter Visitor | |
|---|---|---|---|---|---|---|---|

常見月份

| 1 | 2 | 3 | 4 | 5 | 6 | 7 | 8 | 9 | 10 | 11 | 12 |
|---|---|---|---|---|---|---|---|---|---|---|---|

| 留鳥<br>Resident | 迷鳥<br>Vagrant | 偶見鳥<br>Occasional Visitor |
|---|---|---|

# 冠海雀

(普) guān hǎi què

體長 length：26cm

Japanese Murrelet | *Synthliboramphus wumizusume*

嘴 短帶淺灰色。繁殖期具冠羽，喉黑色，眼後有白色斑。非繁殖時頭灰色，頭頂較深色，眼的前後白色。

Bill short and pale grey in color. In breeding plumage, it has crested head, black throat and white patch behind eyes. Non-breeding plumage has greyish head with darker cap, and white patches in front of and behind the eyes.

1 non-breeding 非繁殖羽：HK Southern waters 香港南面水域；5-May-07, 07 年 5 月 5 日：Lo Kar Man 盧嘉孟

| 春季過境遷徙鳥 Spring Passage Migrant | | | 夏候鳥 Summer Visitor | | | 秋季過境遷徙鳥 Autumn Passage Migrant | | | 冬候鳥 Winter Visitor | | |
|---|---|---|---|---|---|---|---|---|---|---|---|
| 1 | 2 | 3 | 4 | 5 | 6 | 7 | 8 | 9 | 10 | 11 | 12 |

常見月份

| 留鳥 Resident | | | | 迷鳥 Vagrant | | | | 偶見鳥 Occasional Visitor | | | |
|---|---|---|---|---|---|---|---|---|---|---|---|

# 白尾鸌

（普）bái wěi méng
（粵）鸌：音蒙

體長 length：70-82cm

## White-tailed Tropicbird │ *Phaethon lepturus*

**體**型略小但尾羽很長的白色海鳥（體長37厘米並不包括尾羽）。成鳥有白色長尾，身體主要為白色，眼周圍黑色，初級飛羽外側羽毛黑色，翼上覆羽具黑色條紋。幼鳥尾部並無延長，上體具有濃密褐色橫紋。

A smaller seabird with a long white tail streamers (length 37cm excluding the tail). White long tail and body. Dark eye patch. Dark outer primaries and black bar on upper wing coverts. Immature lacks tail streamers, upperparts are heavily spotted brown.

1 juvenile 幼鳥；Po Toi 蒲台；May-08, 08 年 5 月；Pippen Ho 何志剛
2 juvenile 幼鳥；HK Southeastern waters 香港東南水域；4-May-08, 08 年 5 月 4 日；Sung Yik Hei 宋亦希
3 juvenile 幼鳥；HK Southeastern waters 香港東南水域；4-May-08, 08 年 5 月 4 日；Sung Yik Hei 宋亦希
4 juvenile 幼鳥；HK Southeastern waters 香港東南水域；4-May-08, 08 年 5 月 4 日；Christina Chan 陳燕明
5 juvenile 幼鳥；HK Southeastern waters 香港東南水域；May-08, 08 年 5 月；Cherry Wong 黃卓研

| 春季過境遷徙鳥<br>Spring Passage Migrant | | | 夏候鳥<br>Summer Visitor | | | 秋季過境遷徙鳥<br>Autumn Passage Migrant | | | 冬候鳥<br>Winter Visitor | | |
|---|---|---|---|---|---|---|---|---|---|---|---|
| 常見月份 1 | 2 | 3 | 4 | 5 | 6 | 7 | 8 | 9 | 10 | 11 | 12 |
| | 留鳥<br>Resident | | | 迷鳥<br>Vagrant | | | 偶見鳥<br>Occasional Visitor | | | | |

# 黃嘴潛鳥

(普) huáng zuǐ qián niǎo

體長 length：76-91cm

## Yellow-billed Loon | *Gavia adamsii*

**頸** 長而粗壯，上體深灰褐色，臉、喉至頸前部白色，粗長的淡黃色嘴經常指向上方。繁殖羽頭頸黑色。常見於海中游泳。

Long neck thick and strong bill. Upperparts dark greyish brown. Face, throat to front part of neck white. The thick and long yellow bill usually points upwards. Breeding birds have a black head and neck. Pelagic.

[1] juvenile 幼鳥：Sai Kung 西貢：Jan-08, 08 年 1 月：Yu Yat Tung 余日東
[2] juvenile 幼鳥：Sai Kung 西貢：Jan-08, 08 年 1 月：Tam Yip Shing 譚業成
[3] juvenile 幼鳥：Sai Kung 西貢：Jan-08, 08 年 1 月：Pippen Ho 何志剛

| 春季過境遷徙鳥<br>Spring Passage Migrant | | | 夏候鳥<br>Summer Visitor | | | 秋季過境遷徙鳥<br>Autumn Passage Migrant | | | 冬候鳥<br>Winter Visitor | | |
|---|---|---|---|---|---|---|---|---|---|---|---|
| 1 | 2 | 3 | 4 | 5 | 6 | 7 | 8 | 9 | 10 | 11 | 12 |

常見月份

| 留鳥<br>Resident | 迷鳥<br>Vagrant | 偶見鳥<br>Occasional Visitor |
|---|---|---|

# 紅喉潛鳥

(普) hóng hóu qián niǎo

體長 length：53-69cm

## Red-throated Loon | *Gavia stellata*

其他名稱 Other Names：Red-throated Diver

大型潛鳥，嘴尖，末端向上翹。頭、後頸至上體褐灰色，上背羽緣有淡色紋。非繁殖鳥的頸側、面頰、喉和胸白色、無紋。繁殖鳥面頰和頸側灰色，有明顯紅喉。不時潛入水中捕捉食物。本港曾有一個冬季記錄。

Large diver, with pointed bill and the bill-tip slightly pointed upwards. Head, hind neck and upperparts brownish grey, with pale feather fringes on upperparts. The neck turn into grey during breeding season, with a prominent red throat. Often dives under water to catch food.

1 adult 成鳥；HK Southern Waters 香港南部水域；29 Mar-18, 18 年 3 月 29 日；Beetle Cheng 鄭諾銘
2 juvenile 幼鳥；Nam Chung 南涌；13-Mar-05, 05 年 3 月 13 日；Michelle and Peter Wong 江敏兒、黃理沛
3 juvenile 幼鳥；Nam Chung 南涌；13-Mar-05, 05 年 3 月 13 日；Michelle and Peter Wong 江敏兒、黃理沛
4 juvenile 幼鳥；Nam Chung 南涌；20-Feb-05, 05 年 2 月 20 日；Michelle and Peter Wong 江敏兒、黃理沛
5 juvenile 幼鳥；Nam Chung 南涌；20-Feb-05, 05 年 2 月 20 日；Michelle and Peter Wong 江敏兒、黃理沛
6 juvenile 幼鳥；Nam Chung 南涌；13-Mar-05, 05 年 3 月 13 日；Doris Chu 朱詠兒
7 juvenile 幼鳥；Nam Chung 南涌；20-Feb-05, 05 年 2 月 20 日；Michelle and Peter Wong 江敏兒、黃理沛
8 juvenile 幼鳥；Nam Chung 南涌；13-Mar-05, 05 年 3 月 13 日；Michelle and Peter Wong 江敏兒、黃理沛

| | 春季過境遷徙鳥 Spring Passage Migrant | | | 夏候鳥 Summer Visitor | | | 秋季過境遷徙鳥 Autumn Passage Migrant | | | 冬候鳥 Winter Visitor | | |
|---|---|---|---|---|---|---|---|---|---|---|---|---|
| 常見月份 | 1 | 2 | 3 | 4 | 5 | 6 | 7 | 8 | 9 | 10 | 11 | 12 |
| | 留鳥 Resident | | | | 迷鳥 Vagrant | | | | 偶見鳥 Occasional Visitor | | | |

# 短尾鸌

(普)duǎn wěi hú
(粵)鸌：音護

體長 length：40-45cm

## Short-tailed Shearwater | *Ardenna tenuirostris*

中等體型的深褐色海鳥。嘴較白額鸌短，頭部及眼部的顏色較深。翼下較淺色。腳灰黑色，飛行時伸出於尾後。

Middle-sized seabird with overall dark brown colour. Bill is shorter than Streaked Shearwater, with head and area around the eyes are much darker. Underwings paler. Darkish legs appear longer than the tail when pointed backward in flight.

1 HK Southeastern waters 香港東南水域；May-08, 08 年 5 月；Isaac Chan 陳家強
2 HK Southeastern waters 香港東南水域；May-06, 06 年 5 月；Jemi and John Holmes 孔思義、黃亞萍
3 HK Southeastern waters 香港東南水域；2-May-06, 06 年 5 月 2 日；Yu Yat Tung 余日東
4 Po Toi 蒲台；May-08, 08 年 5 月；Pippen Ho 何志剛
5 Po Toi 蒲台；May-08, 08 年 5 月；Pippen Ho 何志剛
6 HK Southern waters 香港南面水域；May-08, 08 年 5 月；Martin Hale 夏敖天

| | 春季過境遷徙鳥<br>Spring Passage Migrant | | | 夏候鳥<br>Summer Visitor | | | 秋季過境遷徙鳥<br>Autumn Passage Migrant | | | 冬候鳥<br>Winter Visitor | | |
|---|---|---|---|---|---|---|---|---|---|---|---|---|
| 常見月份 | 1 | 2 | 3 | 4 | 5 | 6 | 7 | 8 | 9 | 10 | 11 | 12 |
| | 留鳥<br>Resident | | | | 迷鳥<br>Vagrant | | | | 偶見鳥<br>Occasional Visitor | | | |

鸌科
Procellariidae

# 白額鸌 普 bái é hù
粵 鸌：音護

體長 length：45-52cm

Streaked Shearwater │ *Calonectris leucomelas*

1 2

大型海鳥，體型厚重。前額及面部明顯白色，嘴色淡，嘴尖灰色。頭頂、後頸、上翼、上體至尾部深褐色。喉、胸至下體白色，翼下白色，飛羽深褐色，翼下初級覆羽呈褐色條紋。腳粉紅色。

Large seabird, with heavy body. Prominent white forehead and cheek, pale bill with dusky tip. Upperparts dark brown. Underparts mainly white apart from brown flight feathers and brownish streaks on primary coverts. Pinkish legs.

1 HK Southeastern waters 香港東南水域；17-Apr-05, 05 年 4 月 17 日：Michelle and Peter Wong 江敏兒·黃理沛
2 HK Southeastern waters 香港東南水域；17-Apr-05, 05 年 4 月 17 日：Michelle and Peter Wong 江敏兒·黃理沛

| 春季過境遷徙鳥 Spring Passage Migrant | | 夏候鳥 Summer Visitor | | 秋季過境遷徙鳥 Autumn Passage Migrant | | 冬候鳥 Winter Visitor | |
|---|---|---|---|---|---|---|---|
| 常見月份 1 | 2 | 3 | 4 | 5 | 6 | 7 | 8 | 9 | 10 | 11 | 12 |

| 留鳥 Resident | 迷鳥 Vagrant | 偶見鳥 Occasional Visitor |
|---|---|---|

# 黑鸛

(普) hēi guàn
(粵) 鸛：音貫

體長 length：95-100cm

## Black Stork | *Ciconia nigra*

巨 型水鳥，嘴和腳紅色，成鳥上體黑色帶有光澤，下體白色。幼鳥上體褐色。飛行時頸和腳伸直，像十字架。冬天偶有單隻在后海灣一帶出現。

Huge waterbird. Bill and legs are red. Adult has glossy black upperparts and white belly. Juvenile is brown above. Flies with neck and legs outstretched resembling a cross. Singles sometimes found around Deep Bay area in winter.

1 juvenile 幼鳥：Mai Po 米埔：Nov-08, 08 年 11 月：Andy Li 李偉仁
2 juvenile 幼鳥：Mai Po 米埔：Tam Yiu Leung 譚耀良
3 juvenile 幼鳥：Mai Po 米埔：Oct-04, 04 年 10 月：Cherry Wong 黃卓研
4 juvenile 幼鳥：Mai Po 米埔：Oct-04, 04 年 10 月：Pippen Ho 何志剛
5 juvenile 幼鳥：Mai Po 米埔：Dickson Wong 黃志俊

| 春季過境遷徙鳥 Spring Passage Migrant | | | 夏候鳥 Summer Visitor | | | 秋季過境遷徙鳥 Autumn Passage Migrant | | | 冬候鳥 Winter Visitor | | |
|---|---|---|---|---|---|---|---|---|---|---|---|
| 1 | 2 | 3 | 4 | 5 | 6 | 7 | 8 | 9 | 10 | 11 | 12 |

常見月份

| 留鳥 Resident | 迷鳥 Vagrant | 偶見鳥 Occasional Visitor |
|---|---|---|

# 東方白鸛

(普) dōng fāng bái guàn
(粵) 鸛：音貫

體長 length：110-115cm

## Oriental Stork | *Ciconia boyciana*

其他名稱 Other names：Eastern White Stork, Oriental White Stork

[1]

巨型具有長嘴的水鳥，頸和腳長，嘴黑色，腳紅色，眼周皮膚紅色，除飛羽黑色外全身白色，飛行時頸和腳伸直，像十字架。通常只有一至兩隻同時出現。

Huge waterbird with long bill, long neck and long legs. Bill is black and legs are red. The orbital skin is red. White overall except for black flight feathers. Flies with neck and legs outstretched. One or two found in winter.

[1] non-breeding 非繁殖羽；Mai Po 米埔；Jan-04, 04 年 1 月；KK Hui 許光杰
[2] non-breeding 非繁殖羽；Mai Po 米埔；15-Jan-94, 94 年 1 月 15 日；Lo Kar Man 盧嘉孟
[3] non-breeding 非繁殖羽；Mai Po 米埔；Jan-04, 04 年 1 月；Michelle & Peter Wong 江敏兒、黃理沛
[4] non-breeding adult 非繁殖羽成鳥；Mai Po 米埔；Jan-04, 04 年 1 月；Henry Lui 呂德恆
[5] non-breeding adult 非繁殖羽成鳥；Mai Po 米埔；10-Jan-04, 04 年 1 月 10 日；Lo Kar Man 盧嘉孟
[6] Mai Po 米埔；Dec-90, 90 年 10 月；Jemi and John Holmes 孔思義、黃亞萍
[7] non-breeding adult 非繁殖羽成鳥；Mai Po 米埔；10-Jan-04, 04 年 1 月 10 日；Lo Kar Man 盧嘉孟

| | 春季過境遷徙鳥 Spring Passage Migrant | | | 夏候鳥 Summer Visitor | | | 秋季過境遷徙鳥 Autumn Passage Migrant | | | 冬候鳥 Winter Visitor | | |
|---|---|---|---|---|---|---|---|---|---|---|---|---|
| 常見月份 | 1 | 2 | 3 | 4 | 5 | 6 | 7 | 8 | 9 | 10 | 11 | 12 |
| | 留鳥 Resident | | | | 迷鳥 Vagrant | | | | 偶見鳥 Occasional Visitor | | | |

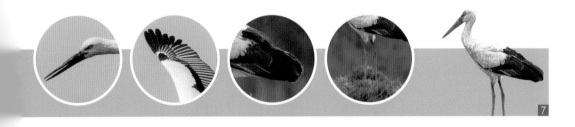

軍艦鳥科
**Fregatidae**

# 黑腹軍艦鳥
hēi fù jūn jiàn niǎo

體長 length：82-105cm

## Great Frigatebird | *Fregata minor*

其他名稱 Other names：小軍艦鳥

1

**翼**長而末端較尖。尾部分叉。雄性成鳥全身深色，雌性成鳥胸及脇部白色。幼鳥及亞成鳥頭部白色。腹部的白色部分可延伸至脇部。

Elongated and sharp-pointed wing; deeply forked tail. Male: totally dark. Female: pale throat, white breast and flank. Juvenile and sub-adult with white head. The whitish colour on upper breast can extend to the flank.

[1] immature, probably male 未成年鳥，很可能是雄鳥；HK Southeastern waters 香港東南水域；Apr-08, 08 年 4 月；Cherry Wong 黃卓研
[2] immature, probably male 未成年鳥，很可能是雄鳥；HK Southeastern waters 香港東南水域；Apr-08, 08 年 4 月；Sammy Sam and Winnie Wong 森美與雲妮
[3] immature, probably male 未成年鳥，很可能是雄鳥；HK Southeastern waters 香港東南水域；Apr-08, 08 年 4 月；Sammy Sam and Winnie Wong 森美與雲妮
[4] immature, probably male 未成年鳥，很可能是雄鳥；HK Southeastern waters 香港東南水域；27-Apr-08, 08 年 4 月 27 日；Sung Yik Hei 宋亦希

| | 春季過境遷徙鳥<br>Spring Passage Migrant | | | 夏候鳥<br>Summer Visitor | | | 秋季過境遷徙鳥<br>Autumn Passage Migrant | | | 冬候鳥<br>Winter Visitor | | |
|---|---|---|---|---|---|---|---|---|---|---|---|---|
| 常見月份 | 1 | 2 | 3 | 4 | 5 | 6 | 7 | 8 | 9 | 10 | 11 | 12 |
| | 留鳥<br>Resident | | | | 迷鳥<br>Vagrant | | | | 偶見鳥<br>Occasional Visitor | | | |

# 白斑軍艦鳥

(普) bái bān jūn jiàn niǎo

體長 length：66-81cm

## Lesser Frigatebird | *Fregata ariel*

擅 長飛行的海鳥，較一般軍艦鳥小。翼黑色，細而尖長，尾開叉很深。成鳥頭、上體和尾黑色。幼鳥頭部棕色，腹部及腋羽白色。

Small Frigatebird. Aerial seabird with slender and pointed black wings. Tail deeply forked. Adult has black glossy head, upperparts, belly and tail. Juveniles have brownish head, belly and axillaries white.

① immature 未成年鳥：Po Toi 蒲台：31-Mar-09, 09 年 3 月 31日：Geoff Welch
② immature 未成年鳥：Po Toi 蒲台：6-May-03, 03 年 5 月 6 日：Geoff Welch
③ immature 未成年鳥：Po Toi 蒲台：31-Mar-09, 09 年 3 月31日：Geoff Welch
④ immature 未成年鳥：Mai Po 米埔：Mar 01, 01 年 3 月：Jemi and John Holmes 孔思義、黃亞萍

| 春季過境遷徙鳥<br>Spring Passage Migrant | | | 夏候鳥<br>Summer Visitor | | | 秋季過境遷徙鳥<br>Autumn Passage Migrant | | | 冬候鳥<br>Winter Visitor | | |
|---|---|---|---|---|---|---|---|---|---|---|---|
| 1 | 2 | 3 | 4 | 5 | 6 | 7 | 8 | 9 | 10 | 11 | 12 |

常見月份

| 留鳥<br>Resident | 迷鳥<br>Vagrant | 偶見鳥<br>Occasional Visitor |
|---|---|---|

# 紅腳鰹鳥

(普) hóng jiǎo jiān niǎo
(粵) 鰹：音堅

體長 length：66-77cm

## Red-footed Booby | *Sula sula*

小型鰹鳥。成鳥腳紅色，嘴淡藍或沾紅。分淡色型和褐色型，前者大致白色而部分飛羽黑色，後者大致淡褐色而翼部及尾部褐色。未成年鳥大致褐色，嘴淡紅色。

Small-sized booby. Adult has red legs. Pale blue bill might tinged with red. Classified into pale and dark morph. Pale morph is mainly white with some black flight feathers. Dark morph is mainly pale brown with brown wings and tail. Sub-adult is mainly brown with pale pink bill.

[1] immature bird 未成年鳥；West Water 西部水域；Jun-10, 10 年 6 月；Jemi and John Holmes 孔思義・黃亞萍
[2] immature bird 未成年鳥；West Water 西部水域；Jun-10, 10 年 6 月；Jemi and John Holmes 孔思義・黃亞萍

| 春季過境遷徙鳥 Spring Passage Migrant | | | | 夏候鳥 Summer Visitor | | 秋季過境遷徙鳥 Autumn Passage Migrant | | 冬候鳥 Winter Visitor | | | |
|---|---|---|---|---|---|---|---|---|---|---|---|
| 1 | 2 | 3 | 4 | 5 | 6 | 7 | 8 | 9 | 10 | 11 | 12 |

常見月份

| 留鳥 Resident | 迷鳥 Vagrant | 偶見鳥 Occasional Visitor |
|---|---|---|

鰹鳥科
**Sulidae**

# 褐鰹鳥

(普)hè jiān niǎo
(粵)鰹：音堅

體長length：64-74cm

## Brown Booby | *Sula leucogaster*

出沒於太平洋及印度洋的熱帶至亞熱帶海域。身軀顏色配搭非常容易辨認。上半身深褐色，下腹白色，嘴及腳淡黃色，幼鳥的腹部為灰褐色。

Normally occurs in tropical and sub-tropical zones of Pacific and Indian Oceans. Very distinctive color combination: dark brown upperparts and white underparts, pale yellow bill and legs. Juvenile has greyish brown belly.

[1] adult 成鳥：Tam Yiu Leung 譚耀良
[2] adult 成鳥：Tam Yiu Leung 譚耀良
[3] Stanley 赤柱：25-Aug-07, 07 年 8 月 25 日：Michelle and Peter Wong 江敏兒、黃理沛

| | 春季過境遷徙鳥 Spring Passage Migrant | | 夏候鳥 Summer Visitor | | 秋季過境遷徙鳥 Autumn Passage Migrant | | 冬候鳥 Winter Visitor | | |
|---|---|---|---|---|---|---|---|---|---|

| 常見月份 | 1 | 2 | 3 | 4 | 5 | 6 | 7 | 8 | 9 | 10 | 11 | 12 |
|---|---|---|---|---|---|---|---|---|---|---|---|---|

| 留鳥 Resident | 迷鳥 Vagrant | 偶見鳥 Occasional Visitor |
|---|---|---|

# 綠背鸕鷀

(普) lù bèi lú cí
(粵) 鸕鷀：音盧慈

體長 length：92cm

## Japanese Cormorant | *Phalacrocorax capillatus*

其他名稱 Other names：暗綠背鸕鷀, Temminck's Cormorant

1

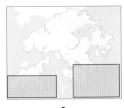

似 普通鸕鷀但兩翼及背部具偏綠色光澤。成鳥臉上黃色裸露皮膚在嘴裂形成一個尖角並包圍眼睛，頦和喉部白色範圍較普通鸕鷀大。尾較短，翼接近身體的後半部而非接近於中央。

Resembles a Great Cormorant. Metal green wings and upperparts. Yellow skin behind the bill forms a point at the gape and also goes around the eye. White colour on chin and throat is much larger than Greater Cormorant. Tail is shorter. Wings are in the rear half of the body rather than central as in Great Comorant.

[1] juvenile 幼鳥；Po Toi 蒲台；Apr-05, 05 年 4 月；Martin Hale 夏敖天
[2] non-breeding adult 非繁殖羽成鳥；Po Toi 蒲台；18-Jan-06, 06 年 1 月 18 日；Geoff Welch

2

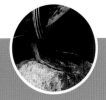

| 春季過境遷徙鳥<br>Spring Passage Migrant | | | 夏候鳥<br>Summer Visitor | | | 秋季過境遷徙鳥<br>Autumn Passage Migrant | | | 冬候鳥<br>Winter Visitor | | |
|---|---|---|---|---|---|---|---|---|---|---|---|
| 1 | 2 | 3 | 4 | 5 | 6 | 7 | 8 | 9 | 10 | 11 | 12 |

常見月份

| 留鳥<br>Resident | 迷鳥<br>Vagrant | 偶見鳥<br>Occasional Visitor |
|---|---|---|

# 普通鸕鷀

(普) pǔ tōng lú cí
(粵) 鸕鷀：音盧慈

其他名稱 Other names：鸕鷀, Cormorant

**Great Cormorant** | *Phalacrocorax carbo*

體長 length：80-100cm

1

大型水鳥。全身黑色，肩和翼帶有銅輝，嘴長，嘴端呈鈎狀，面部裸露皮膚黃色。幼鳥身體羽毛暗褐色，腹部白色。愛群居，飛行時頸部伸長，常集體列隊成一排排或 V 字型飛行，覓食時潛入魚塘或海中捕魚，休息時棲於樹上或岩石上。叫聲像鵝。

Large waterbird. Black all over with glossy wings and shoulders. Bill with hooked tip. The naked cheek is yellow. Juveniles have dark brown plumage and white belly. Gregarious. Flies with neck extended, often in lines or in V-shape. Dives into ponds or the sea to catch fish. Rests on trees or rocks. The voice is a guttural "gwaah, gwaah".

1 breeding 繁殖羽：Tsim Bei Tsui 尖鼻咀；Feb-07, 07 年 2 月；Sammy Sam and Winnie Wong 森美與雲妮
2 non-breeding adult 非繁殖羽成鳥；Mai Po 米埔；Dec-08, 08 年 12 月；James Lam 林文華
3 breeding 繁殖羽；Mai Po 米埔；Mar-08, 08 年 3 月；Kami Hui 許淑君
4 non-breeding adult 非繁殖羽成鳥；Nam Sang Wai 南生圍；Dec-04, 04 年 12 月；Jemi and John Holmes 孔思義、黃亞萍
5 breeding 繁殖羽；Mai Po 米埔；Mar-05, 05 年 3 月；Pippen Ho 何志剛
6 Mai Po 米埔；Dec-06, 06 年 12 月；Henry Lui 呂德恆
7 non-breeding adult 非繁殖羽成鳥；Mai Po 米埔；25-Jan-09, 09 年 1 月 25 日；Owen Chiang 深藍

| | 春季過境遷徙鳥<br>Spring Passage Migrant | | | 夏候鳥<br>Summer Visitor | | | 秋季過境遷徙鳥<br>Autumn Passage Migrant | | 冬候鳥<br>Winter Visitor | | |
|---|---|---|---|---|---|---|---|---|---|---|---|
| 常見月份 | 1 | 2 | 3 | 4 | 5 | 6 | 7 | 8 | 9 | 10 | 11 | 12 |
| | 留鳥<br>Resident | | | | 迷鳥<br>Vagrant | | | | 偶見鳥<br>Occasional Visitor | | |

# 黑頭白鹮

(普) hēi tóu bái huán
(粤) 鹮：音環

體長 length：65-76cm

## Black-headed Ibis | *Threskiornis melanocephalus*

其他名稱 Other names：White Ibis

1

♨

大型水鳥，嘴長而下彎，除頭、頸、向下彎的嘴、以及腳黑色外全身白色，飛行時翼下有粉紅色的裸露皮膚。冬天偶然有單隻或小群在后海灣一帶出沒。1999年11月後再無記錄。

Large-sized waterbird. White overall except for black head, neck, decurved bill and legs. A strip of pink naked skin is visible under the wings in flight. Single individuals or small groups used to be found around Deep Bay in winter. No record after Novemeber 1999.

1 immature 未成年鳥｜Mai Po 米埔｜22-Nov-92, 92 年 11 月 22 日｜Lo Kar Man 盧嘉孟

| 春季過境遷徙鳥<br>Spring Passage Migrant | | | 夏候鳥<br>Summer Visitor | | | 秋季過境遷徙鳥<br>Autumn Passage Migrant | | | 冬候鳥<br>Winter Visitor | | |
|---|---|---|---|---|---|---|---|---|---|---|---|
| 留鳥<br>Resident | | | | 迷鳥<br>Vagrant | | | | 偶見鳥<br>Occasional Visitor | | | |

常見月份

| 1 | 2 | 3 | 4 | 5 | 6 | 7 | 8 | 9 | 10 | 11 | 12 |
|---|---|---|---|---|---|---|---|---|---|---|---|

# 彩䴉

(普) cǎi huán
(粵) 䴉：音環

體長 length：48.5-66cm

Glossy Ibis | *Plegadis falcinellus*

中型帶金屬亮綠色的水鳥，嘴長而下彎。全身遠看像黑色，初級飛羽和翼背較黑。相當罕見，主要於后海灣一帶出沒。

Medium-sized waterbird. Dark glossy-green waterbird with long decurved bill. Appears black from a distance. Primaries and upper wings are nearly black. Very rare, found around Deep Bay area.

1 adult 成鳥：Tai Sang Wai 大生圍：Apr-19, 19 年 4 月：Henry Lui 呂德恆
2 adult 成鳥：Long Valley 塱原：Mar-19, 19 年 3 月：Kwok Tsz Ki 郭子祈
3 non-breeding 非繁殖羽：Mai Po 米埔：May-94, 95 年 5 月：Jemi and John Holmes 孔思義、黃亞萍

| 春季過境遷徙鳥 Spring Passage Migrant | | | 夏候鳥 Summer Visitor | | | 秋季過境遷徙鳥 Autumn Passage Migrant | | | 冬候鳥 Winter Visitor | | |
|---|---|---|---|---|---|---|---|---|---|---|---|
| 1 | 2 | 3 | 4 | 5 | 6 | 7 | 8 | 9 | 10 | 11 | 12 |
| 留鳥 Resident | | | | 迷鳥 Vagrant | | | | 偶見鳥 Occasional Visitor | | | |

常見月份

# 白琵鷺 (普) bǎi pí lù

體長 length：70-95cm

## Eurasian Spoonbill | *Platalea leucorodia*

大型水鳥具有像琵琶形狀的嘴，全身白色，腳和腳趾黑色，嘴長黑色，未成年鳥嘴粉紅色。冬天在后海灣一帶出現，經常有一、兩隻混在相對較矮的黑臉琵鷺群中。

Large waterbird with a long spoon-like bill. White overall. Legs and feet are black. Bill is dark in adults but pink in immatures. Occurs in ones and twos in Deep Bay area in winter, usually among the shorter but more numerous Black-faced Spoonbills.

[1] immature 未成年鳥；Mai Po 米埔；Nov-08, 08 年 11 月；Danny Ho 何國海
[2] immature 未成年鳥；Mai Po 米埔；Dec-08, 08 年 12 月；Joyce Tang 鄧玉蓮
[3] immature 未成年鳥；Mai Po 米埔；Nov-08, 08 年 11 月；Yue Pak Wai 余柏維
[4] immature 未成年鳥；Tin Shui Wai 天水圍；Nov-04, 04 年 11 月；Henry Lui 呂德恆
[5] immature 未成年鳥；Mai Po 米埔；Nov-06, 06 年 11 月；Bill Man 文權溢
[6] immature 未成年鳥；Mai Po 米埔；Dec-08, 08 年 12 月；Joyce Tang 鄧玉蓮
[7] immature 未成年鳥；Mai Po 米埔；Nov-06, 06 年 11 月；Allen Chan 陳志雄
[8] immature 未成年鳥；Mai Po 米埔；Dec-06, 06 年 12 月；Michelle and Peter Wong 江敏兒、黃理沛

| 春季過境遷徙鳥 Spring Passage Migrant | | | | 夏候鳥 Summer Visitor | | 秋季過境遷徙鳥 Autumn Passage Migrant | | | 冬候鳥 Winter Visitor | | |
|---|---|---|---|---|---|---|---|---|---|---|---|
| 留鳥 Resident | | | | 迷鳥 Vagrant | | | | 偶見鳥 Occasional Visitor | | | |

常見月份

| 1 | 2 | 3 | 4 | 5 | 6 | 7 | 8 | 9 | 10 | 11 | 12 |
|---|---|---|---|---|---|---|---|---|---|---|---|

# 黑臉琵鷺

(普)hēi liǎn pí lù

體長 length：60-78.5cm

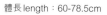

**Black-faced Spoonbill** | *Platalea minor*

大型水鳥，全身白色。嘴長黑色，像琵琶或茶匙。面部有黑色裸露皮膚，體型較白琵鷺小。飛行時頸和腳伸直。繁殖期蓬鬆的冠羽和胸部羽毛沾橙黃色。未成年鳥初級飛羽末端沾黑色，上嘴平滑帶粉紅色，年長鳥上嘴黑及有橫紋。常成群站在淺水中，覓食時將嘴伸入水中左右擺動，休息時常將嘴藏入翅膀中。

Large white waterbird with a long, spoon-like black bill. Distinguished from Eurasian Spoonbill by black facial skin and smaller size. Neck and legs extended in flight. Yellowish crest and breast in breeding plumage. Wing tips of immature are black. The upper surface of bill is smooth and pink in juveniles but black and rippled in older individuals. Usually in groups in shallow water. Sweeps bill sideways in water to catch fish. Buries bill under the wings when resting.

1 breeding 繁殖羽；Mai Po 米埔；22-Mar-04, 04 年 3 月 22 日；Lo Kar Man 盧嘉孟
2 non-breeding adult 非繁殖羽成鳥；Nam Sang Wai 南生圍；Dec-07, 07 年 12 月；Michelle and Peter Wong 江敏兒、黃理沛
3 immature 未成年鳥；Nam Sang Wai 南生圍；Dec-05, 05 年 12 月；Martin Hale 夏敖天
4 Feb-07, 07 年 2 月；Michelle and Peter Wong 江敏兒、黃理沛
5 immature 未成年鳥；Mai Po 米埔；May-08, 08 年 5 月；Cherry Wong 黃卓研
6 non-breeding adult 非繁殖羽成鳥；Mai Po 米埔；Oct-06, 06 年 10 月；James Lam 林文華
7 immature, banded 未成年鳥, 已旗標；Mai Po 米埔；Dec-08, 08 年 12 月；Joyce Tang 鄧玉蓮

| | 春季過境遷徙鳥<br>Spring Passage Migrant | | | 夏候鳥<br>Summer Visitor | | | 秋季過境遷徙鳥<br>Autumn Passage Migrant | | | 冬候鳥<br>Winter Visitor | | |
|---|---|---|---|---|---|---|---|---|---|---|---|---|
| 常見月份 | 1 | 2 | 3 | 4 | 5 | 6 | 7 | 8 | 9 | 10 | 11 | 12 |
| | 留鳥<br>Resident | | | | 迷鳥<br>Vagrant | | | | 偶見鳥<br>Occasional Visitor | | | |

# 大麻鳽

(普) dà má jiān
(粤) 鳽：音軒

體長 length：64-80cm

## Eurasian Bittern | *Botaurus stellaris*

大型鷺鳥，頭頂黑色，身體大致金褐色，身上有細長而濃密的深色斑點，上體有黑斑，飛行時飛羽深色。警戒時站立不動，頭、頸、嘴伸直朝天。常單隻出現，通常只在米埔的蘆葦叢間出沒。

Large ardeid. Cap is black. The body is mainly golden brown with heavy fine dark spots, and black patches on upperparts. Wings are dark in flight. Stands still with head, neck and bill held vertical when alarmed. Usually solitary, restricted to reedbeds of Mai Po.

1 adult 成鳥；Mai Po 米埔；Jan-09, 09 年 1 月；Simon Chan 陳志明
2 adult 成鳥；Mai Po 米埔；Jan-09, 09 年 1 月；Kinni Ho 何建業
3 adult 成鳥；Mai Po 米埔；Nov-06, 06 年 11 月；Michelle and Peter Wong 江敏兒、黃理沛
4 Mai Po 米埔；Jan-02, 02 年 1 月；Stanley Fok 霍棟豪
5 adult 成鳥；Mai Po 米埔；Jan-09, 09 年 1 月；Kinni Ho 何建業
6 adult 成鳥；Mai Po 米埔；Jan-09, 09 年 1 月；Kinni Ho 何建業

| 春季過境遷徙鳥<br>Spring Passage Migrant | | | 夏候鳥<br>Summer Visitor | | 秋季過境遷徙鳥<br>Autumn Passage Migrant | | | 冬候鳥<br>Winter Visitor | | |
|---|---|---|---|---|---|---|---|---|---|---|
| 常見月份 1 | 2 | 3 | 4 | 5 | 6 | 7 | 8 | 9 | 10 | 11 | 12 |

| 留鳥<br>Resident | 迷鳥<br>Vagrant | 偶見鳥<br>Occasional Visitor |
|---|---|---|

# 栗鳽

(普) lì jiān
(粵) 鳽：音軒

體長 length：45-49cm

## Japanese Night Heron | *Gorsachius goisagi*

[1] immature bird 未成年鳥

中 小型鳽。外貌似黑冠鳽，大致栗色，喉、胸至腹部淡色，有深色縱紋，但栗鳽的初級飛羽末端是栗色，頭頂深栗色。

Medium small-sized heron. Resembles Malay Night Heron. Mainly chestnut body. Paler throat, breast and belly with dark streaks. Distinguishes from Malayan Night Heron by chestnut tips to primaries and darker chestnut crown.

[1] immature bird 未成年鳥：Dec-14, 14 年 12 月：Jemi and John Holmes 孔思義、黃亞萍
[2] immature bird 未成年鳥：Sai Kung 西貢；Dec-14, 14 年 12 月：Kinni Ho 何建業
[3] immature bird 未成年鳥：Chung Yun Tak 鍾潤德
[4] immature bird 未成年鳥：Sai Kung 西貢；Dec-14, 14 年 12 月：Kinni Ho 何建業
[5] immature bird 未成年鳥：Dec-14, 14 年 12 月：Jemi and John Holmes 孔思義、黃亞萍

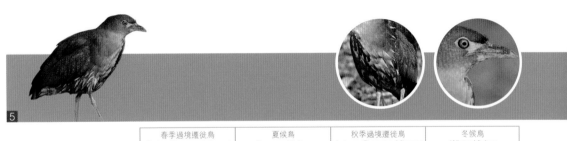

[5]

| 春季過境遷徙鳥<br>Spring Passage Migrant | | | 夏候鳥<br>Summer Visitor | | | 秋季過境遷徙鳥<br>Autumn Passage Migrant | | | 冬候鳥<br>Winter Visitor | | |
|---|---|---|---|---|---|---|---|---|---|---|---|
| 1 | 2 | 3 | 4 | 5 | 6 | 7 | 8 | 9 | 10 | 11 | 12 |

常見月份

| 留鳥<br>Resident | | | 迷鳥<br>Vagrant | | | 偶見鳥<br>Occasional Visitor | | |
|---|---|---|---|---|---|---|---|---|

# 黃葦鳽

(普) huáng wěi jiān
(粵) 鳽：音軒

體長 length：30-40cm

## Yellow Bittern | *Ixobrychus sinensis*

小型鷺鳥，嘴和腳黃色，成鳥頭頂和飛羽黑色，背褐色，翼上覆羽和腹部淺褐色，飛行時和黑色飛羽成對比。幼鳥和池鷺相似，但褐色羽毛上有深色粗縱紋。常單隻出現或小群遷徙。

Small ardeid with yellow bill and legs. Cap and flight feathers are black. Back is brown. Wing coverts and belly are light brown; contrasting with black wings in flight. Juvenile resembles Chinese Pond Heron but has dark broad stripes on brown feathers. Usually solitary but migrates in small groups.

1 adult male 雄成鳥；Po Toi 蒲台；May-07, 07 年 5 月；James Lam 林文華
2 adult male 雄成鳥；Po Toi 蒲台；May-08, 08 年 5 月；Chan Kai Wai 陳佳瑋
3 adult male 雄成鳥；Hong Kong Park 香港公園；Jul-08, 08 年 7 月；Ng I in Yau 吳瑋宥
4 adult male 雄成鳥；Mai Po 米埔；Aug-07, 07 年 8 月；Kinni Ho 何建業
5 1st summer male 第一年夏天雄鳥；Po Toi 蒲台；May-07, 07 年 5 月；Eling Lee 李佩玲
6 1st summer male 第一年夏天雄鳥；Po Toi 蒲台；May-07, 07 年 5 月；Allen Chan 陳志雄
7 1st summer male 第一年夏天雄鳥；Lamma Island 南丫島；May-04, 04 年 5 月；Cherry Wong 黃卓研

| 春季過境遷徙鳥 Spring Passage Migrant | | | 夏候鳥 Summer Visitor | | | 秋季過境遷徙鳥 Autumn Passage Migrant | | | 冬候鳥 Winter Visitor | | |
|---|---|---|---|---|---|---|---|---|---|---|---|
| 1 | 2 | 3 | 4 | 5 | 6 | 7 | 8 | 9 | 10 | 11 | 12 |
| 留鳥 Resident | | | | 迷鳥 Vagrant | | | | 偶見鳥 Occasional Visitor | | | |

常見月份

# 紫背葦�

(普) zǐ bèi wěi jiān
(粵) 鳽：音軒

體長 length：33-42cm

## Von Schrenck's Bittern | *Ixobrychus eurhythmus*

小 型鷺鳥。成鳥頭頂黑色，上體深褐近黑色，腹部淺褐色，中央有一深色紋伸延至喉部，飛行時飛羽深灰色。雌鳥和未成年鳥背和翼有明顯白色斑點，下體有褐色粗縱紋。

Small ardeid. Male has black cap and dark brown upperparts. Belly is light brown with a dark stripe in the middle extending to the throat. Flight feathers are grey in flight. Female and immature have white spots on back and wings, and thick brown stripes on underparts.

1 adult male 雄成鳥：Po Toi 蒲台；May-07, 07 年 5 月；Winnie Wong and Sammy Sam 森美與雲妮
2 adult male 雄成鳥：Po Toi 蒲台；May-07, 07 年 5 月；Michelle and Peter Wong 江敏兒、黃理沛
3 adult male 雄成鳥：Po Toi 蒲台；May 07, 07 年 6 月；Michelle and Peter Wong 江敏兒、黃理沛
4 adult male 雄成鳥：Po Toi 蒲台；May-08, 08 年 5 月；Cherry Wong 黃卓研
5 adult female 雌成鳥：Po Toi 蒲台；Mar-08, 08 年 3 月；Joyce Tang 鄧玉蓮
6 adult male 雄成鳥：Po Toi 蒲台；May-07, 07 年 5 月；Winnie Wong and Sammy Sam 森美與雲妮

| | 春季過境遷徙鳥<br>Spring Passage Migrant | | | 夏候鳥<br>Summer Visitor | | | 秋季過境遷徙鳥<br>Autumn Passage Migrant | | | 冬候鳥<br>Winter Visitor | | |
|---|---|---|---|---|---|---|---|---|---|---|---|---|
| 常見月份 | 1 | 2 | 3 | 4 | 5 | 6 | 7 | 8 | 9 | 10 | 11 | 12 |
| | 留鳥<br>Resident | | | | 迷鳥<br>Vagrant | | | | 偶見鳥<br>Occasional Visitor | | | |

# 栗葦鳽

(普) lì wěi jiān
(粵) 鳽：音軒

體長 length：40-41cm

## Cinnamon Bittern | *Ixobrychus cinnamomeus*

1

**小** 型鷺鳥。成鳥上體均勻紅褐色，飛行時飛羽紅褐色，可和其他
鳽鳥區別。雌鳥上體較暗，下體縱紋較多。幼鳥褐色羽毛上有
粗縱紋和黑白斑點。

Small ardeid with uniform reddish brown upperparts. Distinguished from other
bitterns by reddish brown wings in flight. Female is duller with more streaking
below. Juvenile thickly-striped with light brown spots on brown feathers.

[1] immature 未成年鳥；Long Valley 塱原；Oct-03, 03 年 10 月；Jemi and John Holmes 孔思義、黃亞萍
[2] juvenile 幼鳥；Mai Po 米埔；Sep-07, 07 年 9 月；Michelle and Peter Wong 江敏兒、黃理沛
[3] juvenile 幼鳥；Mai Po 米埔；Sep-07, 07 年 9 月；Allen Chan 陳志雄
[4] juvenile 幼鳥；Mai Po 米埔；Sep-07, 07 年 9 月；Michelle and Peter Wong 江敏兒、黃理沛
[5] juvenile 幼鳥；Mai Po 米埔；Oct-07, 07 年 10 月；Wallace Tse 謝鑑超
[6] adult female 雌成鳥；Mai Po 米埔；Jan-04, 04 年 1 月；Henry Lui 呂德恆
[7] juvenile 幼鳥；Mai Po 米埔；Sep-07, 07 年 9 月；Michelle and Peter Wong 江敏兒、黃理沛

| 春季過境遷徙鳥 Spring Passage Migrant | | | 夏候鳥 Summer Visitor | | | 秋季過境遷徙鳥 Autumn Passage Migrant | | | 冬候鳥 Winter Visitor | | |
|---|---|---|---|---|---|---|---|---|---|---|---|
| 1 | 2 | 3 | 4 | 5 | 6 | 7 | 8 | 9 | 10 | 11 | 12 |

常見月份

| 留鳥 Resident | 迷鳥 Vagrant | 偶見鳥 Occasional Visitor |
|---|---|---|

# 黑鳽

(普) hēi jiān
(粵) 鳽：音軒

體長 length：54-66cm

## Black Bittern | *Ixobrychus flavicollis*

中等體型，全身大致深藍灰色。嘴黃褐色，長而粗。頸側黃色，喉至胸白色具黑色及黃色縱紋。雌鳥上體顏色偏褐，下體白色較多。未成年鳥的上體及翼尖有淡黃褐色羽緣。腳深褐色。

Middle-sized heron, with overall dark greyish blue in color. Bill yellowish brown, long and thick. Sides of the neck yellow, white throat to breast with dark and yellow stripes. Female appears more brownish in upperparts, with fewer stripes on the underparts. Immature birds have yellowish brown feather fringes. Legs dark brown.

1 adult male 雄成鳥；Po Toi 蒲台；May-08, 08 年 5 月；Michelle and Peter Wong 江敏兒、黃理沛
2 adult 成鳥；Po Toi 蒲台；May-07, 07 年 5 月；Geoff Welch
3 adult female 雌成鳥；Po Toi 蒲台；Jul-08, 08 年 7 月；Chan Kai Wai 陳佳瑋
4 adult female 雌成鳥；Po Toi 蒲台；May-08, 08 年 5 月；Ken Fung 馮漢城
5 adult male 雄成鳥；Po Toi 蒲台；May-08, 08 年 5 月；Pippen Ho 何志剛
6 adult male 雄成鳥；Po Toi 蒲台；May-08, 08 年 5 月；Michelle and Peter Wong 江敏兒、黃理沛
7 adult female 雌成鳥；Po Toi 蒲台；May-08, 08 年 5 月；Michelle and Peter Wong 江敏兒、黃理沛

| | 春季過境遷徙鳥<br>Spring Passage Migrant | | 夏候鳥<br>Summer Visitor | | 秋季過境遷徙鳥<br>Autumn Passage Migrant | | 冬候鳥<br>Winter Visitor | | |
|---|---|---|---|---|---|---|---|---|---|
| 常見月份 | 1 | 2 | 3 | 4 | 5 | 6 | 7 | 8 | 9 | 10 | 11 | 12 |
| | 留鳥<br>Resident | | | 迷鳥<br>Vagrant | | | 偶見鳥<br>Occasional Visitor | | |

# 黑冠鳽

普 hēi guān jiān
粵 鳽：音軒

體長 length：45-49cm

## Malayan Night Heron | *Gorsachius melanolophus*

其他名稱 Other names：黑冠麻鷺、黑冠虎斑鳽

體 型粗壯，全身大致栗褐色、深灰色長冠羽蓋着枕部。嘴橄欖色，短而粗，上嘴向下彎。成鳥上體具深色斑點。頰和喉兩側栗褐色，有深色中央紋，並有由黑縱紋組成的中線，伸延至胸部，腹部至尾下覆羽有濃密斑紋。飛行時有明顯深色飛羽和白色翼尖。未成年鳥上體有淡色斑點和淡黃橫斑，下體較淡色。腳橄欖色。在夜間、拂曉及黃昏時份活動。

A stocky chestnut-coloured heron. Dark grey crown and long crest covering the nape. Bill olive in colour, short and thick, with upperbill pointed downward. Adult has dark spots on upperparts. Side of throat and cheeks are chestnut brown, with dark central stripes diffusing to general streaking on belly and vent. Prominent dark flight feathers with white tips. Immature is more spotted and streaked, with olive green legs. Nocturnal and crepuscular.

[1] adult 成鳥：On Po Village 安甫村：Sep-04, 04 年 9 月：Jemi and John Holmes 孔思義、黃亞萍
[2] juvenile 幼鳥：On Po Village 安甫村：Sep-04, 04 年 9 月：Michelle and Peter Wong 江敏兒、黃理沛
[3] juvenile 幼鳥：On Po Village 安甫村：Sep-04, 04 年 9 月：Michelle and Peter Wong 江敏兒、黃理沛
[4] adult 成鳥：On Po Village 安甫村：Sep-04, 04 年 9 月：Michelle and Peter Wong 江敏兒、黃理沛
[5] adult 成鳥：On Po Village 安甫村：Sep-04, 04 年 9 月：Jemi and John Holmes 孔思義、黃亞萍
[6] adult 成鳥：On Po Village 安甫村：Sep-04, 04 年 9 月：Jemi and John Holmes 孔思義、黃亞萍

| 春季過境遷徙鳥<br>Spring Passage Migrant | | 夏候鳥<br>Summer Visitor | | 秋季過境遷徙鳥<br>Autumn Passage Migrant | | 冬候鳥<br>Winter Visitor | |
|---|---|---|---|---|---|---|---|
| 常見月份 1 | 2 | 3 | 4 | 5 | 6 | 7 | 8 | 9 | 10 | 11 | 12 |
| 留鳥<br>Resident | | | 迷鳥<br>Vagrant | | | 偶見鳥<br>Occasional Visitor | |

# 夜鷺 （普）yè lù

體長 length：56-65cm

## Black-crowned Night Heron | *Nycticorax nycticorax*

其他名稱 Other names：Night Heron

1

**中**型鷺鳥，腳黃色，成鳥頭頂和背部深綠色，翼和腹部灰白二色，對比鮮明，繁殖期後枕有兩至三條白色細長飾羽。幼鳥和池鷺很相似，但褐色羽毛上有淺色斑點，飛行時雙翼褐色。

Medium-sized heron with yellow legs. Adult has dark green cap and back contrasting with grey and white wings and belly. Two to three narrow elongated nape feathers appeared during breeding season. Juvenile resembles Chinese Pond Heron but identified by light colour spots on brown feathers and brown wings in flight.

1 adult 成鳥；Hong Kong Park 香港公園；Sep-07, 07 年 9 月；Eling Lee 李佩玲
2 adult 成鳥；Hong Kong Park 香港公園；Sep-07, 07 年 9 月；Bill Man 文權溢
3 juvenile 幼鳥；Mai Po 米埔；Dec-06, 06 年 12 月；Henry Lui 呂德恆
4 adult 成鳥；Tseung Kwan O 將軍澳；Jul-07, 07 年 7 月；Winnie Wong and Sammy Sam 森美與雲妮
5 juvenile and adult 幼鳥和成鳥；Kowloon Park 九龍公園；Aug-08, 08 年 8 月；Aka Ho
6 2nd winter 第二年冬天；Kowloon Park 九龍公園；Jan-03, 03 年 1 月；Cherry Wong 黃卓研

| | 春季過境遷徙鳥<br>Spring Passage Migrant | | | 夏候鳥<br>Summer Visitor | | | 秋季過境遷徙鳥<br>Autumn Passage Migrant | | | 冬候鳥<br>Winter Visitor | | |
|---|---|---|---|---|---|---|---|---|---|---|---|---|
| 常見月份 | 1 | 2 | 3 | 4 | 5 | 6 | 7 | 8 | 9 | 10 | 11 | 12 |
| | 留鳥<br>Resident | | | 迷鳥<br>Vagrant | | | | 偶見鳥<br>Occasional Visitor | | | | |

# 綠鷺 ㊚ㄌㄨˋ ㄌㄨˋ

體長length：35-48cm

## Striated Heron | *Butorides striata*

其他名稱Other names：Little Green Heron

小型鷺鳥，全身灰色，和常見的夜鷺相似但較小，有淺灰色蓑羽。幼鳥上體灰褐色，下體有褐色細紋。愛在樹上築巢。常單隻出現，出沒在沼澤樹叢，冬天時會間中出現在林區溪澗。

Small grey green heron resembling the common Black-crowned Night Heron but smaller, with light grey plumages. Juvenile is dull brown above and heavily streaked below. Usually nests on trees, favours marshy shrubs. Sometimes found in woodland streams in winter.

1 adult 成鳥：Tai Po Kau 大埔滘；Jan-09, 09 年 1 月；Michelle and Peter Wong 江敏兒、黃理沛
2 adult 成鳥：Wun Yiu 碗窰；Cheung Ho Fai 張浩輝
3 juvenile 幼鳥：Mai Po 米埔；Aug-03, 03 年 8 月；Michelle and Peter Wong 江敏兒、黃理沛
4 adult 成鳥：Tai Po Kau 大埔滘；Jan-09, 09 年 1 月；Michelle and Peter Wong 江敏兒、黃理沛
5 1st winter 第一年冬天；Pui O 貝澳；Mar-08, 08 年 3 月；Wallace Tse 謝鑑超

| | 春季過境遷徙鳥<br>Spring Passage Migrant | | | 夏候鳥<br>Summer Visitor | | | 秋季過境遷徙鳥<br>Autumn Passage Migrant | | | 冬候鳥<br>Winter Visitor | | |
|---|---|---|---|---|---|---|---|---|---|---|---|---|
| 常見月份 | 1 | 2 | 3 | 4 | 5 | 6 | 7 | 8 | 9 | 10 | 11 | 12 |
| | | 留鳥<br>Resident | | | | 迷鳥<br>Vagrant | | | | 偶見鳥<br>Occasional Visitor | | |

# 池鷺  (普) chí lù

**Chinese Pond Heron** | *Ardeola bacchus*

體長 length：42-52cm

小 型鷺鳥，嘴長、腳長，腳黃色，靜立時除腹部白色外全身褐色。頭、頸、胸有縱紋，飛行時雙翼白色，十分明顯。繁殖期頭、頸變成深酒紅色，背灰藍色，腹部白色，對比鮮明。叫聲為「閣—閣」聲。

Small heron with long bill and long yellow legs. At rest, appears brown except the white belly, with streaks on head, neck and breast. White wings obvious in flight. In breeding season, head and neck turn wine red while the back becomes greyish blue, contrasting with the white belly. Voice is a "kok-kok".

[1] breeding 繁殖羽：Tsing Yi 青衣；Apr-06, 06 年 4 月：Cherry Wong 黃卓研
[2] non-breeding adult 非繁殖羽成鳥：Chinese University of HK 香港中文大學；Nov-08, 08 年 11 月：Christine and Samuel Ma 馬志榮、蔡美蓮
[3] non-breeding adult 非繁殖羽成鳥：Nam Sang Wai 南生圍；Dec-05, 05 年 12 月：Martin Hale 夏敖天
[4] non-breeding adult 非繁殖羽成鳥：Mai Po 米埔；Oct-05, 05 年 10 月：Henry Lui 呂德恆
[5] 1st winter 第一年冬天：Mai Po 米埔；Nov-06, 06 年 11 月：Andy Kwok 郭匯昌
[6] non-breeding adult 非繁殖羽成鳥：Nam Sang Wai 南生圍；Jan-05, 05 年 1 月：Henry Lui 呂德恆
[7] non-breeding adult 非繁殖羽成鳥：Chinese University of HK 香港中文大學；Dec-05, 05 年 12 月：Cherry Wong 黃卓研
[8] breeding 繁殖羽：Mai Po 米埔；Apr-03, 03 年 4 月：Henry Lui 呂德恆

| | 春季過境遷徙鳥<br>Spring Passage Migrant | | 夏候鳥<br>Summer Visitor | | 秋季過境遷徙鳥<br>Autumn Passage Migrant | | 冬候鳥<br>Winter Visitor | | |
|---|---|---|---|---|---|---|---|---|---|
| 常見月份 | 1 | 2 | 3 | 4 | 5 | 6 | 7 | 8 | 9 | 10 | 11 | 12 |
| | 留鳥<br>Resident | | | 迷鳥<br>Vagrant | | | 偶見鳥<br>Occasional Visitor | | | |

鷺科
Ardeidae

# 牛背鷺 <sup>普</sup>niú bèi lù

體長 length：46-56cm

## Eastern Cattle Egret | *Bubulcus coromandus*

小 型鷺鳥。全身白色，頭部渾圓，腳和腳趾全黑，嘴黃色。嘴和頸較大白鷺短，繁殖時頭、頸和背變成鮮明的橙色，有時腳、嘴和臉部的裸露皮膚發紅。愛群居，時常與黃牛或水牛為伴。

Small, round-headed white egret with black legs and feet. Bill is yellow. Bill and neck are shorter than Great Egret. In breeding season, head, neck and back turn bright orange and the legs, bill and facial skin also turn reddish yellow. Associates with cattle or buffalo. Gregarious.

1 breeding 繁殖羽：Pui O 貝澳；Mar-08, 08 年 3 月；Aka Ho
2 non-breeding adult 非繁殖羽成鳥：Po Toi 蒲台；May-08, 08 年 5 月；Cherry Wong 黃卓研
3 breeding 繁殖羽：Shui Hau 水口；Apr-04, 04 年 4 月；Henry Lui 呂德恒
4 breeding 繁殖羽：Pui O 貝澳；Apr-04, 04 年 4 月；Henry Lui 呂德恒
5 non-breeding adult 非繁殖羽成鳥：Long Valley 塱原；Oct-08, 08 年 10 月；Lau Chu Kwong 劉柱光
6 non-breeding adult 非繁殖羽成鳥：Mai Po 米埔；Oct-05, 05 年 10 月；Henry Lui 呂德恒
7 breeding 繁殖羽：Lut Chau 甩洲；Apr-06, 06 年 4 月；Jemi and John Holmes 孔思義、黃亞萍
8 non-breeding adult 非繁殖羽成鳥：Pui O 貝澳；Mar-08, 08 年 3 月；Isaac Chan 陳家強

| | 春季過境遷徙鳥<br>Spring Passage Migrant | | | 夏候鳥<br>Summer Visitor | | 秋季過境遷徙鳥<br>Autumn Passage Migrant | | | 冬候鳥<br>Winter Visitor | | |
|---|---|---|---|---|---|---|---|---|---|---|---|
| 常見月份 | 1 | 2 | 3 | 4 | 5 | 6 | 7 | 8 | 9 | 10 | 11 | 12 |
| | 留鳥<br>Resident | | | | 迷鳥<br>Vagrant | | | | 偶見鳥<br>Occasional Visitor | | |

# 蒼鷺

普 cāng lù

體長 length：90-98cm

Grey Heron | *Ardea cinerea*

1

大型水鳥。體羽主要是黑、灰、白三色配搭，成鳥羽色對比鮮明，幼鳥則主要是淡灰色。叫聲為「閣—閣」聲。覓食時愛長期呆立水中，當獵物游近時快速啄食。

Large waterbird covered in black, grey and white. Adults have contrasting plumage. Juveniles are mainly grey. Voice is a "kok-kok". Usually stands still in water for long periods, ready to seize prey swimming nearby.

1 non-breeding adult 非繁殖羽成鳥；Mai Po 米埔；Oct-08, 08 年 10 月；Cherry Wong 黃卓研
2 non-breeding adult 非繁殖羽成鳥；Nam Sang Wai 南生圍；Nov-06, 06 年 11 月；Henry Lui 呂德恆
3 juvenile 幼鳥；Pui O 貝澳；Mar-08, 08 年 3 月；Ken Fung 馮漢城
4 non-breeding adult 非繁殖羽成鳥；Mai Po 米埔；Dec-06, 06 年 12 月；Matthew and TH Kwan 關朗曦、關子凱
5 non-breeding adult 非繁殖羽成鳥；Mai Po 米埔；30-Oct-06, 06 年 10 月 30 日；Doris Chu 朱詠兒
6 breeding 繁殖羽；Nam Sang Wai 南生圍；Feb-08, 08 年 2 月；Ng Lin Yau 吳璉宥
7 juvenile 幼鳥；Mai Po 米埔；Oct-06, 06 年 10 月；Cherry Wong 黃卓研

| | 春季過境遷徙鳥<br>Spring Passage Migrant | | | 夏候鳥<br>Summer Visitor | | | 秋季過境遷徙鳥<br>Autumn Passage Migrant | | | 冬候鳥<br>Winter Visitor | | |
|---|---|---|---|---|---|---|---|---|---|---|---|---|
| 常見月份 | 1 | 2 | 3 | 4 | 5 | 6 | 7 | 8 | 9 | 10 | 11 | 12 |
| | 留鳥<br>Resident | | | 迷鳥<br>Vagrant | | | 偶見鳥<br>Occasional Visitor | | | | | |

# 草鷺 (普)cǎo lù

體長 length：78-90cm

## Purple Heron | *Ardea purpurea*

其他名稱 Other names：紫鷺

茻

大型水鳥。外貌和較常見的蒼鷺很相似，但羽毛較深色，主要是紅褐色和灰黑色的配搭。愛單隻躲在沼澤樹叢中。

Large waterbird resembling the more common Grey Heron but appears darker. Reddish brown and greyish black in colour. Usually solitary and secretive, taking cover in marshy undergrowth.

1 adult 成鳥；Mai Po 米埔；Oct-06, 06 年 10 月；Simon Chan 陳志明
2 juvenile 幼鳥；Mai Po 米埔；Nov-05, 05 年 11 月；Pippen Ho 何志剛
3 juvenile 幼鳥；Mai Po 米埔；Oct-06, 06 年 10 月；Frankie Chu 朱錦滿
4 juvenile 幼鳥；Mai Po 米埔；Jan-07, 07 年 1 月；Cherry Wong 黃卓研
5 adult 成鳥；Mai Po 米埔；Oct-07, 07 年 10 月；Kelvin Yam 任德政
6 juvenile 幼鳥；Mai Po 米埔；Nov-05, 05 年 11 月；Henry Lui 呂德恒
7 juvenile 幼鳥；Mai Po 米埔；Jan-07, 07 年 1 月；Cherry Wong 黃卓研

| 春季過境遷徙鳥 Spring Passage Migrant | | | 夏候鳥 Summer Visitor | | | 秋季過境遷徙鳥 Autumn Passage Migrant | | | 冬候鳥 Winter Visitor | | |
|---|---|---|---|---|---|---|---|---|---|---|---|
| 1 | 2 | 3 | 4 | 5 | 6 | 7 | 8 | 9 | 10 | 11 | 12 |
| | | 留鳥 Resident | | | | 迷鳥 Vagrant | | | 偶見鳥 Occasional Visitor | | |

常見月份

# 大白鷺 <sup>曾</sup>dà bái lù

體長 length：80-104cm

## Great Egret | *Ardea alba*

其他名稱 Other names：Great White Heron

大型鷺鳥。全身白色，頸部屈曲的幅度大得像「折斷了」般。腳全黑，嘴黃色。繁殖期嘴會變成黑色。背上長有薄紗般的白色飾羽。叫聲為「閣—閣」聲。

Large white egret. The long neck can appear "kinked" when bent. Feet are black. Bill is yellow, but turns black in breeding season, when gauze-like plumes appear on the back. Voice is a "kok-kok".

[1] breeding 繁殖羽；Mai Po 米埔；25-Feb-07, 07 年 2 月 25 日；Cheung Ho Fai 張浩輝
[2] breeding 繁殖羽；Mai Po 米埔；May-07, 07 年 5 月；Cherry Wong 黃卓研
[3] non-breeding adult 非繁殖羽成鳥；Nam Chung 南涌；Oct-07, 07 年 10 月；Aka Ho
[4] non-breeding adult 非繁殖羽成鳥；Mai Po 米埔；11-Sep-07, 07 年 9 月 11 日；Anita Lee 李雅婷
[5] breeding 繁殖羽；Mai Po 米埔；Apr-06, 06 年 4 月；Tam Yip Shing 譚業成
[6] non-breeding adult 非繁殖羽成鳥；Nam Sang Wai 南生圍；Nov-05, 05 年 11 月；Andy Kwok 郭匯昌
[7] breeding 繁殖羽；Mai Po 米埔；24-Feb-07, 07 年 2 月 24 日；Michelle and Peter Wong 江敏兒、黃理沛

| 春季過境遷徙鳥<br>Spring Passage Migrant | | | 夏候鳥<br>Summer Visitor | | | 秋季過境遷徙鳥<br>Autumn Passage Migrant | | | 冬候鳥<br>Winter Visitor | | |
|---|---|---|---|---|---|---|---|---|---|---|---|
| 1 | 2 | 3 | 4 | 5 | 6 | 7 | 8 | 9 | 10 | 11 | 12 |
| 留鳥<br>Resident | | | 迷鳥<br>Vagrant | | | | 偶見鳥<br>Occasional Visitor | | | | |

常見月份

# 中白鷺 （普）zhōng bái lù

## Intermediate Egret | *Ardea intermedia*

體長 length：56-72cm

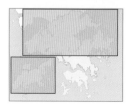

中型鷺鳥。全身白色，腳全黑，嘴黃色，嘴尖常沾黑色。外貌和較常見的大白鷺很相似，但嘴較短，嘴角不像大白鷺般延伸到眼後，頸較粗且屈曲幅度較小。

Medium-sized white egret. Black legs and yellow bill with dusky tip. Resembles the more common Great Egret, but bill shorter than Great Egret and gape line does not extend behind the eye. Neck is thicker and bends to a smaller extent.

[1] non-breeding adult 非繁殖羽成鳥；Mai Po 米埔；4-Dec-05, 05 年 12 月 4 日；Doris Chu 朱詠兒
[2] non brooding adult 非繁殖羽成鳥，Mai Po 米埔，Oct-05, 05 年 10 月；Henry Lui 呂德恒
[3] non-breeding adult moulting into breeding plumage 非繁殖羽成鳥轉換繁殖羽；Mai Po 米埔；Apr-08, 08 年 4 月；Michael Schmitz
[4] breeding 繁殖羽；Mai Po 米埔；Mar-07, 07 年 3 月；James Lam 林文華
[5] non-breeding adult moulting into breeding plumage 非繁殖羽成鳥轉換繁殖羽；Mai Po 米埔；28-Feb-05, 05 年 2 月 28 日；Wong Hok Sze 王學思
[6] non-breeding adult 非繁殖羽成鳥；Mai Po 米埔；Jan-08, 08 年 1 月；Pang Chun Chiu 彭俊超

| 春季過境遷徙鳥 Spring Passage Migrant | | | | 夏候鳥 Summer Visitor | | | 秋季過境遷徙鳥 Autumn Passage Migrant | | 冬候鳥 Winter Visitor | | |
|---|---|---|---|---|---|---|---|---|---|---|---|
| 1 | 2 | 3 | 4 | 5 | 6 | 7 | 8 | 9 | 10 | 11 | 12 |
| 留鳥 Resident | | | | 迷鳥 Vagrant | | | | 偶見鳥 Occasional Visitor | | | |

常見月份

鷺科
**Ardeidae**

# 小白鷺 <sub>普</sub> xiǎo bái lù

體長 length：55-65cm

## Little Egret | *Egretta garzetta*

1

中至小型鷺鳥，是香港最常見的鷺鳥。全身白色，外貌很像大白鷺，但體型較小，嘴全黑，腳黑，腳趾黃色。繁殖期頭上長出兩條細長的飾羽，面頰上黃綠色的裸露皮膚變得鮮明，甚至帶有紅色。叫聲為沙啞的「呀—呀」聲。有羽毛偏灰的深色型，但較為罕見。

Medium to small-sized egret. The most common and widespread egret in Hong Kong. White all over. Resembles Great Egret but is smaller. Bill and legs are black and toes are yellow. During breeding season, two narrow elongated nape feathers appear behind the head and yellowish green facial skin flushes bright red. Voice is a harsh "ah-ah". Dark morph has greyish plumage but very rare to see.

1 non-breeding adult moulting into breeding plumage 非繁殖羽成鳥轉換繁殖羽；Tai O 大澳；Dec-06, 06 年 12 月；Matthew and TH Kwan 關朗曦、關子凱
2 non-breeding adult 非繁殖羽成鳥；Mai Po 米埔；Apr-07, 07 年 4 月；James Lam 林文華
3 breeding 繁殖羽；Nam Chung 南涌；May-08, 08 年 5 月；Ken Fung 馮漢城
4 breeding 繁殖羽；Mai Po 米埔；Mar-08, 08 年 3 月；Kami Hui 許淑君
5 non-breeding adult 非繁殖羽成鳥；Tsuen Wan 荃灣；Jan-05, 05 年 1 月；Cherry Wong 黃卓研
6 breeding 繁殖羽；Mai Po 米埔；Jan-08, 08 年 1 月；Pang Chun Chiu 彭俊超
7 dark morph 深色型；Mai Po 米埔；29-Mar-09, 09 年 3 月 29 日；Kinni Ho 何建業

| | 春季過境遷徙鳥<br>Spring Passage Migrant | | | 夏候鳥<br>Summer Visitor | | | 秋季過境遷徙鳥<br>Autumn Passage Migrant | | | 冬候鳥<br>Winter Visitor | | |
|---|---|---|---|---|---|---|---|---|---|---|---|---|
| 常見月份 | 1 | 2 | 3 | 4 | 5 | 6 | 7 | 8 | 9 | 10 | 11 | 12 |
| | 留鳥<br>Resident | | | 迷鳥<br>Vagrant | | | | 偶見鳥<br>Occasional Visitor | | | | |

鷺科
Ardeidae

# 岩鷺 普yán lù

體長 length：58-66cm

Pacific Reef Heron | *Egretta sacra*

其他名稱 Other names：Eastern Reef Egret, Reef Egret

中至小型鷺鳥。全身深灰色，腳和嘴灰黃色。繁殖期嘴變成黃或橙黃色，後枕有小撮飾羽。常單隻或小群在岩石海岸出現，覓食時愛呆立水邊，當獵物游近時迅速啄食。

Medium to small-sized dark grey egret. Legs and bill are greyish yellow. During breeding season, bill and legs turn orange-yellow. A short tufty crest also appears on the hind nape. Usually single or in small groups along rocky shorelines. Usually stalks in shallow water, pecks at prey that swims nearby.

[1] non-breeding adult 非繁殖羽成鳥；Aberdeen 香港仔；Dec-07, 07 年 12 月；Raymond Cheng 鄭兆文
[2] 1st winter 第一年冬天；Po Toi 蒲台；Apr-07, 07 年 4 月；James Lam 林文華
[3] breeding 繁殖羽；Tai Shue Wan 大樹灣；Jan-09, 09 年 1 月；Aka Ho
[4] juvenile 幼鳥；HK Southwestern waters 香港西南水域；Aug-07, 07 年 8 月；Winnie Wong and Sammy Sam 森美與雲妮
[5] breeding 繁殖羽；HK Southwestern waters 香港西南水域；May-08, 08 年 5 月；Isaac Chan 陳家強
[6] non-breeding adult 非繁殖羽成鳥；Cheung Chau 長洲；Dec-04, 04 年 12 月；Henry Lui 呂德恆
[7] non-breeding adult 非繁殖羽成鳥；Stanley 赤柱；Sep-07, 07 年 9 月；Chan Kai Wai 陳佳瑋

| 春季過境遷徙鳥 Spring Passage Migrant | | | 夏候鳥 Summer Visitor | | | 秋季過境遷徙鳥 Autumn Passage Migrant | | | 冬候鳥 Winter Visitor | | |
|---|---|---|---|---|---|---|---|---|---|---|---|
| 1 | 2 | 3 | 4 | 5 | 6 | 7 | 8 | 9 | 10 | 11 | 12 |
| 留鳥 Resident | | | 迷鳥 Vagrant | | | | | 偶見鳥 Occasional Visitor | | | |

常見月份

# 黃嘴白鷺
(普)huáng zuǐ bái lù

體長 length：65-68cm

## Chinese Egret | *Egretta eulophotes*

其他名稱 Other names：Swinhoe's Egret

中型鷺鳥。全身白色，腳黑而腳趾黃色，外貌和常見的小白鷺相似，但嘴黃或橙黃色，嘴端尖細。繁殖期面頰上裸露皮膚變成鮮藍色，後枕有一束較長的飾羽。常單隻出現，覓食時十分活躍。

Medium-sized pure-white egret with distinctive black legs and yellow toes. Resembles the common Little Egret but bill is yellow to yellowish-orange pointed bill tip. During breeding season, the facial skin turns bright blue, shortish and shaggy plumage appears. Usually seen singly. Very active when feeding.

1 breeding 繁殖羽：Mai Po 米埔：19-May-06, 06 年 5 月 19 日；Owen Chiang 深藍
2 breeding 繁殖羽；Mai Po 米埔；Apr-08, 08 年 4 月；Andy Kwok 郭匯昌
3 breeding 繁殖羽；Mai Po 米埔；Apr-07, 07 年 4 月；Lee Kai Hong 李啟康
4 non-breeding 非繁殖羽；Mirs Bay 大鵬灣；Marcus Ho 何萬邦
5 breeding 繁殖羽；Mai Po 米埔；Apr-08, 08 年 4 月；Michael Schmitz

| | 春季過境遷徙鳥<br>Spring Passage Migrant | | | 夏候鳥<br>Summer Visitor | | | 秋季過境遷徙鳥<br>Autumn Passage Migrant | | | 冬候鳥<br>Winter Visitor | | |
|---|---|---|---|---|---|---|---|---|---|---|---|---|
| 常見月份 | 1 | 2 | 3 | 4 | 5 | 6 | 7 | 8 | 9 | 10 | 11 | 12 |
| | 留鳥<br>Resident | | | | 迷鳥<br>Vagrant | | | | 偶見鳥<br>Occasional Visitor | | | |

# 卷羽鵜鶘

(普) juǎn yǔ tí hú
(粵) 鵜鶘：音提湖

體長 length：160-180cm

## Dalmatian Pelican | *Pelecanus crispus*

巨型水鳥，嘴橙黃色，有很大的囊袋。除初級飛羽末端黑色外，全身大致灰白色，囊袋淡橙色。幼鳥嘴、囊袋、頭和上體均沾有灰色。愛群居。2006年之前每年冬天均有小群在后海灣一帶出現。

Huge waterbird with yellowish orange bill and big gular pouch. Looks greyish white except the black primaries. The gular pouch appears light orange in colour. The bill, gular pouch, head and upperparts of juveniles are tinted grey. Gregarious. Before 2006, a small group could be found in Deep Bay every winter.

1 non-breeding 非繁殖羽；Mai Po 米埔；26-Feb-94, 94 年 2 月 26 日；Lo Kar Man 盧嘉孟
2 breeding 繁殖羽；Mai Po 米埔；Mar-04, 04 年 3 月；Martin Hale 夏敖天
3 breeding 繁殖羽；Mai Po 米埔；Feb-08, 08 年 2 月；Kelvin Yam 任德政
4 juvenile 幼鳥；Mai Po 米埔；Karl Ng 伍耀成
5 breeding 繁殖羽；Mai Po 米埔；Samson So 蘇毅雄
6 breeding 繁殖羽；Nam Sang Wai 南生圍；Mar-05, 05 年 3 月；Pippen Ho 何志剛

| | 春季過境遷徙鳥<br>Spring Passage Migrant | | | 夏候鳥<br>Summer Visitor | | 秋季過境遷徙鳥<br>Autumn Passage Migrant | | | 冬候鳥<br>Winter Visitor | | |
|---|---|---|---|---|---|---|---|---|---|---|---|
| 常見月份 | 1 | 2 | 3 | 4 | 5 | 6 | 7 | 8 | 9 | 10 | 11 | 12 |
| | 留鳥<br>Resident | | | | 迷鳥<br>Vagrant | | | | 偶見鳥<br>Occasional Visitor | | |

# 鶚 (普) è

體長 length：55-58cm

## Western Osprey | *Pandion haliaetus*

其他名稱 Other names：魚鷹

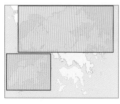

中型猛禽，全身大致褐、白兩色，頭、頸至下體白色，粗黑貫眼紋由嘴基伸延至枕部，胸有黑色橫帶，翼面近黑色。常停在淺水處的木樁上，覓食時會在水面上定點振翅，俯衝而下捕取獵物。

Medium-sized raptor. Mainly brown and white. Head, neck to underparts are white. Broad black band through eye to nape. Black breast band. Upper wings are nearly black. Usually perches on stakes in shallows. Hovers over water and plunge-dives feet-first for prey.

1 adult male 雄成鳥；Mai Po 米埔；Apr-07, 07 年 4 月；Kinni Ho 何建業
2 adult female 雌成鳥；Mai Po 米埔；Sep-04, 04 年 9 月；Henry Lui 呂德恆
3 HK Northeastern Waters 香港東北水域；Oct-03, 03 年 10 月；Henry Lui 呂德恆
4 adult 成鳥；Mai Po 米埔；Mar-08, 08 年 3 月；Cherry Wong 黃卓研
5 juvenile 幼鳥；Mai Po 米埔；Dec-06, 06 年 12 月；Cherry Wong 黃卓研
6 adult female 雌成鳥；Tsim Bei Tsui 尖鼻咀；Lee Hok Fei 李鶴飛

| 春季過境遷徙鳥 Spring Passage Migrant | | | | 夏候鳥 Summer Visitor | | | 秋季過境遷徙鳥 Autumn Passage Migrant | | 冬候鳥 Winter Visitor | | |
|---|---|---|---|---|---|---|---|---|---|---|---|
| 常見月份 | | | | | | | | | | | |
| 1 | 2 | 3 | 4 | 5 | 6 | 7 | 8 | 9 | 10 | 11 | 12 |
| 留鳥 Resident | | | | 迷鳥 Vagrant | | | | 偶見鳥 Occasional Visitor | | | |

# 鳳頭蜂鷹

(普) fèng tóu fēng yīng

體長 length：52-68cm

## Crested Honey Buzzard | *Pernis ptilorhynchus*

其他名稱 Other names：蜂鷹, Oriental Honey Buzzard

[1]

[2]

**體**型較普通鵟大，毛色多變，頭部伸得較長。翼長，次級飛羽特長，尾羽長而末端圓。飛行時，初級和次級飛羽翼底有很多橫紋，可以此辨認。淺色型鳳頭蜂鷹頸部有明顯深色縱紋，而深色型下體則為均勻的深褐色。

Bigger than Common Buzzard with variable plumage, and the head protrudes more. Wings are long, with budging secondaries. Tail feathers long and round-tipped. In flight, it shows many bars on the underside of primary and secondary feathers that is diagnostic. Pale morph individuals show some streaks on necks. Dark morphs have uniform dark brown underparts.

[1] juvenile intermediate morph 中間型幼鳥：Tai Po Kau 大埔滘：29-Nov-08, 08 年 11 月 29 日：Owen Chiang 深藍
[2] adult 成鳥：Nei Lak Shan 彌勒山：Feb-20, 20 年 2 月：Henry Lui 呂德恆
[3] juvenile dark morph 深色型幼鳥：Mai Po 米埔：21-Oct-08, 08 年 10 月 21 日：Sung Yik Hei 宋亦希

[3]

| 春季過境遷徙鳥<br>Spring Passage Migrant | | | 夏候鳥<br>Summer Visitor | | | 秋季過境遷徙鳥<br>Autumn Passage Migrant | | | 冬候鳥<br>Winter Visitor | | |
|---|---|---|---|---|---|---|---|---|---|---|---|
| 1 | 2 | 3 | 4 | 5 | 6 | 7 | 8 | 9 | 10 | 11 | 12 |
| 留鳥<br>Resident | | | | 迷鳥<br>Vagrant | | | | 偶見鳥<br>Occasional Visitor | | | |

常見月份

鷹科
**Accipitridae**

# 黑翅鳶

(普) hēi chì yuān
(粵) 鳶：音淵

體長 length：30-37cm

## Black-winged Kite | *Elanus caeruleus*

其他名稱 Other names：Black-shouldered Kite

全身黑、白、灰色配搭的小型猛禽，頭較闊大，眼紅色，身體羽毛淡灰或白色，肩膀明顯黑色。飛行時可見翼底初級飛羽黑色，時常定點振翅，滑翔時雙翼上舉作深 V 形，尾部方正。

Small raptor, overall black, white and grey. Head is wide and large. Eyes are red. Body is light grey or white. Shoulder patch large and black. Black underwing primaries can be seen in flight. Hovers in air; steep V-shaped wings when gliding. Square tail.

1 adult 成鳥：Tsim Bei Tsui 尖鼻咀；24-Oct-04, 04 年 10 月 24 日；Michelle and Peter Wong 江敏兒、黃理沛
2 juvenile 幼鳥：Martin Hale 夏敖天
3 juvenile 幼鳥：Martin Hale 夏敖天
4 adult 成鳥：Tai O 大澳；Oct-07, 07 年 10 月；Ng Lin Yau 吳璉宥
5 juvenile 幼鳥：Mai Po 米埔；Owen Chiang 深藍
6 Mai Po 米埔；Aug-04, 04 年 8 月；Henry Lui 呂德恆

| | 春季過境遷徙鳥<br>Spring Passage Migrant | | | 夏候鳥<br>Summer Visitor | | | 秋季過境遷徙鳥<br>Autumn Passage Migrant | | | 冬候鳥<br>Winter Visitor | | |
|---|---|---|---|---|---|---|---|---|---|---|---|---|
| 常見月份 | 1 | 2 | 3 | 4 | 5 | 6 | 7 | 8 | 9 | 10 | 11 | 12 |
| | 留鳥<br>Resident | | | | 迷鳥<br>Vagrant | | | | 偶見鳥<br>Occasional Visitor | | | |

鷹科
Accipitridae

# 黑冠鵑隼

(普) hēi guān juān sǔn
(粵) 隼:音準

體長 length:23-35cm

Black Baza | *Aviceda leuphotes*

小型猛禽,全身大致黑、白兩色,有長冠羽;停棲時可見頭、頸、翼及尾黑色,胸部有寬黑橫帶,腹部有黑和黃褐色相間細橫紋。飛行時翼寬而圓,常在樹頂間短距離飛行。常單隻或四至五隻小群出現,遷徙時一群可達二、三十隻。

Small raptor. Mainly black and white, with long crest. Head, neck, wings and tail are black. Broad black band on breast and narrow buff bands on belly visible when perched. Wings broad and round in flight. Usually flies among tree tops for short distances. Singles or family parties of four to five. Migrating flocks may number anoumd 20 to 30.

[1] adult 成鳥:Sai Kung 西貢:Jul-05, 05 年 7 月:Pippen Ho 何志剛
[2] adult 成鳥:Sai Kung 西貢:9-Jul-05, 05 年 7 月 9 日:Michelle and Peter Wong 江敏兒、黃理沛
[3] adult 成鳥:Sai Kung 西貢:9-Jul-05, 05 年 7 月 9 日:Michelle and Peter Wong 江敏兒、黃理沛
[4] adult 成鳥:Sai Kung 西貢:Sai Kung 西貢:Jul 06, 06 年 7 月:Kinni Ho 何建業

| 春季過境遷徙鳥<br>Spring Passage Migrant | | | 夏候鳥<br>Summer Visitor | | | 秋季過境遷徙鳥<br>Autumn Passage Migrant | | | 冬候鳥<br>Winter Visitor | | |
|---|---|---|---|---|---|---|---|---|---|---|---|
| 1 | 2 | 3 | 4 | 5 | 6 | 7 | 8 | 9 | 10 | 11 | 12 |
| 留鳥<br>Resident | | | | 迷鳥<br>Vagrant | | | | 偶見鳥<br>Occasional Visitor | | | |

常見月份

# 禿鷲
(普) tū jiù
(圖) 鷲：音就

體長 length：100-120cm

Cinereous Vulture | *Aegypius monachus*

香 港體型最大的猛禽，羽色為近黑的深褐色。成鳥的後枕光禿，嘴基蠟膜粉紅色。飛行時雙翼平放，相當寬長，翼端飛羽分開像手指，翼內緣呈鋸齒狀，尾呈楔狀。

The largest raptor in Hong Kong. Dark brown (appears black) overall. Adult has bare nape and pink cere at the base of the bill. Extremely long broad wings held flat in flight, with fingers and saw-toothed trailing edges. Tail is wedge-shaped.

1 juvenile 幼鳥；Mai Po 米埔；Geoff Carey 賈知行
2 immature 未成年鳥；Mai Po 米埔；Geoff Carey 賈知行
3 immature 未成年鳥；Mai Po 米埔；Geoff Carey 賈知行
4 immature 未成年鳥；Ma Tso Lung 馬草壟；Jemi and John Holmes 孔思義．黃亞萍

| 春季過境遷徙鳥<br>Spring Passage Migrant | | | 夏候鳥<br>Summer Visitor | | | 秋季過境遷徙鳥<br>Autumn Passage Migrant | | | 冬候鳥<br>Winter Visitor | | |
|---|---|---|---|---|---|---|---|---|---|---|---|
| 1 | 2 | 3 | 4 | 5 | 6 | 7 | 8 | 9 | 10 | 11 | 12 |
| 留鳥<br>Resident | | | | 迷鳥<br>Vagrant | | | | 偶見鳥<br>Occasional Visitor | | | |

常見月份

# 蛇鵰
(普) shé diāo
(粵) 鵰：音丟

體長 length：50-74cm

Crested Serpent Eagle | *Spilornis cheela*

[1]

中型猛禽，飛行時飛羽及尾端有明顯的淺色橫帶。幼鳥羽色較淡，翅膀佈滿黑點和橫紋。站立時可見獨特的冠羽，後枕和腹部有白色斑點。盤旋時翅膀呈淺V型，高空盤旋時發出多節嘯聲，遠處也可以聽到。

Medium-sized raptor. Obvious white bands on wings and tail edge during flight. Juvenile much paler in colour, with spots and strips on wings. Distinctive crest with white spots on nape when perched. V-shaped profile in flight. Whistling calls can be heard from a long distance.

[1] adult 成鳥；Sai Kung 西貢；Dec-08, 08 年 12 月；Andy Kwok 郭匯昌
[2] adult 成鳥；Sai Kung 西貢；Dec-08, 08 年 12 月；Andy Kwok 郭匯昌
[3] Siu Lam 小欖；Jimmy Chim 詹玉明
[4] adult 成鳥；Tai Po Kau 大埔滘；Apr-08, 08 年 4 月；Ng Lin Yau 吳璉宥
[5] adult 成鳥；Siu Lam 小欖；Jimmy Chim 詹玉明
[6] adult 成鳥；Long Valley 塱原；Mar-07, 07 年 3 月；Law Kam Man 羅錦文

| 春季過境遷徙鳥<br>Spring Passage Migrant | | | 夏候鳥<br>Summer Visitor | | | 秋季過境遷徙鳥<br>Autumn Passage Migrant | | | 冬候鳥<br>Winter Visitor | | |
|---|---|---|---|---|---|---|---|---|---|---|---|

常見月份

| 1 | 2 | 3 | 4 | 5 | 6 | 7 | 8 | 9 | 10 | 11 | 12 |
|---|---|---|---|---|---|---|---|---|---|---|---|

| 留鳥<br>Resident | | | 迷鳥<br>Vagrant | | | 偶見鳥<br>Occasional Visitor | | |
|---|---|---|---|---|---|---|---|---|

# 烏鵰

(普) wū diāo
(粵) 鵰：音丟

體長 length：59-71cm

## Greater Spotted Eagle | *Clanga clanga*

大型猛禽。成鳥全身深褐色，爪及嘴黃色。幼鳥全身有白色斑點。相對白肩鵰，飛行中的烏鵰的翅膀較寬，尾部較短圓，頸部較短，頭部看來較小，飛羽不時有殘缺。

Large-sized raptor. Adult is dark brown in colour, with yellow bill and claws. Juvenile has white spots. In flight, it appears smaller than Imperial Eagle but has broader wings and shorter rounded tail. Neck is shorter and head is smaller. Sometimes seen with broken feathers.

[1] adult 成鳥；Mai Po 米埔；Oct-06, 06 年 10 月；Frankie Chu 朱錦滿
[2] juvenile 幼鳥；Mai Po 米埔；Dec-06, 06 年 12 月；Henry Lui 呂德恆
[3] juvenile pale morph 淺色型幼鳥；Mai Po 米埔；27-Dec-04, 04 年 12 月 27 日；Michelle and Peter Wong 江敏兒、黃理沛
[4] adult 成鳥；Mai Po 米埔；30-Oct-06, 06 年 10 月 30 日；Michelle and Peter Wong 江敏兒、黃理沛

| 春季過境遷徙鳥 Spring Passage Migrant | | | | 夏候鳥 Summer Visitor | | 秋季過境遷徙鳥 Autumn Passage Migrant | | | 冬候鳥 Winter Visitor | | |
|---|---|---|---|---|---|---|---|---|---|---|---|
| 1 | 2 | 3 | 4 | 5 | 6 | 7 | 8 | 9 | 10 | 11 | 12 |

常見月份

| 留鳥 Resident | 迷鳥 Vagrant | 偶見鳥 Occasional Visitor |
|---|---|---|

# 草原鵰

(普) cǎo yuán diāo
(粵) 鵰：音丟

體長length：60-81cm

Steppe Eagle | *Aquila nipalensis*

動 作緩慢的鵰類，全身深褐色，或會與體型稍大的白肩鵰混淆。幼鳥淡棕褐色，斑點較白肩鵰少。像白肩鵰一樣，草原鵰翼下大覆羽末端形成明顯白色橫帶，而翼後邊緣及尾端淺色橫帶比白肩鵰更為明顯。

Appears slow-moving. Most likely to be confused with Imperial Eagle, but is slightly smaller and adult is dark brown. Juvenile Steppe is tawny-brown, less spotted than Imperial. Like Imperial Eagle, Stepple Eagle has white tips to greater coverts, but it forms more prominent white line. On Steppe pale trailing edge to wing and tail-band are more distinct than Imperial.

[1] 1st winter 第一年冬天：Mai Po 米埔；Dec-08, 08 年 12 月：Daniel CK Chan 陳志光
[2] Mai Po 米埔；1-Apr-09, 09 年 4 月 1 日：Kinni Ho 何建業
[3] Mai Po 米埔；1-Apr-09, 09 年 4 月 1 日：Cherry Wong 黃卓研
[4] Mai Po 米埔；1-Apr-09, 09 年 4 月 1 日：Cherry Wong 黃卓研

| 春季過境遷徙鳥 Spring Passage Migrant | | | 夏候鳥 Summer Visitor | | | 秋季過境遷徙鳥 Autumn Passage Migrant | | | 冬候鳥 Winter Visitor | | |
|---|---|---|---|---|---|---|---|---|---|---|---|
| 1 | 2 | 3 | 4 | 5 | 6 | 7 | 8 | 9 | 10 | 11 | 12 |
| 留鳥 Resident | | | | 迷鳥 Vagrant | | | | 偶見鳥 Occasional Visitor | | | |

常見月份

# 白肩鵰

(普) bái jiān diāo
(粵) 鵰：音丟

體長 length：72-84cm

## Eastern Imperial Eagle | *Aquila heliaca*

大型猛禽。成鳥體羽深褐色，頭後至枕部淡褐色。幼鳥全身淡褐色，初級飛羽、次級飛羽及尾羽黑色。飛行時，頸長及頭部大而突出，尾羽長而近似長方形。

Large-sized raptor. Adult has dark brown body with pale brown hind crown and nape. Juvenile has light brown body, with black primaries, secondaries, and tail. In flight it shows long neck and big head, with long rectangular tail.

1 adult 成鳥；Mai Po 米埔；Nov-06, 06 年 11 月；Henry Lui 呂德恆
2 juvenile 幼鳥；Mai Po 米埔；20-Dec-03, 03 年 12 月 20 日；Michelle and Peter Wong 江敏兒‧黃理沛
3 juvenile 幼鳥；Mai Po 米埔；Dec-06, 06 年 12 月；Kinni Ho 何建業
4 juvenile 幼鳥；Mai Po 米埔；Dec-06, 06 年 12 月；Cherry Wong 黃卓研
5 adult 成鳥；Mai Po 米埔；25-Dec-06, 06 年 12 月 25 日；Michelle and Peter Wong 江敏兒‧黃理沛

| 春季過境遷徙鳥<br>Spring Passage Migrant | | | | 夏候鳥<br>Summer Visitor | | 秋季過境遷徙鳥<br>Autumn Passage Migrant | | 冬候鳥<br>Winter Visitor | | | |
|---|---|---|---|---|---|---|---|---|---|---|---|
| 1 | 2 | 3 | 4 | 5 | 6 | 7 | 8 | 9 | 10 | 11 | 12 |
| 留鳥<br>Resident | | | | 迷鳥<br>Vagrant | | | | 偶見鳥<br>Occasional Visitor | | | |

常見月份

# 白腹隼鵰

(普) bái fù sǔn diāo
(粵) 隼鵰：音準丟

體長 length：55-67cm

## Bonelli's Eagle | *Aquila fasciata*

其他名稱 Other names：白腹山鵰, Bonelli's Hawk Eagle

中型猛禽。成鳥上體黑褐色，下體灰白而有縱紋；初級和次級飛羽黑色，翼底有白斑。尾羽淡灰白色，末端有深色寬橫帶。幼鳥上身和翼下覆羽淡褐色，飛羽和尾下覆羽淡灰。

Medium-sized raptor. Adult has dark brown upperparts, greyish white lower-parts with stripes. Black primaries and secondary feathers, with white patches on underwings. Long, pale tail with wide terminal dark band. Juvenile has pale brown body and wing lining, light grey flight feathers and undertail.

[1] juvenile moulting into adult plumage 幼鳥轉換成羽；Mai Po 米埔；30-Sep-07, 07 年 9 月 30 日；Michelle and Peter Wong 江敏兒、黃理沛
[2] adult 成鳥；Fei Ngo Shan 飛鵝山；Aug-08, 08 年 8 月；Ng Lin Yau 吳璉宥
[3] juvenile 幼鳥；Mai Po 米埔；KK Hui 許光杰
[4] adult 成鳥；Mai Po 米埔；Captain Wong 黃倫昌
[5] Mai Po 米埔；Cheung Ho Fai 張浩輝

| 春季過境遷徙鳥<br>Spring Passage Migrant | | | 夏候鳥<br>Summer Visitor | | | 秋季過境遷徙鳥<br>Autumn Passage Migrant | | | 冬候鳥<br>Winter Visitor | | |
|---|---|---|---|---|---|---|---|---|---|---|---|
| 1 | 2 | 3 | 4 | 5 | 6 | 7 | 8 | 9 | 10 | 11 | 12 |

常見月份

| 留鳥<br>Resident | 迷鳥<br>Vagrant | 偶見鳥<br>Occasional Visitor |
|---|---|---|

鷹科
**Accipitridae**

# 鳳頭鷹 ⓅＬ fèng tóu yīng

體長 length：30-46cm

Crested Goshawk | *Accipiter trivirgatus*

其他名稱 Other names：鳳頭蒼鷹, Asian Crested Goshawk

1

中型猛禽。雄鳥深灰褐色，胸、腹有褐色橫紋。飛行時翅膀短圓，次級飛羽較長成圓弧狀，白色尾下覆羽突出，有時會抖動雙翼展示。雌鳥和幼鳥羽色偏褐。

Medium-sized raptor. Adult is dark greyish brown with brown stripes on breast and belly. Obviously round wings in flight, bulging secondaries and prominent white undertail coverts, with occasional quivering display. Female and juvenile are brownish in colour.

1 adult 成鳥：Hong Kong Park 香港公園；Dec-03, 03 年 12 月；Isaac Chan 陳家強
2 juvenile 幼鳥：Kwai Chung 葵涌；Owen Chiang 深藍
3 juvenile 幼鳥：Tai Po Kau 大埔滘；Jun-07, 07 年 6 月；Martin Hale 夏敖天
4 adult 成鳥：Siu Lek Yuen 小瀝源；May-08, 08 年 5 月；Ken Fung 馮漢城
5 adult female 雌成鳥：Cheung Chau 長洲；Mar-06, 06 年 3 月；Henry Lui 呂德恆
6 juvenile 幼鳥：Kwai Chung 葵涌；Jul-07, 07 年 7 月；Sammy Sam and Winnie Wong 森美與雲妮
7 adult male 雄成鳥：Hong Kong Park 香港公園；Dec-03, 03 年 12 月；Isaac Chan 陳家強

| | 春季過境遷徙鳥<br>Spring Passage Migrant | | | 夏候鳥<br>Summer Visitor | | | 秋季過境遷徙鳥<br>Autumn Passage Migrant | | | 冬候鳥<br>Winter Visitor | | |
|---|---|---|---|---|---|---|---|---|---|---|---|---|
| 常見月份 | 1 | 2 | 3 | 4 | 5 | 6 | 7 | 8 | 9 | 10 | 11 | 12 |
| | 留鳥<br>Resident | | | | 迷鳥<br>Vagrant | | | | 偶見鳥<br>Occasional Visitor | | | |

# 赤腹鷹 <sup>普</sup> chì fù yīng

體長 length：23-35cm

## Chinese Sparrowhawk | *Accipiter soloensis*

[1]

[2]

小型猛禽。成鳥頭至背部灰色，嘴黑色，蠟膜和腳橙黃色，胸及脇部帶淡紅褐色，尾部深灰色。飛行時翼底白色，翼尖黑色。雄性成鳥虹膜深紅，雌性及未成年鳥虹膜黃色。幼鳥背部毛色偏褐，胸及腹部有褐色橫紋。

Small raptor. Adult has grey head and upperparts, black bill, yellowish orange cere and legs. Breast and flanks appear pale reddish brown, tail dark grey. White under wing and black wing tips in flight. Male has red iris, female and immature have yellow iris. Juvenile has brownish upperparts and brown bars on breast and belly.

[1] adult female 雌成鳥；Po Toi 蒲台；Apr-07, 07 年 4 月；James Lam 林文華
[2] adult male 雄成鳥；Po Toi 蒲台；9-Apr-06, 06 年 4 月 9 日；Michelle and Peter Wong 江敏兒、黃理沛
[3] juvenile male moulting into adult plumage 雄幼鳥轉換成羽；Po Toi 蒲台；May-08, 08 年 5 月；Eling Lee 李佩玲
[4] juvenile 幼鳥；Mai Po 米埔；12-Oct-08, 08 年 10 月 12 日；Sung Yik Hei 宋亦希
[5] adult male 雄成鳥；Po Toi 蒲台；May-07, 07 年 5 月；Helen Chan 陳燕芳
[6] adult female 雌成鳥；Po Toi 蒲台；Apr-07, 07 年 4 月；Wallace Tse 謝鑑超

| | 春季過境遷徙鳥<br>Spring Passage Migrant | | 夏候鳥<br>Summer Visitor | | 秋季過境遷徙鳥<br>Autumn Passage Migrant | | 冬候鳥<br>Winter Visitor | | |
|---|---|---|---|---|---|---|---|---|---|
| 常見月份 | 1 | 2 | 3 | 4 | 5 | 6 | 7 | 8 | 9 | 10 | 11 | 12 |
| | 留鳥<br>Resident | | | 迷鳥<br>Vagrant | | | 偶見鳥<br>Occasional Visitor | | |

# 日本松雀鷹

（普）rì běn sōng què yīng

Japanese Sparrowhawk | *Accipiter gularis*

體長 length：23-30cm

小型猛禽。雄鳥頭部和上體灰色，胸有淺棕色橫紋，尾灰色及有闊橫紋。雌鳥毛色較深。雄性成鳥虹膜深紅，雌性及未成年鳥虹膜黃色。與較常見的松雀鷹比較，日本松雀鷹體型較小，胸、腹毛色較淡，翼尖較長，尾較短，尾上橫紋淺色的比深色的寬。

Small raptor. Male has grey head and upperparts. Light brown strips on breast. Tail is grey in colour with broad bands. Adult male has red iris, female and immature have yellow iris. Female is darker overall. Compared to the more common Besra, Japanese Sparrowhawk is smaller in size, has lighter colour on breast and belly, longer wing tips, shorter while tail, with narrower dark tail bars.

1 adult female 雌成鳥：Tai Mo Shan 大帽山；Tam Yiu Leung 譚耀良
2 female 雌鳥：Cloudy Hill 九龍坑山；Oct-02, 02 年 10 月；Jemi and John Holmes 孔思義、黃亞萍
3 immature 未成年鳥：Long Valley 塱原；13 Nov-16, 16 年 11 月 13 日；Beetle Cheng 鄭諾銘

| 春季過境遷徙鳥<br>Spring Passage Migrant | | | 夏候鳥<br>Summer Visitor | | | 秋季過境遷徙鳥<br>Autumn Passage Migrant | | 冬候鳥<br>Winter Visitor | | |
|---|---|---|---|---|---|---|---|---|---|---|

常見月份

| 1 | 2 | 3 | 4 | 5 | 6 | 7 | 8 | 9 | 10 | 11 | 12 |
|---|---|---|---|---|---|---|---|---|---|---|---|

| 留鳥<br>Resident | 迷鳥<br>Vagrant | 偶見鳥<br>Occasional Visitor |
|---|---|---|

# 松雀鷹

(普) sōng què yīng

體長 length：24-36cm

## Besra | *Accipiter virgatus*

[1] [2] [3] [4]

小 型猛禽。雄鳥深灰褐色，腹部白色並有褐色的粗橫紋。雌鳥顏色較深褐，下腹有明顯橫紋。幼鳥下體縱紋較多。體型比日本松雀鷹稍大，翼較短圓，尾較長，尾羽深淺橫紋寬度相等。

Small raptor. Male is dark greyish brown, white belly with thick brown strips across. Female is darker brown in colour, with more obvious brown strips on breast. Juvenile has more stripes on underparts. Slightly bigger than Japanese Sparrowhawk, with short, round wings and longer tail. Tail bars have even thickness.

[1] adult 成鳥：Sha Tin Pass 沙田㘭：17-Jul-04, 04 年 7 月 17 日；Michelle and Peter Wong 江敏兒 · 黃理沛
[2] juvenile 幼鳥：Cheung Chau 長洲：Apr-05, 05 年 4 月；Henry Lui 呂德恆
[3] juvenile 幼鳥：Mai Po 米埔：Sep-08, 08 年 9 月；Cherry Wong 黃卓研
[4] juvenile 幼鳥：Clear Water Bay 清水灣；Jul-07, 07 年 7 月；Wee Hock Kee 黃福基

| 春季過境遷徙鳥 Spring Passage Migrant | | | 夏候鳥 Summer Visitor | | | 秋季過境遷徙鳥 Autumn Passage Migrant | | | 冬候鳥 Winter Visitor | | |
|---|---|---|---|---|---|---|---|---|---|---|---|
| 1 | 2 | 3 | 4 | 5 | 6 | 7 | 8 | 9 | 10 | 11 | 12 |

常見月份

| 留鳥 Resident | | | | 迷鳥 Vagrant | | | | 偶見鳥 Occasional Visitor | | | |
|---|---|---|---|---|---|---|---|---|---|---|---|

# 雀鷹 (普) què yīng

體長 length：28-40cm

## Eurasian Sparrowhawk | *Accipiter nisus*

其他名稱 Other names：Northern Sparrow Hawk

雄 鳥石灰色的上體，面頰具橙褐斑點，胸口和腹部滿佈橙褐色條紋，尾部有模糊深色的條紋。沒有鬚紋，較長的翼和尾羽。雌鳥體型較大，有較白色的長眉紋和帶較褐的灰上體。

Male has slate-grey upperparts, browish orange wash on chins, light underparts with browish orange bars, tinged color bars on tail. No mesial, longer wings and tail. Female is bigger, with visible white supercilium and pale brown-grey upperparts.

1 adult 成鳥：Long Valley 塱原；Oct-07, 07 年 10 月；Fung Siu Ping 馮少萍
2 adult male 雄成鳥：Shek Kong 石崗；Mar-06, 06 年 3 月；Tam Yip Shing 譚業成
3 adult 成鳥：Tai Lam 大欖；Mar-19, 19 年 3 月；Kwok Tsz Ki 郭子祈
4 adult male 雄成鳥：Shek Kong 石崗；Mar-06, 06 年 3 月；Tam Yip Shing 譚業成
5 juvenile 幼鳥：Mai Po 米埔；Nov-06, 06 年 11 月；Pippen Ho 何志剛
6 juvenile 幼鳥：Mai Po 米埔；Nov-06, 06 年 11 月；Kinni Ho 何建業

| 春季過境遷徙鳥<br>Spring Passage Migrant | | | | 夏候鳥<br>Summer Visitor | | 秋季過境遷徙鳥<br>Autumn Passage Migrant | | 冬候鳥<br>Winter Visitor | | | |
|---|---|---|---|---|---|---|---|---|---|---|---|
| 常見月份 1 | 2 | 3 | 4 | 5 | 6 | 7 | 8 | 9 | 10 | 11 | 12 |
| 留鳥<br>Resident | | | | 迷鳥<br>Vagrant | | | | 偶見鳥<br>Occasional Visitor | | | |

# 蒼鷹
(普) cāng yīng

Northern Goshawk | *Accipiter gentilis*

體長 length：46-63cm

有 白色粗眉紋的大型鷹類。成鳥上體深灰色，下體近白色並有深色幼紋。未成年鳥上體褐色，下體近白色並有褐色粗紋。

Large hawk with thick white supercilia. Adult has dark grey upperparts and white underparts with thin dark streaks. Sub-adult has brown upperparts and whitish underparts with thick brown streaks.

1 adult female 雌成鳥；Mai Po 米埔；Nov-17, 17 年 11 月：Wong Leung Hung 黃良熊
2 adult male 雄成鳥；Mai Po 米埔；Jan-18, 18 年 1 月：Peter Chan 陳佳瑋
3 adult male 雄成鳥；Mai Po 米埔；Jan-18, 18 年 1 月：Peter Chan 陳佳瑋
4 adult male 雄成鳥；Mai Po 米埔；Jan-18, 18 年 1 月：Peter Chan 陳佳瑋

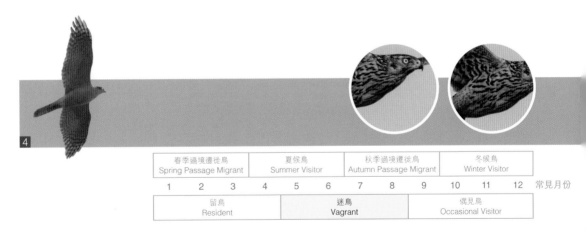

| 春季過境遷徙鳥<br>Spring Passage Migrant | | | 夏候鳥<br>Summer Visitor | | | 秋季過境遷徙鳥<br>Autumn Passage Migrant | | | 冬候鳥<br>Winter Visitor | | |
|---|---|---|---|---|---|---|---|---|---|---|---|
| 1 | 2 | 3 | 4 | 5 | 6 | 7 | 8 | 9 | 10 | 11 | 12 |

常見月份

| 留鳥<br>Resident | 迷鳥<br>Vagrant | 偶見鳥<br>Occasional Visitor |
|---|---|---|

# 白腹鷂

(普) bái fù yáo
(粵) 鷂：音堯

體長 length：47-55cm

## Eastern Marsh Harrier | *Circus spilonotus*

其他名稱 Other names：澤鷂

1

中型猛禽。雄鳥主要黑、白、灰三色，腹部白色，尾羽灰色。正面看可隱約看見頭部有像貓頭鷹的臉盤，雌鳥及未成年鳥主要褐色，羽色多變。經常低飛，翅膀成淺V型。香港錄得的雄性成鳥較少。

Medium-sized raptor. Adult male is mainly black, white and grey. White belly and grey tail. Look from the front, will find it has a facial disc similar to the owls. Female and immatures are mainly brown, with great variation in plumage. Always flies low on V shaped wings. Occasional records of adult males in Hong Kong.

1 juvenile female 雌幼鳥；Mai Po 米埔；Nov-08, 08 年 11 月；Cherry Wong 黃卓研
2 adult male 雄成鳥；Mai Po 米埔；Jan-09, 09 年 1 月；Jemi and John Holmes 孔思義、黃亞萍
3 juvenile female 雌幼鳥；Mai Po 米埔；30-Oct-06, 06 年 10 月 30 日；Michelle and Peter Wong 江敏兒、黃理沛
4 juvenile female 雌幼鳥；Mai Po 米埔；30-Oct-06, 06 年 10 月 30 日；Michelle and Peter Wong 江敏兒、黃理沛
5 juvenile male 雄幼鳥；Mai Po 米埔；Dec-06, 06 年 12 月；Cherry Wong 黃卓研
6 juvenile female 雌幼鳥；Mai Po 米埔；Oct-06, 06 年 10 月；Frankie Chu 朱錦滿

| | 春季過境遷徙鳥 Spring Passage Migrant | | | 夏候鳥 Summer Visitor | | 秋季過境遷徙鳥 Autumn Passage Migrant | | 冬候鳥 Winter Visitor | | |
|---|---|---|---|---|---|---|---|---|---|---|
| 常見月份 | 1 | 2 | 3 | 4 | 5 | 6 | 7 | 8 | 9 | 10 | 11 | 12 |
| | 留鳥 Resident | | | | 迷鳥 Vagrant | | | | 偶見鳥 Occasional Visitor | | |

2

3

4

5

6

# 鵲鷂

(普) què yáo
(粵) 鷂：音堯

體長 length：43-50cm

## Pied Harrier | *Circus melanoleucos*

中型猛禽。雄鳥頭、頸、胸、背部和初級飛羽呈黑色，腹、腰及兩脇白色。飛行時背部有明顯的黑色斑紋伸展至翼上，翼下灰白色，滑翔時翅膀呈 V 形。雌鳥及幼鳥看似白腹鷂，有像貓頭鷹的臉盤。但初級飛羽基部有橫紋。

Medium-sized raptor. Head, neck, breast, upperparts and primary feathers are black. Belly, rump and flanks are white. Obvious black stripes on back extend behind distinctive white upperwing coverts in flight. underwings are pale grey. V-shaped wings when gliding. Female and juvenile resemble Eastern Marsh Harrier with a facial disc similar to the owls, but have bars on the base of primaries.

[1] juvenile 幼鳥：Mai Po 米埔：Oct-07, 07 年 10 月：Yue Pak Wai 余柏維
[2] juvenile 幼鳥：Mai Po 米埔：Oct-07, 07 年 10 月：Yue Pak Wai 余柏維
[3] adult male 雄成鳥：Long Valley 塱原：Lo Kar Man 盧嘉孟

| 春季過境遷徙鳥 Spring Passage Migrant | | | 夏候鳥 Summer Visitor | | | 秋季過境遷徙鳥 Autumn Passage Migrant | | 冬候鳥 Winter Visitor | | | |
|---|---|---|---|---|---|---|---|---|---|---|---|
| 1 | 2 | 3 | 4 | 5 | 6 | 7 | 8 | 9 | 10 | 11 | 12 |

常見月份

| 留鳥 Resident | 迷鳥 Vagrant | 偶見鳥 Occasional Visitor |
|---|---|---|

# 栗鳶

(普) lì yuān
(粵) 鳶：音淵

體長 length：44-52cm

**Brahminy Kite** | *Haliastur indus*

中型猛禽，成鳥顏色對比明顯，十分易認。頭和胸白色，上體、臀及尾栗色，翼尖黑色，對比鮮明。未成年鳥全身大致深褐色，和黑鳶很相似，但尾圓。常近水而居。

Medium-sized raptor. Adult is distinctive by its contrasting colours. White head and breast; chestnut upperparts, vent and tail; black wing tip. Immature is mainly dark brown. Resembles Black Kite but tail is rounded. Lives near water.

1 adult 成鳥：Mai Po 米埔；Oct-89, 89 年 10 月；Jemi and John Holmes 孔思義，黃亞萍

| 春季過境遷徙鳥 Spring Passage Migrant | | | 夏候鳥 Summer Visitor | | | 秋季過境遷徙鳥 Autumn Passage Migrant | | | 冬候鳥 Winter Visitor | | |
|---|---|---|---|---|---|---|---|---|---|---|---|
| 1 | 2 | 3 | 4 | 5 | 6 | 7 | 8 | 9 | 10 | 11 | 12 |

常見月份

| 留鳥 Resident | 迷鳥 Vagrant | 偶見鳥 Occasional Visitor |
|---|---|---|

# 黑鳶

(普)hēi yuān
(粵)鳶：音淵

體長 length：44-66cm

## Black Kite | *Milvus migrans*

其他名稱 Other names：麻鷹, 黑耳鳶, Black-eared Kite

[1]

香港最常見的猛禽，全身大致深褐色，耳羽深色。經常在高空盤旋，初級飛羽分開像手指；尾末端開叉，可以此和其他猛禽區別。間中停在木桿或樹上，叫聲為一聲長嘯後有數節短促的嘯聲。常見於市區上空，在香港繁殖。晚間停棲在林區，昂船洲和馬己仙峽都是主要的夜棲地點。

The commonest raptor in Hong Kong. Mainly dark brown with dark ear coverts. Usually soars high in the sky. Appeared to have "fingers" at wing tip. Distinguished from other raptors by a slightly forked tail. Perches on poles or trees. Voice is a long whine followed by a series of shorter ones. Very common and widespread. Also seen over urban areas. Breeds in Hong Kong. Roosts in woodland at night. Stonecutters Island and Magazine Gap are the main roosting sites.

[1] adult 成鳥：Victoria Peak 太平山；May-07, 07 年 5 月；Joyce Tang 鄧玉蓮
[2] juvenile 幼鳥：Stanley 赤柱；Jul-08, 08 年 7 月；Chan Kai Wai 陳佳瑋
[3] adult 成鳥：Fung Lok Wai 豐樂圍；Jan-06, 06 年 1 月；Henry Lui 呂德恆
[4] Sai Kung 西貢；Sep-07, 07 年 9 月；Chan Kai Wai 陳佳瑋
[5] juvenile 幼鳥：Stanley 赤柱；Nov-07, 07 年 11 月；Chan Kai Wai 陳佳瑋
[6] adult 成鳥：Magazine Gap 馬己仙峽；Apr-04, 04 年 4 月；Henry Lui 呂德恆
[7] adult 成鳥：Tai Po Kau 大埔滘；Jun-04, 04 年 6 月；Cherry Wong 黃卓研
[8] adult 成鳥：Shau Kei Wan 筲箕灣；15-Apr-07, 07 年 4 月 15 日；Ken Fung 馮漢城

| 春季過境遷徙鳥 Spring Passage Migrant | | | 夏候鳥 Summer Visitor | | | 秋季過境遷徙鳥 Autumn Passage Migrant | | | 冬候鳥 Winter Visitor | | |
|---|---|---|---|---|---|---|---|---|---|---|---|
| 1 | 2 | 3 | 4 | 5 | 6 | 7 | 8 | 9 | 10 | 11 | 12 |
| 留鳥 Resident | | | | 迷鳥 Vagrant | | | | 偶見鳥 Occasional Visitor | | | |

常見月份

鷹科
Accipitridae

# 白腹海鵰
(普) bái fù hǎi diāo
(粵) 鵰：音丟

體長 length：75-85cm

## White-bellied Sea Eagle | *Haliaeetus leucogaster*

大型猛禽，飛行時黑白分明，容易辨認。嘴黑色，成鳥頭、頸至下體及尾羽白色，初級飛羽黑色，有楔形尾。滑翔時雙翼稍為舉起成 V 形，間中停棲在樹上，靜立時上體灰白色，叫聲為急促的「ad-ad-ad-ad」響聲。幼鳥全身褐色，和黑鳶相似。在香港繁殖，喜在偏僻海岸的樹上築巢。

Large black-and-white raptor, unmistakable in flight. Bill is black. Adult has white head, neck, underparts and tail. Black primaries and wedge-shaped tail. V-shaped wings when gliding. Sometimes perches in trees. Light grey upperparts when perched. Voice is a quick "ad-ad-ad-ad". Juvenile all brown resembling Black Kite. Breeds in Hong Kong. Prefers nesting on trees near remote sea coasts.

1 adult 成鳥：Aberdeen 香港仔；Jun-05, 05 年 6 月；Pippen Ho 何志剛
2 adult 成鳥：Fung Lok Wai 豐樂圍；Nov-03, 03 年 11 月；Jemi and John Holmes 孔思義、黃亞萍
3 adult 成鳥：Tin Wan 田灣；Jan-08, 08 年 1 月；Chan Kai Wai 陳佳瑋
4 adult 成鳥：Aberdeen 香港仔；17-Apr-05, 05 年 4 月 17 日；Michelle and Peter Wong 江敏兒、黃理沛
5 juvenile 幼鳥：Shum Chung 深涌；Cheung Ho Fai 張浩輝
6 adult 成鳥：Aberdeen 香港仔；Jun-05, 05 年 6 月；Pippen Ho 何志剛

| | 春季過境遷徙鳥<br>Spring Passage Migrant | | 夏候鳥<br>Summer Visitor | | 秋季過境遷徙鳥<br>Autumn Passage Migrant | | 冬候鳥<br>Winter Visitor | |
|---|---|---|---|---|---|---|---|---|
| 常見月份 | 1 | 2 | 3 | 4 | 5 | 6 | 7 | 8 | 9 | 10 | 11 | 12 |
| | 留鳥<br>Resident | | | 迷鳥<br>Vagrant | | | 偶見鳥<br>Occasional Visitor | | |

# 灰臉鵟鷹

(普) huī liǎn kuáng yīng
(圖) 鵟：音狂

體長 length：41-48cm

## Grey-faced Buzzard | *Butastur indicus*

其他名稱 Other names：Grey-faced Buzzard Eagle

修長的中型猛禽，頭頂由淺至上體較深的褐色，灰頰、眼淺色、喉白色，深色鬚紋和下頦紋，胸部佈滿褐和白色的條紋，尾羽有三條深色間條紋。翼下色淡有幼密橫紋，初級飛羽邊緣黑色。

Medium-sized and slim-bodied. Upperparts deep brown. Grey chins, pale eye and white throat, dark sub-moustachial and mesial stripes, white and brown stripes on belly, three brown bars on tail. Pale finely-barred underwing with black primary tips.

1 adult female 雌成鳥 Apr-09, 09 年 4 月：Pang Chun Chiu 彭俊超
2 adult female 雌成鳥 Apr-09, 09 年 4 月：Pang Chun Chiu 彭俊超
3 adult female 雌成鳥：Siu Lek Yuen 小瀝源；Apr-08, 08 年 4 月：Ken Fung 馮漢城
4 adult female 雌成鳥 Apr-09, 09 年 4 月：Pang Chun Chiu 彭俊超
5 adult female 雌成鳥：Kadoorie Farm 嘉道理農場；Apr-05, 05 年 4 月：Cherry Wong 黃卓研

| 常見月份 | 春季過境遷徙鳥 Spring Passage Migrant | | | | 夏候鳥 Summer Visitor | | 秋季過境遷徙鳥 Autumn Passage Migrant | | 冬候鳥 Winter Visitor | | |
|---|---|---|---|---|---|---|---|---|---|---|---|
| | 1 | 2 | 3 | 4 | 5 | 6 | 7 | 8 | 9 | 10 | 11 | 12 |
| | 留鳥 Resident | | | | 迷鳥 Vagrant | | | | 偶見鳥 Occasional Visitor | | |

# 普通鵟

(普) pǔ tōng kuáng
(粵) 鵟：音狂

體長 length：42-54cm

## Eastern Buzzard | *Buteo japonicus*

1

中型猛禽，羽色多變，深淺不一。全身褐色，下體淡黃褐色，有淡褐色縱紋；飛行時翼圓而寬，尾部呈扇形。翼尖黑色，翼底淺色，翼角有黑斑。有時站在枯樹或電燈柱上。

Medium-sized raptor. Plumage variable. Overall brown in colour. Pale yellowish brown underparts with pale brown streaks. In flight it shows round and broad wings, round tail, black primary tips and pale underwing with black patches. Sometimes perches on tree branches and lamp posts.

[1] adult, race *vulpinus* 成鳥, *vulpinus* 亞種：Kam Tin 錦田：Nov- 07, 07 年 11 月：Raymond Cheng 鄭兆文
[2] adult, race *vulpinus* 成鳥, *vulpinus* 亞種：Mai Po 米埔：Owen Chiang 深藍
[3] juvenile 幼鳥：Cheung Chau 長洲：Nov-04, 04 年 11 月：Henry Lui 呂德恆
[4] juvenile 幼鳥：Long Valley 塱原：Dec-06, 06 年 12 月：Martin Hale 夏敖天
[5] juvenile 幼鳥：Mai Po 米埔：Nov-07, 07 年 11 月：James Lam 林文華
[6] juvenile 幼鳥：Lamma Island 南丫島：Dec-05, 05 年 12 月：Harry Li 李炳偉
[7] juvenile 幼鳥：Shek Kong 石崗：Jan-09, 09 年 1 月：Danny Ho 何國海
[8] juvenile 幼鳥：Shek Kong 石崗：Jan-09, 09 年 1 月：Andy Kwok 郭匯昌

| 春季過境遷徙鳥 Spring Passage Migrant | | | | 夏候鳥 Summer Visitor | | 秋季過境遷徙鳥 Autumn Passage Migrant | | | 冬候鳥 Winter Visitor | | |
|---|---|---|---|---|---|---|---|---|---|---|---|
| 1 | 2 | 3 | 4 | 5 | 6 | 7 | 8 | 9 | 10 | 11 | 12 |

常見月份

| 留鳥 Resident | 迷鳥 Vagrant | 偶見鳥 Occasional Visitor |
|---|---|---|

# 領角鴞

普 lǐng jiǎo xiāo
粵 鴞：音囂

體長 length：23-25cm

## Collared Scops Owl | *Otus lettia*

1

小型有長角羽的貓頭鷹，香港常見，主要在夜間活動。虹膜紅褐色，後頸有明顯淡黃褐色領紋，上體深褐色，下體較淡色，背具深色紋。體型較紅角鴞大。叫聲為輕柔的「hoot」，每10秒重複一次。

Small-sized owl with ear tufts. Common in Hong Kong, mainly active at night. Reddish brown iris, pale brownish yellow hind collar. Dark brown upperparts, lighter upperparts with dark streaks. Larger than Oriental Scops Owl. Call a soft "hoot", repeats every 10 seconds.

1 adult 成鳥：Cloudy Hill 九龍坑山；Oct-06, 06 年 10 月；Henry Lui 呂德恆
2 adult 成鳥：Robin's Nest 紅花嶺；Oct-04, 04 年 10 月；Jemi and John Holmes 孔思義、黃亞萍
3 adult 成鳥：Po Toi 蒲台；May-08, 08 年 5 月；Eling Lee 李佩玲
4 adult 成鳥：Hang Tau 坑頭；May-03, 03 年 5 月；Jemi and John Holmes 孔思義、黃亞萍
5 adult 成鳥：Po Toi 蒲台；Sep-05, 05 年 9 月；Allen Chan 陳志雄
6 adult 成鳥：North Point 北角；Dec-07, 07 年 12 月；Owen Chiang 深藍

| | 春季過境遷徙鳥<br>Spring Passage Migrant | | | 夏候鳥<br>Summer Visitor | | | 秋季過境遷徙鳥<br>Autumn Passage Migrant | | | 冬候鳥<br>Winter Visitor | | |
|---|---|---|---|---|---|---|---|---|---|---|---|---|
| 常見月份 | 1 | 2 | 3 | 4 | 5 | 6 | 7 | 8 | 9 | 10 | 11 | 12 |
| | 留鳥<br>Resident | | | | 迷鳥<br>Vagrant | | | | 偶見鳥<br>Occasional Visitor | | | |

 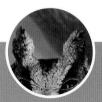 

# 紅角鴞

（普）hóng jiǎo xiāo
（粵）鴞：音囂

體長 length：18-21cm

## Oriental Scops Owl | *Otus sunia*

其他名稱 Other names：Asian Scops Owl

小 型貓頭鷹，主要在夜間活動。毛色多變，有灰色型和紅棕色型兩種。虹膜黃色，肩羽有白色帶，下體有深色縱紋。受驚時會豎起角羽。

Small-sized owl with ear tufts, active at night. Plumage variable, with grey or rufous morph. Yellow iris. White edge on feathers at scapulars forming a line. Heavy dark streaks on underparts. Raises ear-tufts when alarmed.

1 adult, race *stictonotus* rufous morph 棕色型成鳥, *stictonotus* 亞種：Cloudy Hill 九龍坑山；4-Nov-06, 06 年 11 月 4 日；Michelle and Peter Wong 江敏兒、黃理沛

2 adult grey morph, race unknown 灰色型成鳥, 不知名亞種：Cloudy Hill 九龍坑山；Nov-04, 04 年 11 月；Jemi and John Holmes 孔思義、黃亞萍

3 adult grey morph, race unknown 灰色型成鳥, 不知名亞種：Cloudy Hill 九龍坑山；Nov-00, 00 年 11 月；Jemi and John Holmes 孔思義、黃亞萍

4 adult 成鳥：Tsing Yi park 青衣公園；Oct-19, 19年10月；Roman Lo 羅文凱

5 adult, race *stictonotus* rufous morph 棕色型成鳥, *stictonotus* 亞種：Cloudy Hill 九龍坑山；Oct-02, 02 年 10 月；Jemi and John Holmes 孔思義、黃亞萍

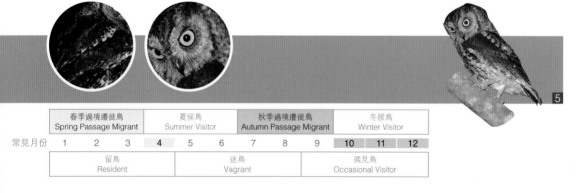

| | 春季過境遷徙鳥<br>Spring Passage Migrant | | | 夏候鳥<br>Summer Visitor | | | 秋季過境遷徙鳥<br>Autumn Passage Migrant | | | 冬候鳥<br>Winter Visitor | | |
|---|---|---|---|---|---|---|---|---|---|---|---|---|
| 常見月份 | 1 | 2 | 3 | 4 | 5 | 6 | 7 | 8 | 9 | 10 | 11 | 12 |
| | 留鳥<br>Resident | | | | 迷鳥<br>Vagrant | | | | 偶見鳥<br>Occasional Visitor | | | |

# 鵰鴞

(普) diāo xiāo
(粵) 鴞：音囂

體長 length：60-75cm

## Eurasian Eagle-Owl | *Bubo bubo*

大 型貓頭鷹，有顯眼的耳羽和臉盤，身體褐色有黑色縱紋，喉及頰淡白色，虹膜黃色。羽毛伸延至腳趾。

Large-sized owl, with prominent ear-tufts and facial disc. Brown in colour with heavy blackish streaks. Whitish cheeks and throat, yellow iris. Feathers extended to toes.

[1] adult 成鳥：Kowloon Peak 飛鵝山；Aug-08, 08 年 8 月；Ng Lin Yau 吳璉宥
[2] adult 成鳥：Lai Chi Kok 荔枝角；30-Nov-07, 07 年 11 月 30 日；Michelle and Peter Wong 江敏兒、黃理沛
[3] adult 成鳥：Kowloon Peak 飛鵝山；Aug-08, 08 年 8 月；Chan Kai Wai 陳佳瑋

| 春季過境遷徙鳥<br>Spring Passage Migrant | | 夏候鳥<br>Summer Visitor | | 秋季過境遷徙鳥<br>Autumn Passage Migrant | | 冬候鳥<br>Winter Visitor | |
|---|---|---|---|---|---|---|---|

| 1 | 2 | 3 | 4 | 5 | 6 | 7 | 8 | 9 | 10 | 11 | 12 | 常見月份 |
|---|---|---|---|---|---|---|---|---|---|---|---|---|

| 留鳥<br>Resident | 迷鳥<br>Vagrant | 偶見鳥<br>Occasional Visitor |
|---|---|---|

# 褐漁鴞

(普) hè yú xiāo
(圖) 鴞：音囂

體長length：48-58cm

## Brown Fish Owl | *Ketupa zeylonensis*

大型紅褐色的貓頭鷹，具有顯眼的角羽，虹膜黃色，全身有深褐色縱紋，翼兩側有較寬闊的白翼帶。與鵰鴞比較，褐漁鴞的長羽較鬆軟，面盤不明顯，腳趾無羽毛。

Large-sized rufous-brown owl, with prominent ear-tufts, yellow iris. Overall body has darkish streaks, and whitish bands appeared on both wings. Compared with Eagle Owls, Brown Fish Owl has softer ear-tufts, unclear facial disc and no feather on toes.

1 adult 成鳥：Pui O 貝澳：Feb-08, 08 年 2 月：Ng Lin Yau 吳璉宥
2 juvenile 幼鳥：Sai Kung 西貢：Jun 09, 09 年 6 月：Wong Hok Sze 王學思
3 adult 成鳥：Pak Nai 白泥：Oct-04, 04 年 10 月：Jemi and John Holmes 孔思義‧黃亞萍
4 adult 成鳥：Pui O 貝澳：Dec-08, 08 年 12 月：Ng Lin Yau 吳璉宥
5 adult 成鳥：Pui O 貝澳：Feb-08, 08 年 2 月：Chan Kai Wai 陳佳瑋

| | 春季過境遷徙鳥<br>Spring Passage Migrant | | | 夏候鳥<br>Summer Visitor | | | 秋季過境遷徙鳥<br>Autumn Passage Migrant | | | 冬候鳥<br>Winter Visitor | | |
|---|---|---|---|---|---|---|---|---|---|---|---|---|
| 常見月份 | 1 | 2 | 3 | 4 | 5 | 6 | 7 | 8 | 9 | 10 | 11 | 12 |
| | 留鳥<br>Resident | | | | 迷鳥<br>Vagrant | | | | 偶見鳥<br>Occasional Visitor | | | |

鴟鴞科
Strigidae

# 褐林鴞

(普) hè lín xiāo
(粵) 鴞：音囂

體長 length：40-55cm

## Brown Wood Owl | *Strix leptogrammica*

大型貓頭鷹。有明顯的深色「眼鏡」和白色眉，無角羽。上體深褐色，下體黃褐色，全身滿佈細密紅褐色的橫斑。

Large-sized owl. Distinctive with conspicuous "spectacles" and white eyebrow, no ear-tufts. Dark chocolate brown upperparts, buffy underparts, with the whole body covered with dense reddish brown strips.

[1] adult 成鳥：Tai Lam 大欖；Oct-19, 19 年 10 月；Matthew Kwan 關朗曦
[2] adult 成鳥：Tai Lam 大欖；Nov-19, 19 年 11 月；Kwok Tsz Ki 郭子祈
[3] Lead Mine Pass 鉛礦坳；Nov-07, 07 年 11 月；Wong Choi On 黃才安

| | 春季過境遷徙鳥<br>Spring Passage Migrant | | | 夏候鳥<br>Summer Visitor | | | 秋季過境遷徙鳥<br>Autumn Passage Migrant | | | 冬候鳥<br>Winter Visitor | | |
|---|---|---|---|---|---|---|---|---|---|---|---|---|
| 常見月份 | 1 | 2 | 3 | 4 | 5 | 6 | 7 | 8 | 9 | 10 | 11 | 12 |
| | 留鳥<br>Resident | | | | 迷鳥<br>Vagrant | | | | 偶見鳥<br>Occasional Visitor | | | |

# 領鵂鶹

- 普 līng xiū liú
- 粵 鵂鶹：音丘留

體長 length：15-17cm

## Collared Owlet | *Glaucidium brodiei*

小型貓頭鷹。虹膜黃色，頭圓沒耳羽，頭頂及後枕褐色而帶白斑，頸後有像眼睛般的黃色領圈。上體灰褐色帶淺橙色斑，下體白色，兩脇有褐色闊縱紋。

Small-sized owl. Yellow irises. Round head without ear tufts. Brown crown and nape tinged with white spots. Yellow collars on the back of its neck look like a pair of eyes. Greyish-brown upperparts tinged with light orange spots. White underparts. Broad brown streaks on flanks.

1 adult 成鳥：Tai Po Kau 大埔滘；Apr-19：Matthew Kwan 關朗曦
2 adult 成鳥：Tai Po Kau 大埔滘；Apr-19：Matthew Kwan 關朗曦
3 adult 成鳥：Tai Po Kau 大埔滘；Apr-19：Ken Fung 馮漢城

| 春季過境遷徙鳥 Spring Passage Migrant | | | 夏候鳥 Summer Visitor | | | 秋季過境遷徙鳥 Autumn Passage Migrant | | | 冬候鳥 Winter Visitor | | |
|---|---|---|---|---|---|---|---|---|---|---|---|
| 1 | 2 | 3 | 4 | 5 | 6 | 7 | 8 | 9 | 10 | 11 | 12 |

常見月份

| 留鳥 Resident | 迷鳥 Vagrant | 偶見鳥 Occasional Visitor |
|---|---|---|

# 斑頭鵂鶹

(普) bān tóu xiū liú
(粵) 鵂鶹：音丘留

體長 length：22-25cm

## Asian Barred Owlet | *Glaucidium cuculoides*

其他名稱 Other names：Barred Owlet

頭部渾圓，虹膜黃色，嘴黃色。頭和上體深褐色而帶濃密淡褐色細橫紋，尾部深色，下體偏白而帶褐色斑。叫聲為急促而漸強的一串「咯咯」聲，有時像水樽斟水時氣泡上升的聲音。在清晨和黃昏活動，常停棲在顯眼的地方。

Round-headed. Yellow iris. Yellow bill. Head and upperparts are dark brown with dense pale browish bars. Tail is dark and underparts are whitish with brown stripes. Call a tight series of "Ka-kup, Ka-kup" rising in volume. Also a prolonged bubbling warble. Active at dawn and dusk. Perches prominently.

[1] adult 成鳥：Kam Tin 錦田；Apr-07, 07 年 4 月：Kinni Ho 何建業
[2] adult 成鳥：Kam Tin 錦田；Apr-07, 07 年 4 月：Martin Hale 夏敖天
[3] adult 成鳥：Kam Tin 錦田；Apr-07, 07 年 4 月：Joyce Tang 鄧玉蓮
[4] juvenile 幼鳥：Kam Tin 錦田；Jun-08, 08 年 6 月：Chan Kai Wai 陳佳瑋
[5] adult 成鳥：Kam Tin 錦田；Jan-09, 09 年 1 月：Cherry Wong 黃卓研
[6] juvenile 幼鳥：Kam Tin 錦田；13-May-07, 07 年 5 月 13 日：Michelle and Peter Wong 江敏兒、黃理沛
[7] adult 成鳥：Kam Tin 錦田；Apr-07, 07 年 4 月：Kinni Ho 何建業
[8] adult 成鳥：Mai Po 米埔；May-18, 18 年 5 月：Henry Lui 呂德恆
[9] adult 成鳥：Shek Kong 石崗；Mar-07, 07 年 3 月：Andy Kwok 郭匯昌

| 春季過境遷徙鳥 Spring Passage Migrant | | | 夏候鳥 Summer Visitor | | | 秋季過境遷徙鳥 Autumn Passage Migrant | | | 冬候鳥 Winter Visitor | | |
|---|---|---|---|---|---|---|---|---|---|---|---|
| 1 | 2 | 3 | 4 | 5 | 6 | 7 | 8 | 9 | 10 | 11 | 12 |
| 留鳥 Resident | | | | 迷鳥 Vagrant | | | | 偶見鳥 Occasional Visitor | | | |

常見月份

鶹鴞科
Strigidae

# 鷹鴞

(普) yīng xiāo
(粵) 鴞：音囂

體長 length：29-33cm

## Northern Boobook | *Ninox japonica*

其他名稱 Other names：Brown Hawk Owl

中型貓頭鷹，頭圓，虹膜黃色。頭和上體暗褐色，下體白色，帶深褐色粗縱紋。尾較長，有淺色橫紋。常站在禿枝上，十分顯眼。

Medium-sized owl. Round head, yellow iris. Head and upperparts dark brown. Underparts white, broadly streaked dark brown. Long tail with pale bars. Perches prominently on bare branches.

1 adult 成鳥；Lai Chi Kok 荔枝角；Nov-08, 08 年 11 月；Pippen Ho 何志剛
2 adult 成鳥；Lai Chi Kok 荔枝角；Nov-07, 07 年 11 月；Ng Lin Yau 吳璉宥
3 adult 成鳥；Lai Chi Kok 荔枝角；Nov-08, 08 年 11 月；Chan Kin Chung Gary 陳建中
4 adult 成鳥；Lai Chi Kok 荔枝角；Nov-08, 08 年 11 月；Freeman Yue 余柏維
5 adult 成鳥；Lai Chi Kok 荔枝角；Nov-08, 08 年 11 月；Eling Lee 李佩玲
6 adult 成鳥；Lai Chi Kok 荔枝角；Nov-08, 08 年 11 月；Ng Lin Yau 吳璉宥

| | 春季過境遷徙鳥 Spring Passage Migrant | | 夏候鳥 Summer Visitor | | 秋季過境遷徙鳥 Autumn Passage Migrant | | 冬候鳥 Winter Visitor | |
|---|---|---|---|---|---|---|---|---|
| 常見月份 | 1 | 2 | 3 | 4 | 5 | 6 | 7 | 8 | 9 | 10 | 11 | 12 |
| | 留鳥 Resident | | | 迷鳥 Vagrant | | | 偶見鳥 Occasional Visitor | |

鴟鴞科
**Strigidae**

# 短耳鴞

(普) duǎn ěr xiāo
(粵) 鴞：音囂

體長 length：37cm

## Short-eared Owl | *Asio flammeus*

1 2

黃褐色具深褐色粗紋的貓頭鷹，頭部細小，角羽短小不明顯，但面盤顯著虹膜黃色。站立時接近水平方向，尾短。飛行時翼底淡色，翼尖深色和有深色腕斑。

A yellowish brown owl with heavy dark browish streaks. Short ear-tufts on small head. Prominent facial disc and yellow iris. Appear horizontal when it stands still. Short tail. In flight, it shows pale underwing, with dark carpal patches and blackish wing tips.

1 adult 成鳥；Mai Po 米埔；Jan-06, 06 年 1 月；Cheung Ho Fai 張浩輝
2 adult 成鳥；Tai Sang Wai 大生圍；May-17, 17 年 5 月；Kwok Tsz Ki 郭子祈
3 adult 成鳥；Lut Chau 甩洲；Dec-19, 19 年 12 月；Roman Lo 羅文凱

3

| 春季過境遷徙鳥<br>Spring Passage Migrant | | | | 夏候鳥<br>Summer Visitor | | | 秋季過境遷徙鳥<br>Autumn Passage Migrant | | | 冬候鳥<br>Winter Visitor | |
|---|---|---|---|---|---|---|---|---|---|---|---|
| 1 | 2 | 3 | 4 | 5 | 6 | 7 | 8 | 9 | 10 | 11 | 12 |

常見月份

| 留鳥<br>Resident | | | 迷鳥<br>Vagrant | | | 偶見鳥<br>Occasional Visitor | | |
|---|---|---|---|---|---|---|---|---|

# 藍胸佛法僧

(普) lán xiōng fó fǎ sēng

European Roller | *Coracias garrulus*

體長 length：31-32cm

除 背部橙褐色外，全身羽毛大致粉藍色，身軀矮壯。嘴黑色，有幼紅褐眼圈，腳黃色。

Mainly pale blue body with brownish-orange back. Chunky body. Black bill. Thin reddish-brown eye rings. Yellow legs.

[1] adult 成鳥：Tuen Mun 屯門：Oct-10, 10 年 10 月；Kinni Ho 何建業

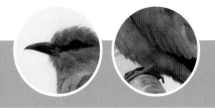

| 春季過境遷徙鳥<br>Spring Passage Migrant | | | 夏候鳥<br>Summer Visitor | | | 秋季過境遷徙鳥<br>Autumn Passage Migrant | | | 冬候鳥<br>Winter Visitor | | |
|---|---|---|---|---|---|---|---|---|---|---|---|
| 1 | 2 | 3 | 4 | 5 | 6 | 7 | 8 | 9 | 10 | 11 | 12 |
| 留鳥<br>Resident | | | | 迷鳥<br>Vagrant | | | | 偶見鳥<br>Occasional Visitor | | | |

常見月份

# 三寶鳥 <sub></sub>sān bǎo niǎo

體長 length：27-32cm

## Oriental Dollarbird | *Eurystomus orientalis*

其他名稱 Other names：佛法僧, Broad-billed Roller

1

**嘴**粗，紅色，嘴端有時沾黑，腳紅色。遠看全身深色，光線充足時可見軀體藍綠色，頭部深色，喉部沾紫藍色。飛行時雙翼寬長，有明顯大白圓斑。愛站在顯眼易見的電線和樹幹上。

Bill red and thick, sometimes bill tip appears black. Red legs. Body appears dark from a distance but is bluish green in good light. Dark head with purplish blue throat. In flight, it shows long broad wings with large round "dollar" - shaped white patches. Perches prominently on bare trees and wires.

1 adult 成鳥：Siu Lek Yuen 小瀝源：May-08, 08 年 5 月：Ken Fung 馮漢城
2 juvenile 幼鳥：Po Toi 蒲台：Oct-08, 08 年 10 月：James Lam 林文華
3 juvenile 幼鳥：Po Toi 蒲台：Oct-08, 08 年 10 月：Ng Lin Yau 吳璉宥
4 juvenile 幼鳥：Po Toi 蒲台：Oct-08, 08 年 10 月：Andy Kwok 郭匯昌
5 adult 成鳥：Po Toi 蒲台：Apr-07, 07 年 4 月：Owen Chiang 深藍
6 juvenile 幼鳥：Po Toi 蒲台：Oct-08, 08 年 10 月：Andy Kwok 郭匯昌
7 juvenile 幼鳥：Po Toi 蒲台：Oct-06, 06 年 10 月：Kinni Ho 何建業

| | 春季過境遷徙鳥<br>Spring Passage Migrant | | | 夏候鳥<br>Summer Visitor | | | 秋季過境遷徙鳥<br>Autumn Passage Migrant | | | 冬候鳥<br>Winter Visitor | | |
|---|---|---|---|---|---|---|---|---|---|---|---|---|
| 常見月份 | 1 | 2 | 3 | 4 | 5 | 6 | 7 | 8 | 9 | 10 | 11 | 12 |
| | 留鳥<br>Resident | | | | 迷鳥<br>Vagrant | | | | 偶見鳥<br>Occasional Visitor | | | |

# 戴勝 <sup>普</sup>dài shèng

## Eurasian Hoopoe | *Upupa epops*

體長 length：19-32cm

有奇特黃褐色冠羽，通常在後枕收起成束，豎起時可見頂端黑白色。嘴黑色，尖長而下彎。頭至胸部黃褐色，上體黑白斑駁，腹至尾下覆羽白色。飛行時像波浪般上下起伏，初級飛羽有闊大白斑，次級飛羽有四條白色橫紋。叫聲像「好寶寶」。通常單隻出現。

Distinctive yellowish brown crest, often carries flat as a tuft behind the nape, shows black and white tips when erects. Bill long and decurved. Yellowish brown from head to breast, white belly and undertail coverts. Folded wings black-and-white. Undulating flight, with obvious white band on primaries and four black-and-white bands on secondaries. Call a "Hoo-poo-poo". Usually appears singly.

1 adult 成鳥；Chinese University of HK 香港中文大學；Feb-05, 05 年 2 月；Allen Chan 陳志雄
2 adult 成鳥；Lamma Island 南丫島；Jan-09, 09 年 1 月；Eling Lee 李佩玲
3 adult 成鳥；Po Toi 蒲台；Jan-08, 08 年 1 月；James Lam 林文華
4 adult 成鳥；Po Toi 蒲台；Dec-07, 07 年 12 月；Fung Siu Ping 馮少萍
5 adult 成鳥；Sai Kung 西貢；Aug-08, 08 年 8 月；Chan Kin Chung Gary 陳建中
6 adult 成鳥；Long Valley 塱原；Jan-08, 08 年 1 月；Ho Kam Wing 何錦榮
7 adult 成鳥；Lamma Island 南丫島；Jan-09, 09 年 1 月；Eling Lee 李佩玲
8 adult 成鳥；Sai Kung 西貢；Aug-08, 08 年 8 月；Chan Kin Chung Gary 陳建中

| | 春季過境遷徙鳥<br>Spring Passage Migrant | | | 夏候鳥<br>Summer Visitor | | | 秋季過境遷徙鳥<br>Autumn Passage Migrant | | | 冬候鳥<br>Winter Visitor | | |
|---|---|---|---|---|---|---|---|---|---|---|---|---|
| 常見月份 | 1 | 2 | 3 | 4 | 5 | 6 | 7 | 8 | 9 | 10 | 11 | 12 |
| | 留鳥<br>Resident | | | | 迷鳥<br>Vagrant | | | | 偶見鳥<br>Occasional Visitor | | | |

# 白胸翡翠

(普) bái xiōng fěi cuì

體長 length：27-28cm

**White-throated Kingfisher** | *Halcyon smyrnensis*

其他名稱 Other names：白胸魚郎, White-breasted Kingfisher

中型翠鳥。頭深褐色，和藍綠色的上體成強烈對比。紅嘴粗而長，喉至胸部大片白色，翼上覆羽有大片深褐色，腹部至尾下覆羽深褐色，腳紅色。飛行時，翼上有大片白斑。叫聲為高音響亮的「傑——傑」笑聲，有時為哀怨而急促的叫聲，音調由高轉低。

Medium-sized kingfisher. Brown head and turquoise blue upperparts. Large dark brown patch on wing coverts. Red bill thick and long, legs red. Large white patch from throat to breast. Belly to undertail coverts dark brown. In flight, it shows a white patch on the wings. Call a laughing "kee-kee" at high pitch, sometimes also a plaintive call with falling pitch.

[1] adult 成鳥；Kowloon Park 九龍公園；Nov-06, 06 年 11 月；Owen Chiang 深藍
[2] juvenile 幼鳥；Mai Po 米埔；23-Sep-07, 07 年 9 月 23 日；Owen Chiang 深藍
[3] juvenile 幼鳥；Tseung Kwan O 將軍澳；Jul-07, 07 年 7 月；Andy Cheung 張玉良
[4] juvenile 幼鳥；Tseung Kwan O 將軍澳；Jul-07, 07 年 7 月；Sammy Sam and Winnie Wong 森美與雲妮
[5] adult 成鳥；Mai Po 米埔；Oct-06, 06 年 10 月；James Lam 林文華
[6] juvenile 幼鳥；Tseung Kwan O 將軍澳；Jul-07, 07 年 7 月；Pippen Ho 何志剛
[7] juvenile 幼鳥；Tseung Kwan O 將軍澳；Jul-07, 07 年 7 月；Chan Kin Chung Gary 陳建中
[8] juvenile 幼鳥；Tseung Kwan O 將軍澳；Jul-07, 07 年 7 月；Owen Chiang 深藍

| | 春季過境遷徙鳥<br>Spring Passage Migrant | | | 夏候鳥<br>Summer Visitor | | | 秋季過境遷徙鳥<br>Autumn Passage Migrant | | | 冬候鳥<br>Winter Visitor | | |
|---|---|---|---|---|---|---|---|---|---|---|---|---|
| 常見月份 | 1 | 2 | 3 | 4 | 5 | 6 | 7 | 8 | 9 | 10 | 11 | 12 |
| | 留鳥<br>Resident | | | | 迷鳥<br>Vagrant | | | | 偶見鳥<br>Occasional Visitor | | | |

# 藍翡翠

(普) lán fěi cuì

體長 length：28cm

## Black-capped Kingfisher | *Halcyon pileata*

其他名稱 Other names：黑頭翡翠

中型翠鳥。頭黑色，頸環至胸部白色，紅嘴粗而長，上體至尾部紫藍色，翼上覆羽黑色。下體橙色。飛行時，翼上有大片白斑。叫聲為高音響亮的一連串急促「傑」笑聲。

Medium-sized kingfisher. Black head, white collar and breast. Red bill thick and long. Upperparts to tail violet with black wing coverts. Shows large white patches on wings in flight. Underparts orange. Call a series of loud and high-pitched laughing "kee".

1 adult 成鳥：Nam Sang Wai 南生圍；Dec-08, 08 年 12 月；Andy Kwok 郭匯昌
2 juvenile 幼鳥：Tai Sang Wai 大生圍；Oct-07, 07 年 10 月；Raymond Cheng 鄭兆文
3 1st winter 第一年冬天：Nam Sang Wai 南生圍；Nov-06, 06 年 11 月；Andy Kwok 郭匯昌
4 adult 成鳥：Mai Po 米埔；Oct-05, 05 年 10 月；Felix Ng 伍昌齡
5 juvenile 幼鳥：Mai Po 米埔；Oct-05, 05 年 10 月；Angus Lau 劉劍明
6 juvenile 幼鳥：Mai Po 米埔；Oct-07, 07 年 10 月；Liu Jian Zhong 劉健忠
7 adult 成鳥：Mai Po 米埔；Oct-05, 05 年 10 月；Felix Ng 伍昌齡
8 juvenile 幼鳥：Mai Po 米埔；10-Nov-06, 06 年 11 月 10 日；Owen Chiang 深藍

| | 春季過境遷徙鳥<br>Spring Passage Migrant | | | 夏候鳥<br>Summer Visitor | | 秋季過境遷徙鳥<br>Autumn Passage Migrant | | | 冬候鳥<br>Winter Visitor | | |
|---|---|---|---|---|---|---|---|---|---|---|---|
| 常見月份 | 1 | 2 | 3 | 4 | 5 | 6 | 7 | 8 | 9 | 10 | 11 | 12 |

| 留鳥<br>Resident | 迷鳥<br>Vagrant | 偶見鳥<br>Occasional Visitor |
|---|---|---|

翠鳥科
Alcedinidae

# 白領翡翠

體長 length：23-25cm

## Collared Kingfisher | *Todiramphus chloris*

1

上體、翼和尾為藍綠色。眼眉、眼下斑、頸環和整個下體白色，過眼線黑色，下嘴有象牙色嘴基。

Bluish green upperparts, wings and tails. White eyebrow, undereye patch, collar and the entire underparts. Blackish eye-stripes and ivory base to lower mandible.

1 adult 成鳥：Mai Po 米埔；Oct-05, 05 年 10 月：Cherry Wong 黃卓研
2 adult 成鳥：Mai Po 米埔；Oct-05, 05 年 10 月：Cherry Wong 黃卓研
3 adult 成鳥：Tai O 大澳；Feb-05, 05 年 2 月：Michelle and Peter Wong 江敏兒、黃理沛

| | 春季過境遷徙鳥 Spring Passage Migrant | | | 夏候鳥 Summer Visitor | | 秋季過境遷徙鳥 Autumn Passage Migrant | | 冬候鳥 Winter Visitor | | |
|---|---|---|---|---|---|---|---|---|---|---|
| 常見月份 | 1 | 2 | 3 | 4 | 5 | 6 | 7 | 8 | 9 | 10 | 11 | 12 |
| | 留鳥 Resident | | | 迷鳥 Vagrant | | | 偶見鳥 Occasional Visitor | | | |

# 普通翠鳥

普 pǔ tōng cuì niǎo

體長 length：16cm

## Common Kingfisher | *Alcedo atthis*

小 型藍色翠鳥，鮮艷奪目，頭的上半部和翅膀翠綠色，並帶淺色斑點，耳羽和下體橙褐色，喉部和耳後近枕處白色，飛行時背部明顯鮮藍色。雌鳥及幼鳥顏色較淡，雌鳥下嘴紅色。貼水面飛行時會發出尖銳的「cheee」叫聲。

Small-sized blue kingfisher. Colourful. Top half of head and wings are greenish blue with pale spots. Ear coverts and underparts orange brown. Throat and areas behind ear coverts are white. In flight, the bright sky-blue back is highly visible. Female and juvenile are duller. Female has reddish lower bill. Call a high-pitched "cheee" in flight.

1 adult male 雄成鳥；Mai Po 米埔；30-Jan-09, 09 年 1 月 30 日；Owen Chiang 深藍
2 adult male 雄成鳥；Mai Po 米埔；Apr-07, 07 年 4 月；Felix Ng 伍昌齡
3 adult female 雌成鳥；Mai Po 米埔；Jan-09, 09 年 1 月；Kami Hui 許淑君
4 adult male 雄成鳥；Kwai Chung 葵涌；Aug-08, 08 年 8 月；Chan Kin Chung Gary 陳建中
5 adult male 雄成鳥；Long Valley 塱原；Nov-07, 07 年 11 月；Isaac Chan 陳家強
6 adult female 雌成鳥；Nam Sun Wai 南生圍；Jan-08, 08 年 1 月；Cherry Wong 黃卓研
7 juvenile male 雄幼鳥；Kwai Chung 葵涌；Aug-08, 08 年 8 月；Chan Kin Chung Gary 陳建中

| | 春季過境遷徙鳥<br>Spring Passage Migrant | | | 夏候鳥<br>Summer Visitor | | 秋季過境遷徙鳥<br>Autumn Passage Migrant | | | 冬候鳥<br>Winter Visitor | | |
|---|---|---|---|---|---|---|---|---|---|---|---|
| 常見月份 | 1 | 2 | 3 | 4 | 5 | 6 | 7 | 8 | 9 | 10 | 11 | 12 |
| | 留鳥<br>Resident | | | | 迷鳥<br>Vagrant | | | 偶見鳥<br>Occasional Visitor | | | |

翠鳥科
**Alcedinidae**

# 三趾翠鳥

(普) sān zhī cuì niǎo

體長 length：14cm

## Oriental Dwarf Kingfisher | *Ceyx erithaca*

多 種艷麗顏色混雜的小型翠鳥。翅膀黑褐色，上背深紫色。嘴、頭上半部分、尾部及腳橙紅色，腰部紫紅色，下體橙色。喉部有白斑，頸兩側有上黑下白的斑塊。

Small kingfisher with many vivid colours. Dark brown wings. Dark purple upper back. Reddish-orange bill, crown, tail and legs. Purplish-red rump. Orange underparts. White chins. A black over white patch on each side of nape.

1 adult female 雌成鳥；Po Toi 蒲台；May-17, 17 年 5 月；Leo Sit 薛國華
2 adult female 雌成鳥；Po Toi 蒲台；May-17, 17 年 5 月；Jemi and John Holmes 孔思義・黃亞萍
3 adult female 雌成鳥；Po Toi 蒲台；May-17, 17 年 5 月；Leo Sit 薛國華

| | 春季過境遷徙鳥<br>Spring Passage Migrant | | | 夏候鳥<br>Summer Visitor | | | 秋季過境遷徙鳥<br>Autumn Passage Migrant | | | 冬候鳥<br>Winter Visitor | | |
|---|---|---|---|---|---|---|---|---|---|---|---|---|
| 常見月份 | 1 | 2 | 3 | 4 | 5 | 6 | 7 | 8 | 9 | 10 | 11 | 12 |
| | 留鳥<br>Resident | | | | 迷鳥<br>Vagrant | | | | 偶見鳥<br>Occasional Visitor | | | |

# 冠魚狗
(普) guān yú gǒu
(粵) 冠：音官

體長 length：41-43cm

## Crested Kingfisher | *Megaceryle lugubris*

大型的翠鳥，冠羽長而蓬鬆，上體黑白條紋幼細，頷和頸側白色，下體白色，雌鳥胸部有黑斑紋，雄鳥具紅棕色斑紋。雌鳥翼下覆羽紅棕色。嘴黑長，端末角色。

Large kingfisher with long shaggy crest, and finely barred upperparts and underparts. White chin and moustachial stripe. White underparts. Blackish streaking on female breast, while rufous on male's. Female has rufous wing lining. Long black bill, horn-coloured tip.

1 adult female 雌成鳥；Nam Chung 南涌；Kwan Po Kuen 關寶權

| 春季過境遷徙鳥<br>Spring Passage Migrant | | | | 夏候鳥<br>Summer Visitor | | | 秋季過境遷徙鳥<br>Autumn Passage Migrant | | | 冬候鳥<br>Winter Visitor | |
|---|---|---|---|---|---|---|---|---|---|---|---|
| 1 | 2 | 3 | 4 | 5 | 6 | 7 | 8 | 9 | 10 | 11 | 12 |

常見月份

| 留鳥<br>Resident | 迷鳥<br>Vagrant | 偶見鳥<br>Occasional Visitor |
|---|---|---|

翠鳥科
**Alcedinidae**

# 斑魚狗

(普) bān yú gǒu

Pied Kingfisher | *Ceryle rudis*

體長 length：25cm

其他名稱 Other names：斑點魚郎

1

中型翠鳥，黑白斑駁。嘴黑色，下體白色，雄鳥有上寬下窄的兩條黑胸帶，而雌鳥和幼鳥只有一條不完整的胸帶。飛行時常發出顫抖叫聲，經常離水面十餘米定點振翅，一發現獵物便直插水中。

Medium-sized black-and-white kingfisher. Black bill. Underparts mainly white. Male has two black breast bands while female and juvenile have two black round chest patches. Loud twittering call in flight. Often hovers at about 10m above water and plunge-dives to catch prey.

1 adult 成鳥：Lut Chau 甩洲；3-Apr-06, 06 年 4 月 3 日；Owen Chiang 深藍
2 adult male 雄成鳥：Mai Po 米埔；Oct-06, 06 年 10 月；Cherry Wong 黃卓研
3 adult female 雌成鳥：Dec-06, 06 年 12 月；Henry Lui 呂德恆
4 adult female 雌成鳥：Nam Chung 南涌；Mike Luk 陸一朝
5 adult female 雌成鳥：Nam Chung 南涌；Mike Luk 陸一朝
6 adult male 雄成鳥：Mai Po 米埔；Oct-08, 08 年 10 月；Lee Kai Hong 李啟康
7 adult male 雄成鳥：Lut Chau 甩洲；Feb-06, 06 年 2 月；Pippen Ho 何志剛

| 春季過境遷徙鳥<br>Spring Passage Migrant | | | 夏候鳥<br>Summer Visitor | | | 秋季過境遷徙鳥<br>Autumn Passage Migrant | | | 冬候鳥<br>Winter Visitor | | |
|---|---|---|---|---|---|---|---|---|---|---|---|
| 1 | 2 | 3 | 4 | 5 | 6 | 7 | 8 | 9 | 10 | 11 | 12 |
| 留鳥<br>Resident | | | | 迷鳥<br>Vagrant | | | | 偶見鳥<br>Occasional Visitor | | | |

常見月份

蜂虎科
Meropidae

# 栗喉蜂虎 ⑱lì hóu fēng hǔ

體長 length：29cm

## Blue-tailed Bee-eater | *Merops philippinus*

黑色嘴尖長而下彎，有黑色貫眼紋，喉部由黃轉深褐色，上體和下體主要為綠色，頭至背部沾橙，尾粉藍色，中間兩條尾羽特長。遷徙時會在香港成群出現。

Black bill pointed, long and decurved. Black eye-stripes, colour of throat changes from yellow to dark brown. Upperparts and underparts are mainly green in colour. Head to mantle tinted orange. Tail whitish blue with two elongated feathers in the middle. Usually recorded in migrating flocks in Hong Kong.

1 Mai Po 米埔：Apr-06, 06 年 4 月；Henry Lui 呂德恆
2 Mai Po 米埔：Oct-02, 02 年 10 月；Francis Chu 崔汝棠
3 adult 成鳥：Beetle Cheng 鄭諾銘
4 juvenile 幼鳥：Mai Po 米埔；Oct-07, 07 年 10 月；James Lam 林文華

| 春季過境遷徙鳥<br>Spring Passage Migrant | | | 夏候鳥<br>Summer Visitor | | 秋季過境遷徙鳥<br>Autumn Passage Migrant | | | 冬候鳥<br>Winter Visitor | | |
|---|---|---|---|---|---|---|---|---|---|---|
| 常見月份 | 1 | 2 | 3 | 4 | 5 | 6 | 7 | 8 | 9 | 10 | 11 | 12 |

| 留鳥<br>Resident | 迷鳥<br>Vagrant | 偶見鳥<br>Occasional Visitor |
|---|---|---|

# 藍喉蜂虎
(普) lán hóu fēng hǔ

體長 length：21cm

## Blue-throated Bee-eater | *Merops viridis*

其他名稱 Other names：栗頭蜂虎, Chestnut-headed Bee-eater

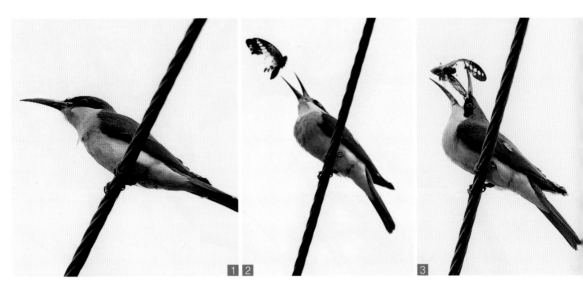

1 2

3

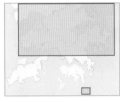

黑 色嘴尖長而下彎，喉藍色，有黑色貫眼紋，頭頂至上背深褐色，下背至尾部藍色，尾長，下體淺綠色。遷徙時會在香港成小群出現。在秋天可見還未長出幼長線尾的幼鳥。

Black bill pointed, long and decurved. Blue throat, black eye-stripe. Crown to mantle dark brown. Back to tail blue. Long tail. Underparts light green. Usually seen in small migrating flocks in Hong Kong. In autumn young birds lack central tail streamers.

1 Tsim Bei Tsui 尖鼻咀：Oct-02, 02 年 10 月；Lee Hok Fei 李鶴飛
2 Tsim Bei Tsui 尖鼻咀：Oct-02, 02 年 10 月；Lee Hok Fei 李鶴飛
3 Tsim Bei Tsui 尖鼻咀：Oct-02, 02 年 10 月；Lee Hok Fei 李鶴飛
4 Tsim Bei Tsui 尖鼻咀：Oct-02, 02 年 10 月；Lee Hok Fei 李鶴飛

4

| 春季過境遷徙鳥 Spring Passage Migrant | | | 夏候鳥 Summer Visitor | | | 秋季過境遷徙鳥 Autumn Passage Migrant | | | 冬候鳥 Winter Visitor | | |
|---|---|---|---|---|---|---|---|---|---|---|---|
| 1 | 2 | 3 | 4 | 5 | 6 | 7 | 8 | 9 | 10 | 11 | 12 |

常見月份

| 留鳥 Resident | 迷鳥 Vagrant | 偶見鳥 Occasional Visitor |
|---|---|---|

# 大擬啄木鳥 (普) dà nǐ zhúo mù niǎo

體長 length：32-35cm

## Great Barbet | *Psilopogon virens*

嘴粗，黃色，腳淡色。頭部深色，上背深褐色，翼、下背至尾部綠色，下體淡黃，有深褐色粗縱紋，尾下覆羽紅色。常在春天鳴叫，叫聲像噪鵑，但較為哀怨，不斷重複。

Bill yellow and thick, pale legs. Dark head, upperparts dark brown. Wings and underparts to tail green, underparts pale yellow with dark brown thick streaks. Undertail coverts red. Call like Koel but more plaintive, continuously repeating in spring.

[1] adult 成鳥：Tai Po Kau 大埔滘；6 Jan-17, 17 年 1 月 6 日；Beetle Cheng 鄭諾銘
[2] Shing Mun 城門水塘；Apr-03, 03 年 4 月；Cherry Wong 黃卓研
[3] juvenile 幼鳥；Shing Mun 城門水塘；Jul-07, 07 年 7 月；Allen Chan 陳志雄
[4] juvenile 幼鳥；Shing Mun 城門水塘；Jul-07, 07 年 7 月；Allen Chan 陳志雄

| 春季過境遷徙鳥 Spring Passage Migrant | | | 夏候鳥 Summer Visitor | | | 秋季過境遷徙鳥 Autumn Passage Migrant | | | 冬候鳥 Winter Visitor | | |
|---|---|---|---|---|---|---|---|---|---|---|---|
| 1 | 2 | 3 | 4 | 5 | 6 | 7 | 8 | 9 | 10 | 11 | 12 |
| 留鳥 Resident | | | | 迷鳥 Vagrant | | | | 偶見鳥 Occasional Visitor | | | |

常見月份

# 黑眉擬啄木鳥
(普) hēi méi nǐ zhúo mù niǎo

體長 length：20-21.5cm

**Chinese Barbet** | *Psilopogon faber*

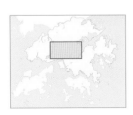

頭部色彩豐富的小型擬啄木鳥。嘴黑色，頭頂黑色，面頰粉藍色，喉部由前至後分別有黃色、粉藍色、紅色。身軀大致青綠色，上體較深色。叫聲為急促響亮的一串「谷谷」聲，遠處可聞。

Small barbet with colourful head. Black bill. Black crown with pale blue face. Throat goes from yellow to pale blue and then red from front to back. Mainly bright green body with darker upperparts. Give a rapid and loud series of "ko-ko" call which can be heard from long distance.

1 adult 成鳥：Tai Po Kau 大埔滘；May-16, 16 年 5 月；Matthew Kwan 關朗曦
2 adult 成鳥：Tai Po Kau 大埔滘；Mar-16, 16 年 3 月；Ng Sze On 吳思安

| 春季過境遷徙鳥<br>Spring Passage Migrant | | | 夏候鳥<br>Summer Visitor | | | 秋季過境遷徙鳥<br>Autumn Passage Migrant | | | 冬候鳥<br>Winter Visitor | | |
|---|---|---|---|---|---|---|---|---|---|---|---|
| 1 | 2 | 3 | 4 | 5 | 6 | 7 | 8 | 9 | 10 | 11 | 12 |
| 留鳥<br>Resident | | | | 迷鳥<br>Vagrant | | | | 偶見鳥<br>Occasional Visitor | | | |

常見月份

# 蟻鴷

（普）yǐ liè
（粵）鴷：音列

體長 length：16-17cm

## Eurasian Wryneck | *Jynx torquilla*

其他名稱 Other names：Wryneck

身體具保護色，主要呈灰褐色，滿佈斑點，腹部顏色較淡。嘴尖而黃色，有深色貫眼紋；中央冠紋黑色，伸延至背部。形態古怪，貌似一片枯葉，能將頸部大角度扭曲。

Camouflaged body brownish grey and mottled, with paler belly. Bill yellow and pointed. Dark eye-stripes. Dark central crown stripe extends to the back. Odd appearance like a dead leaf, often twists the neck to large angles.

1 Lai Chi Kok 荔枝角；Jan-08, 08 年 1 月；Sammy Sam and Winnie Wong 森美與雲妮
2 Lai Chi Kok 荔枝角；Jan-08, 08 年 1 月；Andy Cheung 張玉良
3 Kowloon Park 九龍公園；Dec-06, 06 年 12 月；Owen Chiang 深藍
4 Nam Sang Wai 南生圍；Nov-06, 06 年 11 月；Andy Kwok 郭匯昌
5 Lai Chi Kok 荔枝角；Jan-08, 08 年 1 月；Pippen Ho 何志剛
6 Lai Chi Kok 荔枝角；Jan-08, 08 年 1 月；Pippen Ho 何志剛
7 Lai Chi Kok 荔枝角；Jan-08, 08 年 1 月；Eling Lee 李佩玲
8 HK Wetland Park 香港濕地公園；Nov-06, 06 年 11 月；Kitty Koo 古愛嫻

| 春季過境遷徙鳥 Spring Passage Migrant | | | | 夏候鳥 Summer Visitor | | 秋季過境遷徙鳥 Autumn Passage Migrant | | 冬候鳥 Winter Visitor | | | |
|---|---|---|---|---|---|---|---|---|---|---|---|
| 1 | 2 | 3 | 4 | 5 | 6 | 7 | 8 | 9 | 10 | 11 | 12 |
| 留鳥 Resident | | | | 迷鳥 Vagrant | | | | 偶見鳥 Occasional Visitor | | | |

常見月份

# 斑姬啄木鳥

(普) bān jī zhuó mù niǎo

體長 length：10cm

## Speckled Piculet | *Picumnus innominatus*

其他名稱 Other names：斑啄木鳥, 姬啄木鳥, Spotted Piculet

小型啄木鳥，頭頂橙褐色，翼偏青色，下體偏白，有顯眼黑色斑點。嘴和腳黑色，臉部有黑白紋。叫聲為高音的「的—的—」聲，多在小樹下部覓食，偏好竹林。

Small-sized woodpecker. Brownish orange crown, greenish wings. Whitish underparts with obvious black spots. Black bill and legs. Face with black and white stripes. Call a high pitch "ti- ti-". Usually feeds at lower part of small tree. Prefers bamboo woodland.

1 adult 成鳥：Tai Lam 大欖；Apr-19, 19 年 4 月：John Clough
2 adult 成鳥：Tai Lam 大欖；Apr-19, 19 年 4 月：John Clough
3 male 雄鳥：Ng Tung Chai 梧桐寨；Oct-07, 07 年 10 月：Cheng Nok Ming 鄭諾銘

| 春季過境遷徙鳥 Spring Passage Migrant | | | 夏候鳥 Summer Visitor | | | 秋季過境遷徙鳥 Autumn Passage Migrant | | | 冬候鳥 Winter Visitor | | |
|---|---|---|---|---|---|---|---|---|---|---|---|
| 1 | 2 | 3 | 4 | 5 | 6 | 7 | 8 | 9 | 10 | 11 | 12 |
| 留鳥 Resident | | | | 迷鳥 Vagrant | | | | 偶見鳥 Occasional Visitor | | | |

常見月份

啄木鳥科
Picidae

# 黃嘴栗啄木鳥

(普) huáng zuǐ lì zhuó mù niǎo

體長 length：30cm

## Bay Woodpecker | *Blythipicus pyrrhotis*

其他名稱 Other names：Red-eared Bay Woodpecker, 黃嘴噪啄木鳥

上 體紅褐色，具黑色橫斑，雄鳥頸側及枕紅色。嘴淡綠黃色，腳黑色。叫聲為響亮的下降「嘩—嘩—嘩—嘩」。

Upperparts rufous brown with dark bars. Male has red patch at nape and side of neck. Bill greenish yellow, black legs. Call loud descending "wah-wah-wah-wah".

1 male 雄鳥：Tai Po Kau 大埔滘；5-Jan-07, 07 年 1 月 5 日；Wong Choi On 黃才安
2 female 雌鳥：Tai Po Kau 大埔滘；2-Jan-09, 09 年 1 月 2 日；Louis Cheung 張勇
3 female 雌鳥：Tai Po Kau 大埔滘；Mar-09, 09 年 3 月；Christina Chan 陳燕明
4 female 雌鳥：Tai Po Kau 大埔滘；2-Jan-09, 09 年 1 月 2 日；Louis Cheung 張勇
5 male 雄鳥：Tai Po Kau 大埔滘；4-Apr-09, 09 年 4 月 4 日；Michelle and Peter Wong 江敏兒、黃理沛
6 female 雌鳥：Tai Po Kau 大埔滘；2-Jan-09, 09 年 1 月 2 日；Louis Cheung 張勇

| | 春季過境遷徙鳥<br>Spring Passage Migrant | | | 夏候鳥<br>Summer Visitor | | | 秋季過境遷徙鳥<br>Autumn Passage Migrant | | | 冬候鳥<br>Winter Visitor | | |
|---|---|---|---|---|---|---|---|---|---|---|---|---|
| 常見月份 | 1 | 2 | 3 | 4 | 5 | 6 | 7 | 8 | 9 | 10 | 11 | 12 |
| | 留鳥<br>Resident | | | | 迷鳥<br>Vagrant | | | | 偶見鳥<br>Occasional Visitor | | | |

Michelle & Peter Wong

# 紅隼

普 hóng sǔn
粵 隼：音準

體長 length：27-35cm

## Common Kestrel | *Falco tinnunculus*

[1]

小 型猛禽。成鳥背部紅棕色，尾羽末端有黑色橫帶。雄鳥頭、腰及尾部灰色，雌鳥及幼鳥則為紅棕色。飛行時經常定點振翅，扇展尾部，再俯衝捕捉獵物。

Small-sized raptor. Adult has rufous upperparts, with black terminal bar on long tail. Male has grey head, rump and tail and that of female bird are rufous in colour. Often hovers, fanning tail, then diving to catch prey.

[1] juvenile 幼鳥：Shek Kong 石崗；Mar-07, 07 年 3 月；Sammy Sam and Winnie Wong 森美與雲妮
[2] Kam Tin 錦田；Oct-07, 07 年 10 月；Raymond Cheng 鄭兆文
[3] female 雌鳥：Sai Kung 西貢；Dec-08, 08 年 12 月；Andy Kwok 郭匯昌
[4] juvenile / famale 幼鳥 / 雌鳥：Clear Water Bay 清水灣；Apr-06, 06 年 4 月；Wee Hock Kee 黃福基

| | 春季過境遷徙鳥 Spring Passage Migrant | | | 夏候鳥 Summer Visitor | | 秋季過境遷徙鳥 Autumn Passage Migrant | | 冬候鳥 Winter Visitor | | | |
|---|---|---|---|---|---|---|---|---|---|---|---|
| 常見月份 | 1 | 2 | 3 | 4 | 5 | 6 | 7 | 8 | 9 | 10 | 11 | 12 |
| | 留鳥 Resident | | | | 迷鳥 Vagrant | | | 偶見鳥 Occasional Visitor | | | | |

隼科
Falconidae

# 阿穆爾隼

(普) yà mù ěr sǔn
(粵) 隼：音準

體長 length：28-30cm

Amur Falcon | *Falco amurensis*

[1]

最 易辨認是雌雄鳥都有栗紅色的眼圈和腳。雄鳥全身灰色，尾下覆羽和腿紅或栗紅色，翼下覆羽黑白強烈比對。雌鳥上身是灰色帶黑色條紋，下體是白底滿佈黑色斑點及條紋，腿和脅乳白，翼下覆羽白底滿佈黑色斑點及條紋。

All plumage has the reddish chestnut eye-rings and legs. Male has grey body, red or reddish chestnut thighs and vent, black and white colour at underwing coverts making big contrast in field observation. Female upperparts are grey in colour with black streaks, and white underparts are black spots and streaks. Buff white thigh and vent, grey underwing coverts full of black spots and bars.

[1] juvenile 幼鳥：Tsim Bei Tsui 尖鼻咀；22-Oct-04, 04 年 10 月 22 日：Michelle and Peter Wong 江敏兒、黃理沛
[2] juvenile 幼鳥：Tsim Bei Tsui 尖鼻咀；22-Oct-04, 04 年 10 月 22 日：Michelle and Peter Wong 江敏兒、黃理沛
[3] juvenile 幼鳥：Tsim Bei Tsui 尖鼻咀；22-Oct-04, 04 年 10 月 22 日：Michelle and Peter Wong 江敏兒、黃理沛
[4] adult 成鳥：Tsim Bei Tsui 尖鼻咀；Oct-16, 16 年 10 月：Kwok Tsz Ki 郭子祈
[5] Female adult 雌成鳥：Tsim Bei Tsui 尖鼻咀；Oct-16, 16 年 10 月：Kwok Tsz Ki 郭子祈

| 春季過境遷徙鳥 Spring Passage Migrant | | | 夏候鳥 Summer Visitor | | | 秋季過境遷徙鳥 Autumn Passage Migrant | | 冬候鳥 Winter Visitor | | | |
|---|---|---|---|---|---|---|---|---|---|---|---|
| 1 | 2 | 3 | 4 | 5 | 6 | 7 | 8 | 9 | 10 | 11 | 12 |

常見月份

| 留鳥 Resident | 迷鳥 Vagrant | 偶見鳥 Occasional Visitor |
|---|---|---|

# 遊隼

(普) yóu sǔn
(粵) 隼：音準

體長 length：35-51cm

## Peregrine Falcon | *Falco peregrinus*

1

2

中 型猛禽，軀體粗壯。成鳥上體灰色，下體白色並有深色細橫紋，臉部有深色粗頰紋。幼鳥偏褐色，胸部有褐色縱紋。

Medium-sized raptor. Stout body. Adult has grey upperparts, white belly with dark narrow stripes and a dark, thick moustachial stripe. Juvenile is brownish in colour with brown streaks on belly.

1 adult male, race *pereginator* 雄成鳥, *pereginator* 亞種；Nam Sang Wai 南生圍；Jan-08, 08 年 1 月；Andy Kwok 郭匯昌
2 juvenile, race *japonensis* 幼鳥, *japonensis* 亞種；Mai Po 米埔；Dec-06, 06 年 12 月；Kinni Ho 何建業
3 juvenile 幼鳥；Mai Po 米埔；Jan-08, 08 年 1 月；Pang Chun Chiu 彭俊超
4 Dueling 對決；Mai Po 米埔；24-Mar-07, 07 年 3 月 24 日；Michelle and Peter Wong 江敏兒、黃理沛
5 juvenile, race *japonensis* 幼鳥, *japonensis* 亞種；Mai Po 米埔；Nov-06, 06 年 11 月；Lee Hok Fei 李鶴飛
6 adult, race *pereginator* 成鳥, *pereginator* 亞種；Mirs Bay 大鵬灣；Jul-08, 08 年 7 月；Isaac Chan 陳家強
7 immature 未成年鳥；Beetle Cheng 鄭諾銘

| | 春季過境遷徙鳥<br>Spring Passage Migrant | | | 夏候鳥<br>Summer Visitor | | 秋季過境遷徙鳥<br>Autumn Passage Migrant | | 冬候鳥<br>Winter Visitor | | | |
|---|---|---|---|---|---|---|---|---|---|---|---|
| 常見月份 | 1 | 2 | 3 | 4 | 5 | 6 | 7 | 8 | 9 | 10 | 11 | 12 |
| | 留鳥<br>Resident | | | 迷鳥<br>Vagrant | | | 偶見鳥<br>Occasional Visitor | | | |

# 燕隼

(普) yàn sǔn
(粵) 隼：音準

體長 length：28-36cm

## Eurasian Hobby | *Falco subbuteo*

[1]

[2]

飛　時似雨燕，翼狹長且尖。上體暗灰色。下體淡黃白色，有黑褐色縱紋。成鳥下腹及尾下覆羽栗色。

Appears swift-like in flight, with long pointed wings. It has dark grey upperparts and yellowish white underparts with dark brownish stripes. Adult has chestnut lower abdomen and undertail converts.

[1] adult 成鳥：Po Toi 蒲台；9-Oct-05, 05 年 10 月 9 日；Michelle and Peter Wong 江敏兒、黃理沛
[2] juvenile 幼鳥：Po Toi 蒲台；Sep-08, 08 年 9 月；Cherry Wong 黃卓研
[3] juvenile 幼鳥：Tsim Bei Tsui 尖鼻咀；Oct-04, 04 年 10 月；Henry Lui 呂德恆

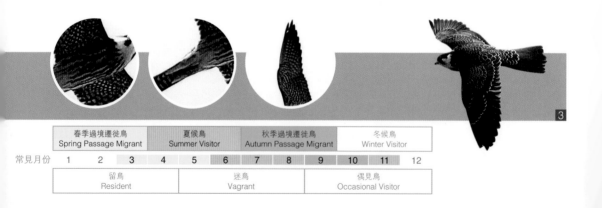

[3]

| 春季過境遷徙鳥 Spring Passage Migrant | | | 夏候鳥 Summer Visitor | | 秋季過境遷徙鳥 Autumn Passage Migrant | | | 冬候鳥 Winter Visitor | | |
|---|---|---|---|---|---|---|---|---|---|---|

常見月份
| 1 | 2 | 3 | 4 | 5 | 6 | 7 | 8 | 9 | 10 | 11 | 12 |
|---|---|---|---|---|---|---|---|---|---|---|---|

| 留鳥 Resident | 迷鳥 Vagrant | 偶見鳥 Occasional Visitor |
|---|---|---|

# 參考資料
# References

Sonobe, K. and Usui, S. (ed.) 1993. *A Field Guide to the Waterbirds of Asia*. Wild Bird Society of Japan, Tokyo.

Wang, S, Zheng G. and Wang, Q. (ed.) 1998. *China Red Data Book of Endangered Animals: Aves*. Science Press, Beijing.

Carey, G. J., Chalmers, M. L., Diskin, D. A., Kennerley, P. R., Leader, P. J., Leven, M. R., Lewthwaite, R. W., Melville, D. S., Turnbull, M., and Young, L. 2001. *The Avifauna of Hong Kong*. Hong Kong Bird Watching Society, Hong Kong.

Hayman P., Marchant J. and Prater T. 1995. *Shorebirds. An identification Guide to the Waders of the World*. Christopher Helm (Publishers) Ltd.

Hong Kong Bird Watching Society. 1957-2004. *Hong Kong Bird Reports (1957-2004)* – annual report publish by Hong Kong Bird Watching Society.

Viney, C., Phillips, K. and Lam, C.Y., 2005. *Birds of Hong Kong and South China*. Government Printer. Hong Kong.

MacKinnon, M., and Phillipps, K. 2000. *A Field Guide to the Birds of China*. Oxford University Press, UK

Hong Kong Bird Watching Society 2005. *A Photographic Guide to the Birds of Hong Kong* Wan Li Book Co Ltd.

King, B.F., Woodcock, M.W. and Dickinaon E.C. 1975 Collins Field Guide Birds of South-East Asia Harper Collins Publishers

聶延秋　　2019　　中國鳥類識別手冊(第二版)　　中國林業出版社

林超英　　2000　　大埔滘觀鳥樂　　天地圖書有限公司、郊野公園之友、香港大學地理及地質學系

馬嘉慧、馮寶基　1999　觀鳥─從后海灣開始　長春社

馬嘉慧、馮寶基、蘇毅雄　2001　觀鳥─從城市開始　香港觀鳥會

約翰‧馬敬能、卡倫‧菲利普斯、何芬奇　2000　中國鳥類野外手冊　湖南教育出版社

王嘉雄等　1991　台灣野鳥圖鑑　台灣野鳥資訊社

許維樞等　1996　中國野鳥圖鑑　翠鳥文化出版社

鄭光美　2002　世界鳥類分類與分布名錄　科學出版社

尹璉、費嘉倫、林超英　2006　香港及華南鳥類　政府新聞處

香港觀鳥會　2005　香港鳥類攝影圖鑑　萬里機構

# 常用網頁
# Websites

Hong Kong Bird Watching Society 香港觀鳥會
*www.hkbws.org.hk*

Hong Kong Observatory 香港天文台網頁
Hong Kong Tide table 香港潮汐表
*http://www.hko.gov.hk/tc/tide/marine/realtide.htm?s=TBT&t=CHART* (中文 Chinese)
*http://www.hko.gov.hk/en/tide/marine/realtide.htm?s=TBT&t=CHART* (英文 English)

# 中文鳥名索引
# Index by Chinese Names

# 英文鳥名索引
# Index by English Common Names

# 學名索引
# Index by Scientific Names

**觀鳥系列01：香港觀鳥全圖鑑** A Photographic Guide to the Birds of Hong Kong

| | |
|---|---|
| 著者 | Author |
| 香港觀鳥會 | The Hong Kong Bird Watching Society |
| 策劃 | Project Co-ordinator |
| 謝妙華 | Pheona Tse |
| 責任編輯 | Project Editor |
| 簡詠怡、陳芷欣 | Karen Kan, Kitty Chan |
| 裝幀設計 | Design |
| 鍾啟善 | Nora Chung |
| 排版 | Typography |
| 劉葉青 | Rosemary Liu |
| 出版者 | Publisher |
| 萬里機構出版有限公司 | Wan Li Book Company Limited |
| 香港北角英皇道499號北角工業大廈20樓 | 20/F, North Point Industrial Building, 499 King's Road, North Point, Hong Kong |
| 電話 | Tel: 2564 7511 |
| 傳真 | Fax: 2565 5539 |
| 電郵 | Email: info@wanlibk.com |
| 網址 | http://www.wanlibk.com |
| | http://www.facebook.com/wanlibk |
| 發行者 | Distributor |
| 香港聯合書刊物流有限公司 | SUP Publishing Logistics (HK) Ltd. |
| 香港荃灣德士古道220-248號荃灣工業中心16樓 | 16/F, Tsuen Wan Industrial Centre, 220-248 Texaco Road, Tsuen Wan, NT., Hong Kong |
| 電話 | Tel: 2150 2100 |
| 傳真 | Fax: 2407 3062 |
| 電郵 | E-mail: info@suplogistics.com.hk |
| 網址 | http://www.suplogistics.com.hk |
| 承印者 | Printer |
| 美雅印刷製本有限公司 | Elegance Printing & Book Binding Co., Ltd. |
| 香港九龍觀塘榮業街6號海濱工業大廈4樓A室 | Block A, 4/F, Hoi Bun Industrial Building, 6 Wing Yip Street, Kwun Tong, Kln., Hong Kong |
| 出版日期 | Publishing Date |
| 二〇二〇年六月第一次印刷 | First print in June 2020 |
| 二〇二一年二月第二次印刷 | Second print in February 2021 |
| 規格 | Specifications |
| 特 16 開（240mm×171mm） | 16K (240mm×171mm) |

ISBN 978-962-14-7238-0

---

**雀鳥視頻** Video of Birds

| | |
|---|---|
| 創作 | Creator |
| 文緝明 | Man Chup Ming |
| 製作 | Production |
| 香港觀鳥會 | The Hong Kong Bird Watching Society |
| 鳴謝 | Acknowledgements |
| 文緝明先生 | Mr Man Chup Ming |
| 黃亞萍小姐 | Ms Jemi AP Wong |
| 孔思義先生 | Mr John Holmes |